Microcomputers in Engineering and Science

INTERNATIONAL COMPUTER SCIENCE SERIES

Consulting editors **A D McGettrick**
University of Strathclyde

J van Leeuwen
University of Utrecht

Microcomputers in Engineering and Science

J. Ffynlo Craine
Graham R. Martin

ADDISON-WESLEY PUBLISHING COMPANY

Wokingham, England · Reading, Massachusetts · Menlo Park, California
Don Mills, Ontario · Amsterdam · Sydney · Singapore · Tokyo · Mexico City · Bogota
Santiago · San Juan

© 1985 Addison-Wesley Publishers Limited
© 1985 Addison-Wesley Publishing Company, Inc.

All rights reserved. No part of this publication may be reproduced, stored in a retrieval system, or transmitted in any form or by any means, electronic, mechanical, photocopying, recording, or otherwise, with prior written permission of the publisher.

The programs presented in this book have been included for their instructional value. They have been tested with care but are not guaranteed for any particular purpose. The publisher does not offer any warranties or representations, nor does it accept any liabilities with respect to the programs.

Cover design by John Gibbs.
Typeset by H. Charlesworth & Co. Ltd. directly from magnetic tape supplied by the authors using **troff**, the UNIX text-processing system.
Printed in Finland by Werner Söderström Osakeyhtiö. Member of Finnprint.

British Library Cataloguing in Publication Data
Craine, J.F.
  Microcomputers in engineering and science.—
  (International computer science series)
  1. Science—Data processing. 2. microcomputers
  3. Engineering—Data processing
  I. Title. II. Martin, G.R. II. Series
  502'.8'5404   Q183.9
  ISBN 0-201-14217-1

Library of Congress Cataloging-in-Publication Data
Craine, J.F.
  Microcomputers in engineering and science.
  (International computer science series)
  Bibliography: p.
  Includes index.
    1. Microcomputers. 2. Microcomputers—Programming.
  3. Computer interfaces. 4. Engineering—Data processing.
  5. Science—Data processing. I. Martin, G.R.
  II. Title. III. Series.
  TK7888.3.C73 1985   502.8'5416   85–18490
  ISBN 0-201-14217-1

ABCDEF 898765

PREFACE

The material in this book is based upon many years of experience, both designing microcomputer-based systems and lecturing on microcomputers and their applications. These lectures have been directed at practising engineers from many disciplines as well as undergraduates reading engineering and computer science at the University of Warwick. Microcomputers are small, robust, and cheap enough to be used as components in a wide range of systems; industrial controllers and laboratory instrumentation are just two applications where mechanical, electromechanical, and non-programmable electronic control systems are being replaced by microcomputers.

As the use of mainframe computers became widespread, engineers and scientists learnt how to program computers to solve purely numerical problems. Today, however, the use of microcomputers as integral parts of larger real-time systems represents a wide departure from this more 'conventional' use of computers. To design and apply microcomputer-based systems in control and instrumentation requires an understanding not only of the microcomputer and its programming, but also of the remainder of the system. Too many books on microcomputers and microprocessors gloss over the problems of interfacing and the real world signals with which they are involved. Likewise, books on engineering control and instrumentation usually neglect the microcomputer itself. In this book we have attempted to redress the balance and include coverage of all the topics essential to the microcomputer applications engineer or the scientific laboratory user. Where our objective of covering all this material in a single volume has meant that it has been impossible fully to do justice to a topic, the text provides references to other books for those readers wanting a deeper insight.

After a brief history of the use and development of microcomputers, the book proceeds to explain how information is represented and processed within a microcomputer. Despite its many shortcomings, BASIC is introduced as a simple programming language because it has become virtually the *lingua franca* of the microcomputing fraternity. A more modern language, Pascal, is then described and its advantages over BASIC, such as the way in which it encourages a structured programming approach, are explained.

The third chapter explains the internal architecture of the microcomputer, and as each section is introduced, the instructions associated with its use are described. Programming examples illustrate how the constructs available in high level programming languages can be implemented in assembly language. The explanation of microprocessor operation has been made as independent of microprocessor type as possible, but where examples are used, 6809 assembly language has been employed because its particularly regular instruction set makes it attractive for this purpose. Peculiarities of the instruction sets of some other microprocessors have been explained where appropriate, and Appendix I provides a comparison of five commonly used types.

Chapter 4 introduces input/output devices and various programming techniques for their operation. The interfacing of microcomputers is of central importance when they are embedded within larger systems. For this reason two further chapters are devoted to digital and analog interfacing, with examples covering commonly encountered problems. The importance of transducers in the control and instrumentation of physical systems is often overlooked, and in this book a whole chapter has been devoted to actuators and sensors for commonly encountered quantities such as position and temperature. As systems become larger, the communication of information between microcomputers becomes progressively more important, and Chapter 8 is dedicated to the subject of data communication. The final chapter deals with the problems encountered in the development and maintenance of microcomputer-based systems. This is another area which is often given scant cover, yet topics such as estimating the amount of effort which a system will require for its development, and design for testability, are of extreme importance in microcomputer-based design.

This book assumes that the reader has only a very basic knowledge of electrical theory such as Ohm's law and simple networks. It is aimed not only at students reading more obviously 'microcomputer related' courses such as computer science and electronic engineering, but at students reading any of the physical sciences or branches of engineering at university or polytechnic, practising engineers, and even the computer hobbyist who is interested in the serious application of the microcomputer.

The publishers wish to thank the following for permission to reproduce material: Digital Equipment Co. Ltd for Figure A.3 and examples from the LSI11 Instruction Set (pp. 425-9); INMOS Ltd for Figure 8.13; Motorola Inc. for programming models of the 6809 and 68000 microprocessors, and examples from their Instruction Sets (pp. 425-9); Rockwell International for Figures 4.19 and A.1 and examples from the 6502 Instruction Set (pp. 425-9); and Zilog (UK) Ltd for Figure A.2 and examples from the Z80 Instruction Set (pp. 425-9).

Finally, the authors wish to record their thanks to Derek Chetwynd for his help in the writing of Chapter 7, and to Andrew McGettrick and Jan van Leeuwen for their valuable comments which caused the book to expand considerably from its originally planned size.

Warwick
May 1985

J.F. Craine
G.R. Martin

CONTENTS

Preface v

Chapter 1 Introducing the Microcomputer 1

    1.1   Microcomputers 1
    1.2   The silicon chip 2
    1.3   From computer to microcomputer 3
    1.4   Systems using microprocessors 4
    1.5   Using a microcomputer 7
    1.6   Communicating with a computer 9
    1.7   Information types 10
    1.8   A closer look at data representation 12
    1.9   Representing data in a computer 17
    1.10 Microcomputer circuits 21
    1.11 Some real applications of the microcomputer 29
    1.12 Questions 30

Chapter 2 Processing Information 31

    2.1   Programming languages 31
    2.2   Algorithm design 32
    2.3   BASIC, a simple programming language 34
    2.4   Extensions to 'Minimal BASIC' 48
    2.5   A second language: Pascal 53
    2.6   Design examples 65
    2.7   Questions 77

Chapter 3 Inside the Microprocessor 78

    3.1   Processing information 79
    3.2   Assignment statements 83
    3.3   Addressing modes 88
    3.4   Data types 94
    3.5   Data handling in other microcomputers 99
    3.6   The processor status register 103
    3.7   Controlling program flow 105
    3.8   Stacks 126
    3.9   Subroutines 128
    3.10 Multiplication and division 131
    3.11 Input/output instructions 134
    3.12 Linking programs in different languages 137
    3.13 Design examples 137
    3.14 Questions 144

Chapter 4 Input/Output Techniques — 146

- 4.1 The structure of an interface — 146
- 4.2 Single-bit input/output — 152
- 4.3 Handshaking — 154
- 4.4 Interrupts — 162
- 4.5 Counting — 170
- 4.6 Direct memory access — 175
- 4.7 A comparison of input/output devices — 178
- 4.8 Design example — 180
- 4.9 Questions — 183

Chapter 5 Digital Interfacing — 186

- 5.1 Transistors — 186
- 5.2 Circuits and signals — 191
- 5.3 Interconnecting logic circuits — 194
- 5.4 Interface circuits for digital inputs — 197
- 5.5 Switch inputs — 205
- 5.6 Interface circuits for digital outputs — 211
- 5.7 Multiplexed outputs — 221
- 5.8 Design examples — 224
- 5.9 Questions — 232

Chapter 6 Analog Interfacing — 234

- 6.1 The structure of an analog interface — 234
- 6.2 The operational amplifier — 242
- 6.3 Digital-to-analog conversion — 249
- 6.4 Analog-to-digital conversion — 258
- 6.5 Analog input circuits — 267
- 6.6 Design example — 269
- 6.7 Questions — 275

Chapter 7 Transducers: Sensors and Actuators — 277

- 7.1 Resistive transducers — 277
- 7.2 Temperature — 281
- 7.3 Light — 284
- 7.4 Position, displacement, and velocity — 288
- 7.5 Acceleration and force — 300
- 7.6 Pressure and flow — 304
- 7.7 Peripheral equipment — 306
- 7.8 Machine vision and speech systems — 318
- 7.9 Design example — 321
- 7.10 Questions — 325

Chapter 8 Machine Communications — 327

- 8.1 Error detection and correction — 327
- 8.2 Serial communications — 331

8.3	Communication between several machines	346
8.4	Standards: handshakes and protocols	352
8.5	Local area networks	358
8.6	Network security	366
8.7	Design example	368
8.8	Questions	378

Chapter 9 System Design and Development — 380

9.1	Designing microcomputer-based systems	380
9.2	Hardware	383
9.3	Software	387
9.4	Software development tools	395
9.5	Test equipment	410
9.6	Design for testability and fault localisation	414

Appendix I A Comparison of Instruction Sets — 421

I.1	Programming models	421
I.2	Single operand data instructions	424
I.3	Dual operand data instructions	425
I.4	Program control instructions	427
I.5	Branch instructions	427
I.6	Manipulating the flag bits	428

Appendix II Symbols — 430

Bibliography — 432

Index — 437

Chapter 1 INTRODUCING THE MICROCOMPUTER

1.1 Microcomputers

The development of the microcomputer during the 1970s brought about a revolution in engineering design. The industrial revolution at the turn of the nineteenth century heralded the development of machines which could replace physical drudgery by mechanical means. Apart from a few exceptions, however, these machines required manual supervision because the problem of controlling this mechanical power was not at all straightforward.

Many types of automatic control system have appeared during the twentieth century, based upon electronic, mechanical, hydraulic and fluidic principles. In each case the design techniques have been similar because each component of the system usually contributes a single well-defined function to the system's behaviour.

The microcomputer represents a fundamentally different approach to the design of a system. Its physical form is quite simple and reliable, consisting of a few general purpose elements which can be programmed to make the system function as required. It is the controlling program which must be designed to give the system the required behaviour, and which will contain 'components' and 'subassemblies' just like any other kind of engineering. The program, or **software**, is just as much a part of the engineered system as the physical **hardware** but it is much less susceptible to failure, provided that it is designed properly.

The idea of programmed systems is not new; electronic computers have been in existence for many decades. However, it has taken the development of the large scale integrated circuit — the **silicon chip** — to produce computers which are cheap, rugged, and reliable enough to be incorporated into engineering designs as components. The techniques of software design are well known to computer scientists and it is not surprising that the principles of good software design and 'software engineering' are essentially those of good engineering design. We shall see in later chapters of this book that engineering design using software allows systems to be designed more easily than using more conventional techniques.

It is the combination of developments in electronic device technology with those in computer technology which has enabled the microcomputer to be produced, and the next two sections will show how these technologies have 'converged' to produce the microelectronic industry which we see today.

1.2 The silicon chip

The basic 'building brick' of modern electronic circuits is the transistor in which an input signal is used to control a larger output signal. The earliest transistors were made in the late 1940s from crystals of a material called germanium, but by the end of the 1950s silicon, the material which is still used for the great majority of electronic devices, was beginning to be used in transistor manufacture. One advantage of silicon is that it allows many of the production steps to be carried out using photographic techniques, and it was soon realised that these techniques could be developed to allow more than one transistor to be fabricated on the same crystal chip of pure silicon. Methods were then discovered for producing other types of electronic component such as resistors, capacitors and diodes on the same chip as the transistors and, importantly, the connections between these components could also be included on the same chip to produce an **integrated circuit**. The fact that the patterns which make up the components of the circuit and the connections between them are all defined photographically means that integrated circuits can be mass produced with very little human labour; there is very little assembly work involved. Conventional circuit assembly techniques are expensive and can be prone to wiring errors.

Integrated circuits were relatively expensive at first and their use was limited to applications where their small size and weight were essential, such as in aircraft and spacecraft. Here another advantage of integrated circuits, that of improved reliability, was important. Many of the failures in conventional electronic circuits occur in the connections between components rather than in the components themselves, and by reducing the number of components the reliability of the system can be substantially increased.

Much of the manufacturing cost of an integrated circuit lies in the packaging of the chip to make a usable electronic component, so that by including as many elements as possible in each integrated circuit the number of packages and connections can be minimised and the cost and reliability optimised. These considerations have led to a steady increase in attainable complexity over the years, with the number of elements which could be incorporated in a single chip almost doubling annually throughout the 1960s and 1970s as manufacturing techniques improved. At the same time, these manufacturing improvements have enabled price reductions to be made, and the price of a given type of integrated circuit often falls by about 30% per annum for the first few years after its introduction.

The main disadvantages of increasing complexity lie in the associated increase in design cost, and the danger that increased specialisation of the circuit will result in a limited range of applications and hence sales. This was already becoming a problem by 1970 and manufacturers became very active in trying to discover large-volume markets for integrated circuits and to define integrated circuit functions which would satisfy as many applications as possible. For example, a counting circuit could be used in an electronic clock or a digital voltmeter, but to fabricate all the electronics for each system on a single chip would require two new designs, each more expensive to produce than that of the simpler but more versatile counter circuit.

Modern integrated circuits (i.c.s) are designed using computer-aided techniques to perform specified functions rather than to have a specified circuit. The economics of i.c. manufacture are such that their design and development are expensive, while production is relatively cheap once the high cost of the manufacturing plant has been recovered. Ideally, an integrated circuit should have a very large number of potential applications so that the development costs can be spread over a large production run. This can be achieved by designing i.c.s which are **programmable** to a greater or lesser degree. The integrated circuits used in mains-driven electronic clocks, for example, can often be programmed to operate on 50 Hz (European) or 60 Hz (American) mains frequencies, and to display the time according to either the 12-hour or 24-hour clock system. In this case four different modes of operation can be obtained from a single design of circuit without incurring the development and inventory costs of four different designs.

1.3 From computer to microcomputer

The development of electronic technology has been the basis of the development of the computer, although the earliest computer antedates electronics by many decades. In fact the choice of candidate for 'first computer' depends very much upon what we choose to regard as a computer; for example one calculating aid, the abacus, has been known since antiquity.

Blaise Pascal invented the first real calculating machine in 1642; this was a purely mechanical device consisting of a set of geared wheels arranged so that a complete revolution of any wheel rotated the wheel immediately to its left through one-tenth of a revolution. In the first half of the nineteenth century Babbage designed programmable calculating 'engines' using mechanical techniques, although computers as we know them today were impractical because of the precision engineering required to construct them. The twentieth century saw the arrival of electromechanical computers, which were developed for specialised military tasks such as code-breaking and gun-aiming. However, these lacked speed and they were soon superseded by fully electronic designs using thermionic valves.

After the Second World War the computer was developed for civilian use as a large and expensive arithmetic-performing machine for business and scientific purposes. The power consumption and unreliability of valves limited the usefulness of computers until valves were ousted by the introduction of the transistor, and new information storage techniques appeared which allowed smaller and more powerful computers to be produced.

The 1960s saw the introduction of the **minicomputer** which was small and cheap enough at a few thousands of pounds — or dollars — to be produced in relatively large numbers for industrial control and laboratory instrumentation purposes. The minicomputer became yet cheaper when integrated circuits were introduced. As these circuits became more complex the number of integrated circuits required to construct a functioning computer fell, until simple minicomputers using only one or two printed circuit boards became possible. These minicomputers enabled significant reductions to be made in the time needed to design, develop, and commission many projects requiring large amounts of electronic logic.

Unlike fixed electronic logic circuits, a computer is a 'do-anything' machine which can be programmed to perform any of a wide range of tasks. The same **physical hardware** can be used in a number of different applications with **software (programs)** defining the behaviour of the system in each case.

More recent developments in integrated circuit technology have led to the introduction of **microcomputers;** small computers fabricated using relatively few integrated circuit components. In fact an entire microcomputer can be made as a single chip. At the heart of any computer is a **Central Processing Unit** or **CPU**, and the corresponding heart of the microcomputer is the **microprocessor** or **MPU** (MicroProcessor Unit), which is simply a CPU implemented on a silicon chip. Its processing power is greater than that of its giant predecessors and yet it is cheap and robust enough to be treated as simply another engineering component.

The microcomputer was conceived as a device which could be programmed in a very flexible fashion to give almost any desired behaviour by means of a list of electronic instructions. Using a microcomputer involves programming skill in producing these lists of instructions as well as more conventional electronic and mechanical design techniques. As its name suggests, the microcomputer is organised in much the same way as a conventional computer; indeed, it may be regarded as the 'natural' outcome of the 'evolution' of the computer from its earliest days.

1.4 Systems using microprocessors

Electronic systems are used for handling information in the most general sense; this information might be a telephone conversation, instrument readings or a company's accounts, but in each case the same main types of operation are involved: the processing, storage and transmission of information. In conventional electronic design these operations are combined at the functional level: for example a counter, whether electronic or mechanical, stores the current count and increments it by one as required. A system such as an electronic clock which employs counters has its storage and processing capabilities spread throughout the system because each counter is able to store and process numbers.

Present day microprocessor based systems depart from this conventional approach by separating the three functions of processing, storage, and transmission into different sections of the system. This partitioning into three main functions was devised by von Neumann during the 1940s, and was not conceived especially for microcomputers. Almost every computer ever made has been designed with this structure, and despite the enormous range in their physical forms, they have all been of essentially the same basic design.

In a microprocessor based system the processing will be performed in the **microprocessor** itself, the storage will be by means of **memory** circuits and the communication of information into and out of the system will be by means of special **Input/Output (I/O)** circuits. It would be impossible to identify a particular piece of hardware which performed the counting in a microprocessor based clock because the time would be stored in the memory and incremented at regular intervals by the microprocessor. However, the software which defined the system's behaviour would contain sections that performed as counters. This apparently rather abstract approach to the **architecture** of the microprocessor and its associated circuits allows it to be very flexible in use, since the system is defined almost entirely in software. The design process is largely one of **software engineering**, and the similar problems of construction and maintenance which occur in conventional engineering are encountered when producing software.

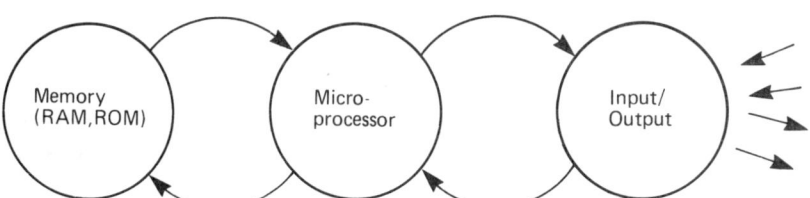

Figure 1.1 Three sections of a typical microcomputer

Figure 1.1 illustrates how these three sections within a microcomputer are connected in terms of the communication of information within the machine. The system is controlled by the microprocessor which supervises the transfer of information between itself and the memory and input/output sections. The external connections relate to the rest (that is, the non-computer part) of the engineering system.

Although only one storage section has been shown in the diagram, in practice two distinct types of memory ('ROM' and 'RAM') are used, as will be seen in section 1.10.5. In each case the word 'memory' is rather inappropriate since a computer memory is more like a filing cabinet in concept; information is stored in a set of numbered 'boxes' and it is referenced by the serial number of the 'box' in question.

The microprocessor processes data under the control of the program, controlling the flow of information to and from memory and input/output devices. Some input/output devices are general purpose types while others are designed for controlling special hardware such as disc drives or controlling information transmission to other computers. Most types of I/O device are programmable to some extent, allowing different modes of operation, while some actually contain special purpose microprocessors to permit quite complex operations to be carried out without directly involving the main microprocessor.

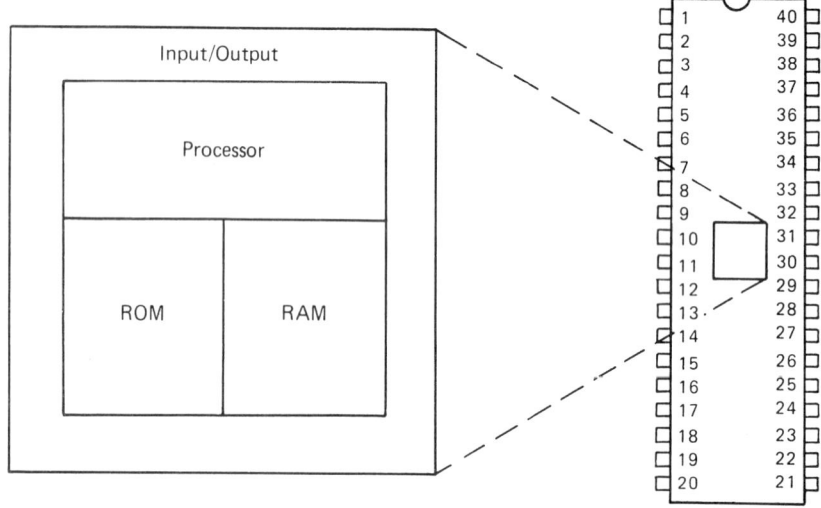

Figure 1.2 Sections of a single-chip microcomputer

The microprocessor, memory, and input/output circuit may all be contained on the same integrated circuit provided that the application does not require too much program or data storage. This is usually the case in low-cost applications such as the **controllers** used in microwave ovens and automatic washing machines. The use of a single package allows considerable cost savings to be made when articles are manufactured in large quantities. As technology develops, more and more powerful processors and larger and larger amounts of memory are being incorporated into **single chip microcomputers** with resulting savings in assembly costs in the final product. For the foreseeable future, however, it will continue to be necessary to interconnect a number of integrated circuits to make a microcomputer whenever larger amounts of storage or input/output are required.

Another major engineering application of microcomputers is in process control. Here the presence of the microcomputer is usually more apparent to the user because provision is normally made for programming the microcomputer for the particular application. In process control applications the benefits of fitting the entire system on to a single chip are usually outweighed by the high design cost involved, because this sort of equipment is produced in smaller quantities. Moreover, process controllers are usually more complicated so that it is more difficult to make them as single integrated circuits. Two approaches are possible; the controller can be implemented as a general purpose microcomputer rather like a more robust version of a hobby computer, or as a 'packaged' system, like that shown in Figure 1.3, designed for replacing controllers based upon older technologies such as electromagnetic relays. In the former case the system would probably be programmed in a conventional programming language such as the ones to be introduced in the next chapter, while in the other case a special-purpose language might be used, for example one which allowed the function of the controller to be described in terms of relay interconnections. In either case programs can be stored in RAM, which allows them to be altered to suit changes in application, but this makes the overall system vulnerable to loss of power unless batteries are used to ensure continuity of supply. Alternatively programs can be stored in ROM, in which case they virtually become part of the electronic 'hardware' and are often referred to as **firmware**.

Figure 1.3 Purpose-designed process controller

More sophisticated process controllers require minicomputers for their implementation, although the use of large scale integrated circuits 'blurs' the distinction between mini- and microcomputers. Products and process controllers of various kinds represent the majority of present-day microcomputer applications, the exact figures depending upon one's interpretation of the word 'product'. Virtually all engineering and scientific uses of microcomputers can be assigned to one or other of these categories.

1.5 Using a microcomputer

Having discussed the range of applications and principles of operation of the microcomputer, it is pertinent to ask what advantages and disadvantages result from using microcomputer technology in preference to more conventional techniques such as electromechanical and hardwired electronic logic.

1.5.1 The advantages

The first advantage has already been mentioned; the large-scale integration of electronic systems has reduced the number of components which are used, leading to an increase in the overall reliability of the system and a reduction in assembly costs. The decrease in size which results from large-scale integration means that equipment based upon microcomputers is usually much smaller, lighter, and more robust than that using older technologies.

Microcomputers can be made in large quantities because they are general purpose devices, and this leads to a much lower unit cost when compared with more conventional methods of producing systems with similar complexity. This standardisation can be extended to the printed circuits on which the integrated circuits are mounted, and very sophisticated microcomputer systems can be purchased as single printed circuit boards at quite reasonable cost. The use of standard components also offers the possibility of standard test fixtures for use in fault diagnosis.

It is the program which defines the function of a microcomputer based system and usually it is in this program that most of the system design is carried out. Many concepts which are familiar to the engineer are found in software engineering, such as the need for modular design and the need to design for **testability** and **maintainability**, topics which will be covered later in this book. In fact, the presence of a microcomputer in a system can ease the problem of fault finding if the possibility of such faults has been anticipated by the designer, since the microcomputer may be used to provide diagnostic information or even to diagnose the precise nature of the fault. However, time dependent faults are less easily detected, and may become apparent only when the equipment has been operating for many hours.

1.5.2 Some disadvantages

The microcomputer might appear at this point to have been presented as something of an engineer's panacea, but unfortunately there are also some disadvantages which arise from using a microcomputer.

The first of these disadvantages is mainly of interest in purely electronic systems; microcomputer logic is slower than hardwired electronic logic by a factor of perhaps 100. In many applications, for example those which have mechanical interfaces, this is not important because the speed of response is limited by external factors rather than the speed of the microcomputer itself.

The problem of fault finding in a malfunctioning microcomputer system is more serious. The fault could be in the electronic hardware, for example a 'dirty joint' in a soldered connection, or it might be due to a programming error, and faults which involve interactions between hardware and software are especially inscrutable. Conventional fault-finding techniques and instruments such as oscilloscopes are frequently useless in such cases, but fortunately methods for locating and identifying faults in equipment have been developed, and some of these will be described in Chapter 9.

A major factor in the use of any engineering technology is the availability of suitable expertise, and this is no less true in the case of the microcomputer. In the past most engineers have been unfamiliar with computer techniques while computer staff have known little of engineering practice, but the arrival of the microcomputer means that each must begin to learn something of the other's problems. The microcomputer presents new demands — or challenges, depending upon one's attitude!

1.6 Communicating with a computer

Conventional computers are intended to be used as tools: they process information provided to them by the user and return it when processed in a form which he or she can understand. This need to communicate with a machine so different from ourselves brings with it the need for a 'language' which both computer and user can understand. This language requires a vocabulary and a grammar just like our own languages, and using a suitable computer language the computer can be programmed to carry out a series of tasks automatically.

The form of the language and its vocabulary depends upon the purpose for which the computer is being used. A user might load a game into a hobby computer with a command such as

    LOAD "DRAGONS"

which is sufficiently like standard English to be easily intelligible to a casual user. However, the game called 'DRAGONS' must itself have been programmed at some time in a language suitable for the purpose of writing games.

In engineering, computers are used not only for running programs which accept information from the user and present the results in a form understandable to humans: computer-aided design and stock control programs, but for running programs which monitor and control engineering systems. In the latter group, applications require special purpose interfaces which allow the microcomputer to interact with its environment in **real time**, so that the activity in the program keeps pace with events outside the computer. Microcomputers are often used in such real-time applications because of their low cost and robustness, and in this case they behave as components of a system rather than as tools. Frequently the identity of the microcomputer is concealed within the system as a whole, and they are referred to as **embedded systems**.

Writing real-time programs is more difficult than most other types of programming. Often the programs must be written using another microcomputer more suited as a program development 'tool'. The first level of communication with larger computers and microcomputers such as might be used for this purpose is provided by an **operating system**. This is a program that allows the user to load programs which have previously been written, to modify them, to translate them into the form used by the computer itself, to run them, to save them for later use, and to allow the computer to carry out any other task which it may be called upon to perform.

Another language may be used to write the actual program to be used for a particular application, or the operating system and programming language may be the same thing, as is the case in simpler hobby computers. Where embedded systems are being developed, another language will almost certainly be used, because languages such as BASIC, beloved of small computer users, detested by many other computer users, and introduced in the next chapter, are not particularly suited to programming embedded systems.

Languages are used to convey meaning. The information carried by human languages can be varied and even rather vague in nature. The languages used by computers are much more precise, and before continuing to discuss programming in the next two chapters, the nature of the information which we require a computer to handle and the way in which it is represented must be discussed.

1.7 Information types

Although it has already been explained that computers exist to manipulate data in various ways, the various forms which this information can take have not yet been discussed. These forms can be divided into two broad categories.

Information of the first type can take one of only a fixed number of states. It can represent a quantity, for example the number of items which have been produced in a batch by a machine, or a pattern such as the settings of some switches on its control panel. This kind of information is often referred to as being **digital** because each value which it can take may be assigned an identifying number. Other examples of this type of information are provided by the names of the twelve months of the year and the number of different ratios which may be selected in a gearbox. In digital computers the signals representing the information being handled can likewise take only a finite number of values or **states**.

The second type of information is continuously variable and can therefore take an infinite number of possible values. Temperature and pressure are just two of the many examples of this type of information which are encountered in engineering and scientific applications. These continuously variable quantities are called **analog** quantities, because in principle they can be represented electronically only by means of a voltage or current which can also vary continuously. Microcomputers are essentially small digital computers, and analog signals must be **digitised** before they can be processed. This important task of converting between analog and digital forms will be discussed in Chapter 6.

There are many ways in which information can be represented within a microcomputer. These are known as **data types** and part of the job of the programmer is to select the correct data type and to ensure that only meaningful operations can be carried out on objects of each type.

1.7.1 Representing integer data

The first of these data types is known simply as **integer**. It allows a finite range of integer numbers to be represented. These numbers can be identifying numbers, for example telephone numbers, as well as quantities such as counts of items. The internal representation of such numbers is the same whether they represent quantities or not. The difference lies in the way in which the computer is programmed to handle them. This is similar to the way in which the same set of digits can represent a quantity such as the distance indicated on a vehicle's mileage counter, or an identification such as a telephone number.

1.7.2 Non-numerical data

Many types of non-numerical quantity can take one of only a limited number of values. The current month of the year, for example, can have one of twelve possible values which are given names from January to December. The four suits in a pack of cards — clubs, diamonds, hearts and spades — provide another example of this type of information. In either case the function of the names or symbols is to enable the months or cards to be distinguished.

These could each be replaced by an identifying number, making it easier for a microcomputer which would identify each name or symbol by an integer. Thus a range of four (integer) numbers will allow the suit of a card to be identified (1 for clubs, 2 for diamonds, and so on), and a range of twelve numbers will permit any month to be represented. Although this use of numbers instead of names is a nuisance, it is not an inconvenience which has been conceived especially for the use of computers. Numbers have long been used to identify months as a form of 'shorthand'.

1.7.3 Representing characters

The digits, letters, and other symbols which are used for communicating with computers represent a special case of non-numerical data. The alphabet, for example, contains 26 letters, each of which exists in two forms, and most computer keyboards allow at least 64 different symbols to be generated.

In principle, any code could be used to represent any symbol, but international standards have been agreed in order to allow communication between different computers. The actual patterns which are used to represent each of these symbols are the same ones which are used elsewhere to represent numerical values; this does not matter, because the program will treat the pattern either as a value or as a symbol according to the context. From the human point of view, it is easier to think of the pattern representing the letter 'A' as the thing which produces an 'A' when printed or displayed, rather than as the corresponding number 65.

1.7.4 Representing real data

Ideally, the numbers which are used to represent analog quantities in digital machines would have an infinite number of possible values like the quantities which they represent. This is the property of **real numbers** in mathematics, which are fundamentally different from integers because in general they require an infinite number of decimal places to allow them to be represented exactly. Computers can handle real numbers, but only to some predetermined accuracy, so that there will be an uncertainty associated with any result of a computation involving real numbers. Most microcomputers represent real numbers with a precision of about six or eight significant figures, which is something of a compromise. Increased accuracy can be obtained at the cost of an increase in storage requirements within the computer and a reduction in the speed at which calculations can be carried out. There is also a maximum and minimum value which can be represented using any system.

1.8 A closer look at data representation

Microcomputers can be programmed using either high level languages (HLLs) or an assembly language. Chapter 2 will be devoted to HLLs, which allow programs to be written in a form that should be independent of the machine on which they run. For this reason they usually conceal the way in which data is stored within the microcomputer, and the programmer need only be concerned with the fact that a limited number of data types is available. Assembly languages are specific to the type of microcomputer in use and will be discussed in Chapter 3.

1.8.1 Representing numerical data

We count in terms of units, tens, hundreds, and higher powers of ten, probably because we normally have a total of ten fingers and thumbs. There is nothing special about the number ten, however, and if we had had a different number of fingers and thumbs — digits — we might have found ourselves counting in terms of eights or twelves, or some other **number base** or **radix**. However, before exploring some of the other number systems which might prove useful to the user of microcomputers, let us take another look at the **decimal number system** in everyday use.

If we write down a number such as 4731, we assume without thinking that it is a 'shorthand' form for saying 'four thousands, seven hundreds, three tens and one unit'. This method of just writing down the multiplying digits without the corresponding powers of the radix is called **positional notation**. Both the order in which the digits are written and their values are important; the base to which the number is represented is usually written as a subscript at the end of the number unless the number is decimal or the base is indicated in some other way. In our example we have assumed that the digits are used to multiply appropriate powers of ten

$$4731 = 4 \times 1000 + 7 \times 100 + 3 \times 10 + 1 \tag{1.1}$$

In positional notation the position of a digit within a number determines its importance or 'weight'. The left-most digit is called the **most significant digit (MSD)**, and the right-most digit is called the **least significant digit (LSD)**.

1.8.2 Binary notation

It is much easier to design electronic circuits which use only two possible states rather than the ten used in decimal notation. These states arise from the electronic switches within the circuit being **on** or **off**, conducting or non-conducting. Similarly, the signals used within the circuit each have only two possible voltage levels: **low** or **high**.

It is convenient for a computer to use a number system based upon the number two, rather than ten with which we are so much more familiar, because each of the signals within the circuit is confined to one of only two possible values: 0 and 1. It does not matter which symbol is assigned to each value; in fact it is impossible to tell which is used by the internal circuitry of a microcomputer. However, it is common for **low** to be represented by the symbol **0** and **high** by **1** in input/output circuits.

Each digit in a binary number is called a **bit**. This word is a rather neat contraction of the words 'BInary' and 'digiT', and the 'bit' also provides us with a unit of information used in information-handling systems. Positional notation is used in binary numbers in the same way as with decimal numbers; thus the binary number

$$1011_2 = 1 \times 8 + 0 \times 4 + 1 \times 2 + 1 \times 1 = 11_{10} \tag{1.2}$$

The left-most bit is referred to as the most significant bit (**MSB**) and the right-most bit as the least significant bit (**LSB**).

Counting in binary is easy apart from the fact that the numbers soon become quite long, as may be seen in Table 1.1. The tenth power of two is 1024, and it is often convenient to 'round off' this quantity to 1000. This is given the abbreviation 'K'; thus

$$2^{16} = 65536 = 64 \times 1024 = 64K$$

Machines, whether they are mechanical counters or computers, contain **registers**: circuits or mechanisms which can store a fixed number of digits. This means that 'leading zeros' must be inserted at the left of a number to make it a fixed length. A decimal counter with five digits will start at the number 00000 and eventually reach 99999. Similarly, an 8-bit binary counter will start at 00000000 and reach 11111111 after 255 (decimal) pulses have been counted. This means that the range of numbers that can be represented in a register of a given length (number of bits) is limited.

In fact the interpretation which is placed upon the bit pattern in a register is largely at the discretion of the programmer. The number system which

Table 1.1 Natural binary

Decimal	Binary
0	0
1	1
2	10
3	11
4	100
5	101
6	110
7	111
8	1000
9	1001
10	1010

was illustrated above was the series of **natural** unsigned binary numbers, the series of integers starting at zero. This does not allow negative or fractional numbers to be represented, and more sophisticated representations are required. First, however, some alternative ways of writing down binary information will be described.

1.8.3 Words, bytes and nibbles

A set of bits processed as a single entity by a computer is called a computer **word**, and this word consists of a group of binary digits in the same way that an ordinary word consists of a group of letters. In principle the number of bits in a word is arbitrary, but in virtually all modern microcomputer designs this number is a multiple of eight.

A block of eight bits represents a useful amount of information because it can contain 256 different patterns of bits, enough to have a distinct pattern for each of the different printed symbols on this page. Because it is such a useful size, a block of eight bits is given the special name **byte**, pronounced 'bite'. Sometimes a smaller unit is used — the **nibble**! A nibble (sometimes spelt 'nybble') is a block of four bits, enough to represent 16 combinations. For example the digits 0 to 9 can each be represented using a different 4-bit nibble.

A word of information in a microcomputer can thus be subdivided into bytes, thence into nibbles, and at the lowest level, into bits:

```
word:        1101101000111011

bytes:             1011010  00111011

nibbles:        1101   1010   0011   1011

bits:     1 1 0 1 1 0 1 0 0 0 1 1 1 0 1 1
```

1.8.4 Other number bases

Although binary numbers are ideal from the point of view of hardware design, they tend to be very difficult to remember. For this reason other number bases are often used as a kind of shorthand to make working with binary numbers easier. Two are in common use, 8 and 16.

Numbers represented to base 8 are called **octal** numbers, and the range of digits allowable is from 0 to 7. The number 576, for example, represents

$$576_8 = 5 \times 64 + 7 \times 8 + 6 = 320 + 56 + 6 = 382_{10} \tag{1.3}$$

Eight permutations are possible using three bits, and each of these can be represented by an octal digit

$$
\begin{array}{ll}
000 .. 0 & 100 .. 4 \\
001 .. 1 & 101 .. 5 \\
010 .. 2 & 110 .. 6 \\
011 .. 3 & 111 .. 7
\end{array}
$$

All that is necessary to convert a binary number to octal representation is to group the digits of the number into threes

$$
\begin{array}{rllllll}
1101101000111011 & = 1 & 101 & 101 & 000 & 111 & 011 & \text{binary} \\
& = 1 & 5 & 5 & 0 & 7 & 3 & \text{octal}
\end{array}
$$

Note that the octal number actually represents a binary pattern, and is used simply to aid memory.

A more compact way of representing binary numbers is to use the base 16. These are called **hexadecimal** numbers (often abbreviated to **hex** for short) and their range of digits is from 0 to 15. Since only ten numerical symbols (the digits 0 to 9) are available, another six are required to represent the digits with values from 10 to 15. These are provided by the letters A to F. The trick for converting binary to hexadecimal form is to group the digits in fours instead of threes:

$$
\begin{array}{llll}
0000 .. 0 & 0100 .. 4 & 1000 .. 8 & 1100 .. C \\
0001 .. 1 & 0101 .. 5 & 1001 .. 9 & 1101 .. D \\
0010 .. 2 & 0110 .. 6 & 1010 .. A & 1110 .. E \\
0011 .. 3 & 0111 .. 7 & 1011 .. B & 1111 .. F
\end{array}
$$

$$
\begin{array}{rllll}
1101101000111011 & = 1101 & 1010 & 0011 & 1011 & \text{binary} \\
& = D & A & 3 & B & \text{hexadecimal}
\end{array}
$$

Hexadecimal notation offers the advantage that it allows each nibble to be represented by a single symbol.

1.8.5 Converting decimal numbers to binary form

Any decimal number can be converted to another base by a simple algorithm. The number is divided repeatedly by the base to which it is required to convert and the remainder of each division is noted. The successive remainders make up the digits of the number to the new base. For example, to convert the decimal number 382 to binary, the method proceeds as

Table 1.2 Conversion to binary

	Quotient	Remainder
382/2 =	191	+ 0
191/2 =	95	+ 1
95/2 =	47	+ 1
47/2 =	23	+ 1
23/2 =	11	+ 1
11/2 =	5	+ 1
5/2 =	2	+ 1
2/2 =	1	+ 0
1/2 =	0	+ 1

382 decimal = 101111110 binary

shown in Table 1.2.

This method is rather longwinded but it can be speeded up by converting to octal or hexadecimal form directly, by dividing by 8 or 16 instead of by 2:

	Quotient	Remainder
382/8 =	47	+ 6
47/8 =	5	+ 7
5/8 =	0	+ 5

382 decimal = 576 octal

	Quotient	Remainder
382/16 =	23	+ 14 (use 'E')
23/16 =	1	+ 7
1/16 =	0	+ 1

382 decimal = 17E hexadecimal

To check that each approach produces the same binary number, write each answer as a binary number:

$$576 \text{ octal} = 101\ 111\ 110 \text{ binary}$$
$$17E \text{ hex} = 1\ 0111\ 1110 \text{ binary}$$

The base to which a number is expressed is often written as a subscript to avoid ambiguity. Thus

$$101111110_2 = 576_8 = 382_{10} = 17E_{16} \tag{1.4}$$

1.9 Representing data in a computer

The 'natural' binary number system allows only non-negative integers to be represented. However, the binary pattern within a word can be interpreted in various ways and other conventions will be introduced in this section which allow a wide range of data types to be stored.

1.9.1 Negative numbers

Negative numbers are represented in decimal notation by a minus sign in front of the corresponding positive number. An equivalent method is sometimes used to represent binary numbers, except that the sign is denoted by a single bit which is 0 for positive numbers and 1 for negative numbers. However, this **sign-magnitude representation** presents some problems in the design of microcomputers (for example, there are two ways of representing zero, $+0$ and -0), and a different method is preferred for use in microcomputers.

In **two's complement representation** the left-most bit in a word is interpreted as a sign, and the way in which two's complement notation works can be explained by considering a rather improbable analogy.

Suppose that a driver were to look at the distance travelled as he reversed a brand new car. The mileage indicator would show the successive numbers 99999, 99998, and so on. This does not mean that the car has travelled 99999 miles of course, because the indicator must be interpreted differently. 99999 really means -1, 99998 means -2, and so on. The rule for converting a ten's complement number to its true meaning is quite simply to subtract it from 100000 (one more than the maximum displayable number 99999).

The same principle is used with two's complement numbers. If an 8-bit counter counts back from 00000000 it will show the successive numbers 11111111, 11111110, and so on. Again the rule is simple; subtract the number from the binary number 100000000 which is one more than the maximum number that the counter can hold. In fact there is an easy way to do this; starting with the original binary number, change each 1 to a 0 and each 0 to a 1 to create a pattern of bits called the **one's complement**, and then add one to the result to obtain the required two's complement number. The most negative number that can be represented using two's complement notation and eight bits is 10000000 (decimal -128), while the most positive number possible (decimal $+127$) is 01111111.

Note that the way in which the bit pattern is interpreted depends upon the convention being used; if the left-most bit were used to represent 128 (two to the power seven) rather than a sign, 01111111 would still mean 127, but the pattern 10000000 would be interpreted as 128.

1.9.2 Fractional numbers

In decimal numbers, digits to the left of the decimal point are assumed to be multiplied by successive positive powers of ten, while those to the right of the point are assumed to be multiplied by successive negative powers. Thus the number 39.37 means

$$39.37 = 3 \times 10 + 9 \times 1 + 3/10 + 7/100 \tag{1.5}$$

Binary fractions can be constructed in the same way, where digits to the right of the binary point are assumed to be multiplied by successive negative powers of two. The binary fraction 101.011 thus represents

$$101.011 = 1 \times 4 + 0 \times 2 + 1 \times 1 + 0/2 + 1/4 + 1/8 \tag{1.6}$$

In computers the binary point cannot be represented directly but is assumed to be at a certain place in the sequence of digits. Sometimes this place is fixed, for example, at the centre of each group of eight bits. Systems like this are somewhat restricted in their range of applicability, and so-called **fixed point** fractional number representations are not widely used.

Alternatively, the position of the binary point can be allowed to move with respect to the digits. The result is something like the 'scientific notation' method of writing down large or small numbers:

$$0.0003937 = 3.937 \times 10^{-4} \tag{1.7a}$$

$$39370 = 3.937 \times 10^4 \tag{1.7b}$$

In each case the effective position of the decimal point has been moved by a 'scaling factor' written with the number 3.937. This can be done by representing a quantity with a pair of binary numbers; a **mantissa** which contains the sign and the first few digits of the number, to whatever accuracy is needed, and an **exponent** which contains the scaling factor as a power of 2 (or 16 in many systems). This type of representation is called **floating point**.

1.9.3 Binary coded decimal numbers

The conversion between binary and decimal notation, while not difficult, sometimes represents a major part of the computation involved in relatively simple applications which require a lot of decimal information to be entered and displayed, such as calculators. The problem can be circumvented, however, by representing numbers in decimal form internally, without taking the trouble to convert them fully to binary. The result is a form of representation where each decimal digit is represented by a four bit number in the range 0 to 9, but the binary code for each digit is stored without further conversion. This increase in convenience is achieved at the cost of efficiency both in terms of storage (a single byte can contain one of only 100 values instead of 256 values) and in terms of program speed, since the use of BCD can sometimes lead to slower program execution. The number 2768 would be represented in **Binary Coded Decimal (BCD)** thus

2=0010, 7=0111, 6=0110, 8=1000

Hence 2768 is represented as 0010 0111 0110 1000

1.9.4 Characters and strings

Another type of information which computers are frequently called upon to handle is in the form of text, which consists of sequences or **strings** of symbols or **characters**. Microcomputers therefore need some means of representing characters internally. In principle any pattern of bits could be assigned to any character but fortunately a standard has been evolved which sees universal adoption among microcomputer systems. The full standard allows for the use of 128 different codes, 95 of which correspond to different symbols such as numerals, letters of the alphabet, and punctuation marks. The remainder are used for a range of control tasks such as positioning the characters on the output medium, supervising communication links and remotely switching equipment off and on. This standard, known as the 'American Standard Code for Information Interchange', or **ASCII**, is listed in part in Table 1.3, which shows the hexadecimal equivalents of the printable characters.

Table 1.3 Printable ASCII characters

Second Digit	First Hexadecimal Digit					
	2	3	4	5	6	7
0	space	0	@	P	`	p
1	!	1	A	Q	a	q
2	"	2	B	R	b	r
3	#	3	C	S	c	s
4	$	4	D	T	d	t
5	%	5	E	U	e	u
6	&	6	F	V	f	v
7	'	7	G	W	g	w
8	(	8	H	X	h	x
9	)	9	I	Y	i	y
A	*	:	J	Z	j	z
B	+	;	K	[	k	{
C	,	<	L	\	l	|
D	-	=	M	]	m	}
E	.	>	N	^	n	~
F	/	?	O	_	o	delete

ASCII is not the only system in use with computers, and a standard code is unnecessary if the designer can be certain that the computer will never be used with other equipment. Although virtually all microcomputers use the ASCII code, other codes are still used in larger computers, for example the Extended Binary Coded Decimal Interchange Code (EBCDIC).

1.9.5 Logical operations

As stated earlier, one advantage of using a binary representation of information is that it allows 'propositions' in formal logic to be expressed concisely. Shannon pointed out in 1938 that switching circuits could be used to evaluate logic statements. In other words, the dualities of on or off, 1 or 0, could be made to stand for the duality of **true** or **false**. It is conventional when using computers to employ a binary **1** to represent **true** and binary **0** to represent **false**.

Three logical operators, **OR**, **AND**, and **Exclusive OR** ('EOR' or 'XOR') combine two binary digits and produce a single digit as a result. The fourth operator, **NOT,** which complements a single binary digit, will be discussed first.

The NOT operation
The effect of the operation **NOT** is to reverse the validity of a statement; if something is not true, it is false, while if it is not false it is true

$$\text{NOT } 1 = 0 \qquad \text{NOT } 0 = 1$$
i.e. NOT true = false   NOT false = true

The AND operation
The operation **AND** is defined by the statement 'If A and B are both true then the result is true, otherwise it is false'. The table below shows the effect of 'ANDing' two quantities; remember that 1 represents true and 0 represents false

$$\begin{array}{ll} 0 \text{ AND } 0 = 0 & \text{false AND false} = \text{false} \\ 0 \text{ AND } 1 = 0 & \text{false AND true} = \text{false} \\ 1 \text{ AND } 0 = 0 & \text{true AND false} = \text{false} \\ 1 \text{ AND } 1 = 1 & \text{true AND true} = \text{true} \end{array}$$

The OR operation
This is defined by the statement 'If A or B, or both, are true then the result is true, otherwise it is false'. If two binary digits are combined using the **OR** operator ('ORed'), there will be two possible outcomes

$$\begin{array}{ll} 0 \text{ OR } 0 = 0 & \text{false OR false} = \text{false} \\ 0 \text{ OR } 1 = 1 & \text{false OR true} = \text{true} \\ 1 \text{ OR } 0 = 1 & \text{true OR false} = \text{true} \\ 1 \text{ OR } 1 = 1 & \text{true OR true} = \text{true} \end{array}$$

Interestingly, but not obviously, the OR function can be expressed using only AND and NOT

$$A \text{ OR } B = \text{NOT (NOT } A \text{ AND NOT } B) \tag{1.8}$$

1.9.5

The Exclusive OR operation

The **Exclusive OR** operation corresponds to the way in which the word 'or' is used in everyday English. In this case the outcome is true if one statement is true, or the other is true, but not if both, or neither are true. In fact this operation is not always explicitly available in high level languages, but it can be programmed using the other operations already introduced

$$A \text{ EOR } B = (A \text{ AND NOT } B) \text{ OR } (B \text{ AND NOT } A) \tag{1.9}$$

The four possible ways in which two logical propositions can be combined using 'Exclusive OR' are

```
0 EOR 0 = 0    false EOR false = false
0 EOR 1 = 1    false EOR true  = true
1 EOR 0 = 1    true  EOR false = true
1 EOR 1 = 0    true  EOR true  = false
```

1.10 Microcomputer circuits

The circuits which make up the microcomputer operate with voltage and current signals which interact according to the rules of the operations just described. The behaviour of simple circuits using switches in series or parallel can be described using AND and OR; in Figure 1.4, a current flows in lamp 1 if switch 'a' and switch 'b' are closed, and in lamp 2 if switch 'c' or switch 'd' is closed.

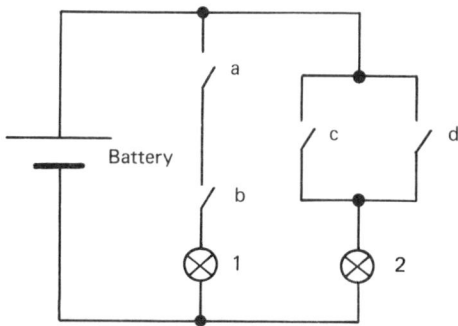

Figure 1.4 Switch logic

These statements can be rephrased as

(Condition for Lamp 1 on) = (Switch 'a' closed) AND (Switch 'b' closed)
(Condition for Lamp 2 on) = (Switch 'c' closed) OR (Switch 'd' closed)

and if the statements in brackets are replaced with symbols a form of algebra results:

$$L_1 = A \text{ AND } B \tag{1.10}$$

$$L_2 = C \text{ OR } D \tag{1.11}$$

This is **boolean algebra**, named after the nineteenth century logician George Boole. Each of the logical operations AND, OR, EOR, and NOT is given a special symbol in boolean algebra, just as the basic arithmetic operations such as addition and subtraction are represented by special symbols. The operation 'AND' is represented by '.', which could be confused with the symbol for multiplication. It is important to bear in mind that boolean algebra operates upon logical statements, not arithmetic quantities. Similarly, 'OR' is denoted by '+', 'Exclusive OR' by '$\oplus$' and 'NOT' by placing a bar above the symbol or expression upon which it operates. Equations 1.10 and 1.11, which described the switch logic, thus become

$$L_1 = A \cdot B \tag{1.10a}$$

$$L_2 = C + D \tag{1.11a}$$

Similarly, the expressions for **OR** and **Exclusive OR** given in equations 1.8 and 1.9 can be rewritten as

$$A + B = (\overline{A \cdot B}) \tag{1.8a}$$

$$A \oplus B = (A \cdot \overline{B}) + (B \cdot \overline{A}) \tag{1.9a}$$

Boolean algebra provides the microcomputer designer and programmer with a 'shorthand' form for describing the operations which the hardware or software is to carry out. Modern computers and microcomputers use transistors as switches, but these transistors are connected to form logical building bricks known as **gates**. A logic gate is an electronic circuit with one or more inputs and one output. The voltages on the inputs and outputs of these gates can each take one of only two values, and the gates are each designed so that the relationships between their input and output levels are determined by a simple boolean expression.

The function of the physical 'hardware' of a microcomputer can be illustrated in terms of a circuit diagram showing gates and their interconnections. Several conventions have been used for indicating the functions of the different types of gates, but the one in most widespread use is based upon the American standard MIL-STD-806B. This is the convention which will be used in the remainder of this book; it indicates the function of each type of gate by means of the outline of the symbol used, as may be seen in Figure 1.5. The key points to note are the use of a straight edge for AND, a curved edge for OR, and a double curved edge for Exclusive OR. The logical operation NOT is denoted by a small 'bubble' where a connection meets a gate.

This book is not primarily concerned with the design of the electronic hardware which makes up microcomputers, because it is usually much more sensible to buy a microcomputer rather than to try and build one. However, a short discussion of the way in which gates are used may lead to an understanding of the basic principles used by microcomputer hardware. The entire circuit of the microcomputer can be expressed in terms of boolean

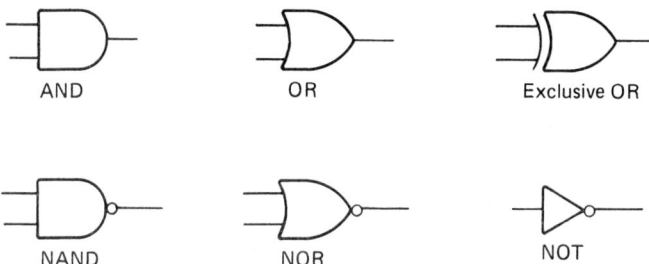

Figure 1.5 Symbols for logic gates

algebra because it is a machine designed to carry out logical operations. This means that its function can be expressed in terms of gates.

A circuit made up of gates will carry out a set of logical operations upon its inputs to produce one or more output signals, and in this way quite complicated operations can be performed. For example, the addition of binary numbers may be carried out by a set of circuits each handling the addition of corresponding binary digits a_i and b_i in the two numbers A and B. Each of these circuits also requires a 'carry' input c_i and produces two output signals: a 'sum' signal s_i and a 'carry' output c_{i+1}.

An analysis of the logic required by each of these circuits shows the boolean expressions for the two outputs to be

$$s_i = a_i \oplus b_i \oplus c_i \tag{1.12}$$

$$c_{i+1} = (a_i.b_i) + (b_i.c_i) + (c_i.a_i) \tag{1.13}$$

These can be implemented using gates as shown in Figure 1.6.

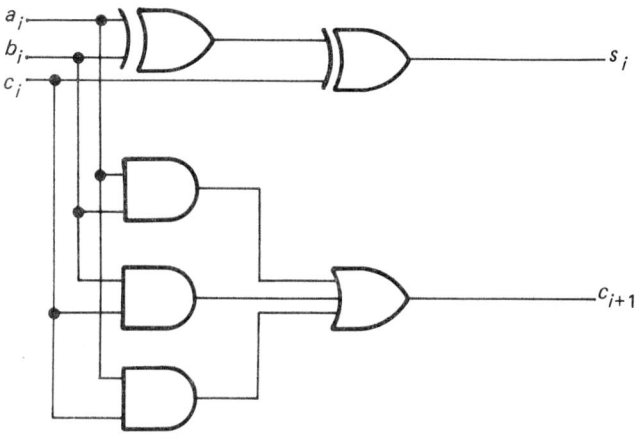

Figure 1.6 Logic for adder circuit

The circuits used in microcomputers are much more complicated than that used in this example, and complexities of thousands or tens of thousands of gates are normal. Not only can all the logic equations used in the design of the circuit of a microcomputer be expressed in terms of the four logical operations already discussed, but each of these four types of operation can be expressed in terms of a single operation called **NAND**.

The NAND operation consists of the operation AND followed by NOT. Although it may appear unnatural, it is easier to produce using integrated circuit electronic techniques. For example, the NOT function requires a NAND gate with a single input, while AND can be obtained if a NAND gate with the requisite number of inputs is followed by another NAND gate acting as a NOT gate. The relationship between OR and exclusive OR and the operations AND and NOT has already been shown.

Although the output voltage of a logic gate depends only upon the current input voltages, gates can be interconnected to form circuits in which the output state depends not only upon the current input signal but past values as well. Consider the circuit of Figure 1.7, which shows two NAND gates connected in a 'cross-coupled' configuration, that is, with the output of each connected to an input of the other.

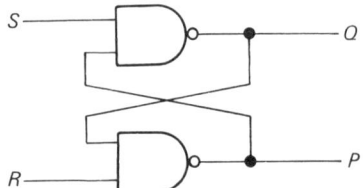

Figure 1.7 Cross-coupled NAND gates

The behaviour of the circuit can be described in boolean algebra as follows:

$$P = \text{NAND}(Q,R) = \overline{Q.R} \tag{1.14}$$

$$Q = \text{NAND}(P,S) = \overline{P.S} \tag{1.15}$$

Eliminating P between the two equations gives

$$Q = \overline{\overline{Q.R}.S} \tag{1.16}$$

which can be manipulated using boolean algebra to give

$$Q = Q.R + \overline{S} \tag{1.17}$$

When both R and S have the value 1, this equation reduces to

$$Q = Q \tag{1.18}$$

which indicates that Q will retain whichever value it currently has. When S has the value 0, Q will be forced to logic 1, while if R has the value 0, Q will be forced to logic 0. The cross-coupled pair of gates provides a simple means of storing information within a microcomputer.

The more complicated circuit of Figure 1.8 provides the designer with an even more useful building brick. The left-most gate inverts the data on input D to give $\bar{D}$, and the pair of gates which follow give output signals described by the boolean expressions

$$R = \overline{\bar{D}.G} = D + \bar{G} \qquad (1.19)$$

$$S = \overline{D.G} = \bar{D} + \bar{G} \qquad (1.20)$$

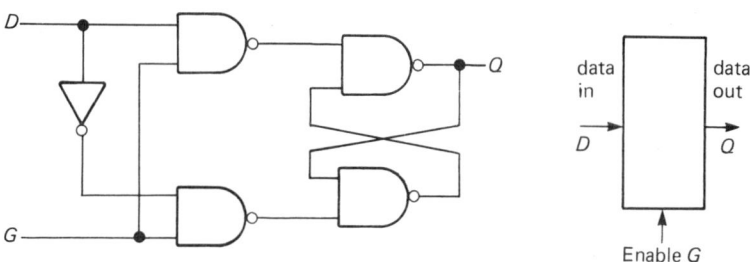

Figure 1.8 Latch circuit and symbol

When these expressions for R and S are substituted into the boolean equation describing the cross-coupled gates and the resulting equation is simplified, the result

$$Q = Q.\bar{G} + D.G \qquad (1.21)$$

shows that while $G=0$, the output Q remains the same irrespective of the input state, but when $G=1$, Q is the same as the input state D. The input G is the **enable input** which determines whether or not information on the **data input** D can be loaded. This circuit is a data **latch** and it provides a means for storing a single bit of information.

The **NOR** operation, consisting of OR followed by NOT, is just as versatile as the NAND operation which has been discussed in the last few pages; in principle, logic circuits of any complexity can be constructed using only NAND or NOR gates.

1.10.1 Registers

This latch circuit is capable of storing a single bit of information, and by connecting a number of similar circuits in parallel with a common enable line a **register** can be produced. Registers are used to hold information within a microcomputer, and they should be thought of as 'boxes' each of which can hold a fixed amount of information, rather than as assemblies of gates. Because an eight-bit register consists of a set of eight latches, it can store any of the 256 patterns which can be represented using eight bits, but only patterns of eight bits, never seven or nine.

Registers make up the bulk of any microcomputer's circuitry and are conventionally represented in diagrams as rectangular boxes with numbers to indicate the register 'length', the number of bits which it contains. Although there are other conventions for numbering the bit positions within registers, the one which is generally preferred is shown in Figure 1.9. The right-most bit is numbered 0, the one immediately to its left is numbered 1, and so on to the left-most bit which for an N-bit register will be bit number $N-1$.

7	6	5	4	3	2	1	0
0	1	1	0	1	0	1	0

Figure 1.9 Bit positions in an 8-bit register

The microcomputer's memory circuit consists of a set of registers which are used simply to store information. Other registers have special functions concerned with inputting and outputting information, or with the internal operation of the microprocessor section of the system.

The ability of the latch to 'remember' a state provides the microcomputer circuit with the ability to carry out sequences of operations. The precise sequence which a microcomputer follows is determined by the program which it contains. Despite its complexity, a microcomputer can be made cheaply because it is manufactured in large quantities, and then 'customised' to a particular application by means of the program.

1.10.2 Buses

Information is communicated between registers along 'highways' or **buses** which can each be regarded as paths with a defined 'width'. In physical terms a bus consists of a number of wires (or metallic tracks on a printed circuit or integrated circuit) and the width is determined by the number of wires used. Note that although it is normal to talk in terms of 'transferring' data from one register to another as if it were some kind of physical material, the action is actually one of copying information and the source is not affected.

System diagrams represent buses as broad arrows, the direction of the arrow indicating the direction in which information can flow. Buses in which the data can be transferred in only one direction are called **unidirectional buses**. Information can pass in either direction along a **bidirectional bus**, which is indicated in diagrams by placing an arrowhead on both ends of the bus. The width of a bus is normally shown by means of a line drawn across it with a number indicating the width against it as illustrated in Figure 1.10.

1.10.3 Addresses

The fact that there is a large number of registers in a microcomputer system means that there must be some method for selecting which one it needs at any instant. This is done by assigning a unique number to each register

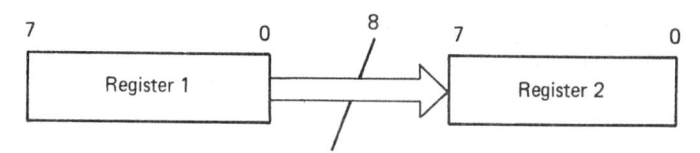

Figure 1.10 8-bit bus connecting two 8-bit registers

called its **address** which is carried on an **address bus**. The system contains circuitry which decodes the electronic signals representing addresses to activate the particular register which is being referenced. In order to distinguish between the registers used within the microprocessor itself and those within the memory circuit, the latter are often referred to as **memory locations**.

The address can be thought of as a sort of map reference in a one dimensional **address space**. The circuits making up a microcomputer each contain sets of registers occupying different locations within this address space and an **address map** can be drawn to show the range of addresses used by each circuit. In some microcomputers the registers in the input/output devices can be addressed in a different way from memory locations, and are described as being located in a separate input/output address space. Some special purpose microcomputers have separate address spaces for program and data, which reduces their flexibility but allows faster operation.

The microprocessor itself contains a number of registers with special functions, and these are often handled differently from the other registers in the system. In most microcomputers they do not have addresses, but in some cases (e.g. the 8048) they are situated at one end of the address space.

1.10.4 The control bus

Further information is needed in addition to the address, to indicate whether information is flowing to or from the register or location, and also the precise time at which the transfer of information is to take place. This control information is carried by the **control bus** and although the precise details of the signals differ from microprocessor to microprocessor, the function of the control bus is the same in every case: to supervise the transfer of information between registers via the data bus.

1.10.5 Microcomputer structure

As was discussed in section 1.4, any electronic system can be thought of as carrying out three main functions upon data: processing, input/output, and storage. In a microcomputer these functions are performed by the microprocessor, input/output, and memory sections respectively, although practical microcomputers use two different kinds of memory.

In each case the word 'memory' is rather inappropriate since a computer memory is more like a filing cabinet in concept; information is stored in a set of numbered 'boxes' and it is referenced by the serial number of the 'box' in question.

One type of memory is 'read-write' or **Random-Access Memory (RAM)** into which data can be **written** and from which data can be **read** again when needed. This data can be read back from the memory in any sequence desired, and not necessarily the same order in which it was written, hence the expression 'random' access memory. The process of reading data from a modern memory circuit is **non-destructive**, so that although a copy of the stored data can be obtained from the memory, the copy kept within the memory is not affected. Most RAMs are **volatile**, which means that any data which has been written into them will be lost if their power supply is switched off or disconnected at any time. So-called **static RAM** circuits will retain data indefinitely provided that the power supply is maintained, but **dynamic RAM** circuits also need to be **refreshed** periodically by means of a special sequence of signals to prevent data from being lost.

Microcomputers also use another type of **Read-Only Memory (ROM)** which is used to hold fixed patterns of information which cannot be affected by the microprocessor; these patterns are not lost when power is removed and are normally used to hold the program which defines the behaviour of a microcomputer-based system. Read only memories can be read 'randomly' like RAMs but unlike RAMs they cannot be used to store variable information. Some ROMs have their data patterns put in during manufacture by a 'mask' or stencil containing the appropriate pattern, and are known as **masked ROMs**, while others are programmable by the user by means of special equipment and are called **programmable ROMs**. The most widely used programmable ROMs are erasable by means of special ultraviolet lamps and are referred to as **EPROMs**, short for **Erasable Programmable Read Only Memories**. Other, newer, types of device can be erased electrically without the need for ultraviolet light. Different physical principles are used by different types, but in each case the information takes much longer to store than to retrieve. Two names are used: **Electrically Erasable Programmable Read Only Memories, EEPROMs** and **Electrically Alterable Read Only Memories, EAROMs**. Unfortunately this proliferation of electronic devices has produced a large number of acronyms such as these and the list of them, although not endless, grows with the passing of the years!

The various sections of a microcomputer may all be incorporated into one integrated circuit or spread over a large number of circuits; their physical form is largely irrelevant to their principles of operation. They are interconnected by the three buses as shown in Figure 1.11. A fourth bus which carries the power to all the circuits is omitted from block diagrams for simplicity.

In normal operation it is the microprocessor that controls the system, generating the address and control signals which allow it to communicate with the input/output and memory sections. The function of the microprocessor is, as its name implies, to process information. It is capable of

1.10.5

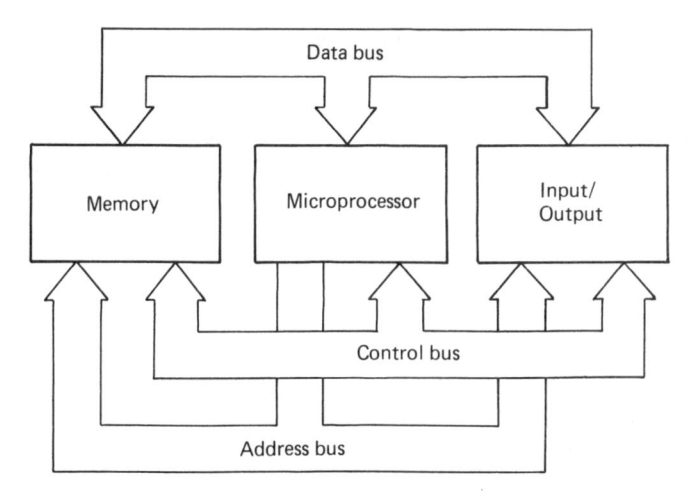

Figure 1.11 The three main buses within a microcomputer

carrying out a wide range of arithmetic and logical operations such as addition and logical 'AND', which it carries out upon all the bits in a register at the same time.

The maximum number n of bits which are processed at the same time is called the **wordlength** of the microprocessor and one with a wordlength of n bits is called an n-bit microprocessor. This wordlength is a factor determining the processing 'power' of a microprocessor. Machines with larger wordlengths are usually more 'powerful' because they are able to process more information at the same time. The earliest microprocessor designs were limited to lengths of only four bits, but this is inconvenient for most applications and longer wordlengths of 8, 16, or even 32 bits are now used.

In most microprocessors the width of the data bus is the same as the wordlength, although this is not necessarily the case. The width of the address bus determines the maximum number of locations which can be selected or 'addressed'. 8-bit microcomputers usually have address bus widths of at least 12 bits, and 16-bit address buses are virtually the rule. The number of different addresses which can be generated using an n-bit address bus is equal to 2 raised to the power n; thus a 16-bit bus allows 65536 different addresses to be represented. Microcomputers with longer wordlengths have 24-bit or even wider address buses and may access over 16 million different locations.

1.11 Some real applications of the microcomputer

In order to illustrate the way in which microcomputers are used in engineering and science, two design examples will be used throughout this book. The first is a simple controller which can be used in applications such as heating water, in which the temperature would be controlled automatically. The second is an instrument for 'capturing' waveforms from laboratory transducers and analysing them.

Each of these systems takes information from the 'outside world' and analyses it. In each case interfaces need to be designed and the microcomputer must be programmed to carry out the appropriate function. The relevant aspects of each system will be examined at the end of each chapter to show the way in which the techniques introduced in that chapter can be used.

1.11.1 Requirements

Later, in Chapter 9, we shall see that any project must commence with a detailed set of requirements which the equipment to be designed must satisfy. From these requirements will be derived a set of specifications for the equipment which will provide the basis of the design. However, the case studies are intended to be of a tutorial nature, the stages of producing formal lists of requirements and specifications will be omitted.

1.12 Questions

1. Convert the decimal numbers 10, 106, and 1066 to their binary equivalents.
2. Convert the binary numbers 10011010, 101000011101, and 1101111010101101 to hexadecimal notation.
3. What are the decimal equivalents of the binary numbers in Question 2?
4. How would the decimal numbers $+5$, $+15$, and -15 be represented using 8-bit two's complement notation?
5. What is the sequence of hexadecimal code needed to represent the letters making up the word 'DRAGON' using ASCII code?
6. Draw a logic circuit to implement the exclusive OR function using the result

    $$A + B = \overline{(\overline{A} . \overline{B})}$$

7. Show that the exclusive OR function can be carried out by means of a circuit consisting solely of NAND gates according to the boolean equation

    $$A \oplus B = \overline{\overline{A.B} . \overline{\overline{A}.B}}$$

    How many NAND gates are needed?
8. Approximately how many gates are needed to implement an 8-bit register?

Chapter 2 PROCESSING INFORMATION

2.1 Programming languages

The function of a computer is to process information according to a **program** or list of operations. Computer programs consist of lists of instructions which must be expressed in a language that is understood by both programmer and computer. Ideally the language used to communicate with a machine should be as 'natural' as possible, but there are many reasons why ordinary English, for example, cannot be used. Human languages have evolved over the course of time to allow people to communicate with each other, and often they do not contain words which would allow us to converse concisely with machines. The richness of meanings associated with the words that we do use in our everyday conversations also allows these languages to be imprecise and ambiguous. This is not usually a problem for us because ordinarily we can resolve any ambiguities without difficulty and because the results of making a mistake tend to be amusing rather than serious. However there are some circumstances such as in legal documents or instruction manuals where the precise meaning of words can be very important.

Special languages with very restricted vocabularies and grammars have been invented for computer programming. In the early days of computing all the languages used were of a type which described what was happening inside the computer in detail but which tended to obscure the overall purpose of the program. Languages like this are called **assembly languages** and they are still used for programming microcomputers. Chapter 3 will explain assembly languages in much more detail because they provide us with an insight into the internal functioning of a microcomputer. However, they are now referred to as being **low level languages** because the trend over the years has been towards languages which appear much more like a human language. These **high level languages** make programming easier by allowing the programmer to describe what the program is to do without having to make reference to the internal operation of the computer itself.

Unfortunately a 'Tower of Babel' has grown with the development of computers, and hundreds of different languages are now in use. Often it is impossible to run programs written for one computer on another one because of this 'language barrier'. Even if two machines do use nominally the same language, they frequently use different 'dialects' which can make them incompatible! One reason for the proliferation of computer languages is that computers are used in a number of specialised fields; the requirements of an engineer programming a machine tool are very different from those of an accountant. The most widely used languages are those which allow mathematics and sequences of logical decisions to be expressed. In this chapter BASIC and Pascal, two languages commonly used with microcomputers, will be introduced and used to illustrate the most important aspects of high level languages.

2.2 Algorithm design

Before a program can be written it must be planned as a series of steps which, when they are followed, will produce the overall effect required. The process is one of breaking a problem into smaller and smaller pieces until a detailed strategy for solving the problem is produced. As an example, consider a vending machine whose task is to dispense drinks. This task can be divided up into a number of smaller tasks: accepting the money, finding out what kind of drink the customer wants, dispensing the drink, and so on. Each of these subtasks can be further divided up into yet smaller ones, such as dropping a cup into position to accept the drink and turning on and off valves to control the flow of hot and cold water.

The program is designed to meet a set of specifications which are derived from a list of requirements that define what that program should do. In this way the production of software is like the production of a wide range of other commodities, and the term **software engineering** is often used to cover the various aspects of software production. The last two chapters of this book look in more detail at the problems to be encountered in designing and implementing microcomputer-based systems.

Usually this process of dividing up the task results in a complicated list of interrelating steps, and various techniques have been devised to make the visualisation of the structure of a program easier. One of the most widely used methods of drawing the operation of a program is the **flowchart** which is used to show the operations to be carried out by the machine and the way in which the program passes from one task to the next.

A flowchart consists of a number of 'boxes' of various shapes which show the tasks to be performed and arrows which indicate the sequence in which they are carried out. The flowcharting standard provides for many different shapes of 'box' symbols (Bycer,1975) for different types of task, but the important ones from the point of view of sketching the flow of a program are the rectangular box used for information processing and the diamond-shaped box which is used to indicate decision points within the program. Figure 2.1 shows a flowchart which might be used when starting the design of a machine to make a cup of instant coffee. It must put in coffee powder, add sugar and powdered milk if they are required, and then add the correct amount of water. The actions to be carried out are shown with rectangular boxes, while the decisions to be made are in diamond-shaped ones. The operation shown in each of the rectangular boxes will need to be analysed in more detail before a program can be written to perform these tasks.

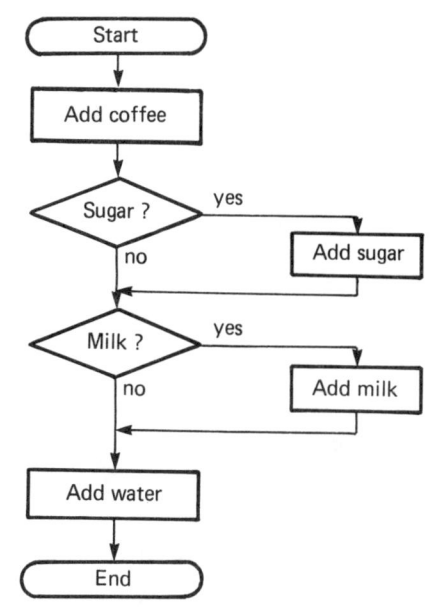

Figure 2.1 Making a cup of coffee

The flowchart is only useful to clarify thinking about the problem, and it is most useful when the flow of control from one part of the program to another is complicated by having many alternative paths as the consequence of many decisions. Flowcharting neither adequately models the program nor supplies any information that would not be found in the program itself. One of the aspects of the 'art' of programming is to produce programs whose operation can be understood readily, and flowcharting is only one of a number of graphical techniques for allowing a programmer to understand the work better.

The **state transition diagram** is another program design aid, and shows the relationship between the various states which a system can enter. This treats the system as a 'finite state machine', a machine which can exist in one of a finite number of possible states and which can switch from one state to another according to a set of rules. This type of description is useful for simple systems where there is little decision-making involved. Figure 2.2 shows a very simple four-state model of the drinks vending machine which could be the starting point for a design.

Later, in Chapter 9, another technique for illustrating the structure of a program will be explained.

Only when the details of this design strategy have been carefully worked out should the program itself be written in a computer language. This

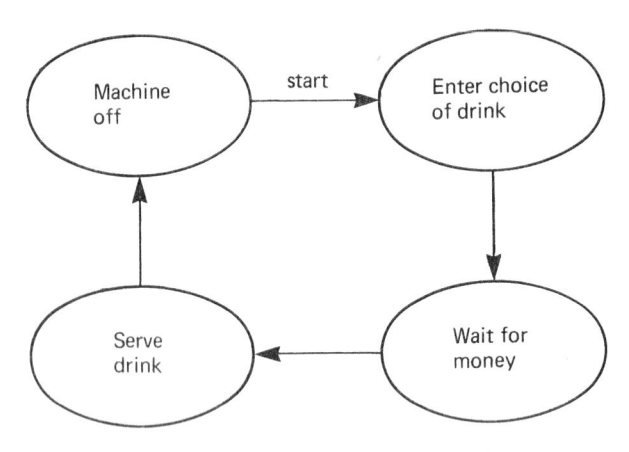

Figure 2.2 State transition diagram

program will be expressed in terms of a sequence of **statements** which are used to define the information to be used, to manipulate this data, and to organise the way in which the program 'flows' from one step to the next.

2.3 BASIC, a simple programming language

BASIC, which is short for 'Beginner's All-purpose Symbolic Instruction Code', was developed during the 1960s as a language for teaching the principles of computer programming, and it provides a useful starting point for a discussion of programming languages. The advent of the microcomputer has seen the widespread use of BASIC as a programming language because it is relatively easy to learn and is adequate for most simple applications. It does, however, have some shortcomings which will be discussed later on.

One problem with BASIC is that it appears in a number of 'dialects', which although they differ only slightly, limit the ability to 'transport' a program written in BASIC from one microcomputer to another. In this section we shall describe 'Minimal BASIC' as defined by the American National Standards Institute (ANSI, 1978).

Each BASIC statement appears on a separate numbered line, and starts with a distinctive keyword such as DATA, LET, PRINT, or GOTO. A statement may merely state something, or it may instruct the computer to do something, in which case it is known as an **executable statement** or an executable instruction. Notice that all the text of a BASIC program is normally written in capital letters, a hangover from the days when computer keyboards and printers had upper case letters only. BASIC ignores spaces in mathematical expressions (and in most other parts of a program as well) thus they can be included to make a program look tidier.

2.3.1 Constants, variables, and expressions

Data within a program can be constant or variable. A **constant** is represented by a number which may be preceded by a minus sign and which may include a decimal point. A simple numeric **variable** is represented in BASIC either by a single letter or by a letter followed by a single digit, and it can be given a value in various ways as the program runs.

These variables and constants can be used to construct mathematical expressions. The symbols used for the four basic arithmetic operations are the same in most computer languages

        +   addition
        −   subtraction
        *   multiplication
        /   division

Brackets can also be used in the conventional way to alter the order in which operations are carried out. BASIC allows the 'uparrow' or '^' symbol to be used to indicate exponentiation ('raise to the power of').

Assignment: the LET *statement*

The most common way in which a variable acquires a value is by means of an **assignment statement**, which according to the 'Minimal BASIC' specification should begin with the keyword LET. In fact this word is optional in many versions of BASIC. The examples of LET statements below show how they can be used to instruct the microcomputer to evaluate a mathematical expression. When each LET statement is executed, the microcomputer will carry out the appropriate calculation

```
10 LET X = 5
20 LET C = 2 * 3.14159 * R
30 LET Y = A*X^2 + B*X + C
40 LET Y = (A*X+B)*X +C
50 LET Z = Z + 1
```

The last example looks rather strange, and is included to underline the fact that assignment statements are not the same as algebraic equations even though they both use 'equals' signs. This statement should be read as meaning 'take the current value of variable Z, add 1 to it, and give this new value to Z'. This emphasises, incidentally, that the value of Z may change.

2.3.2 Input/output

Two special commands in BASIC allow data to be entered into a program and displayed by it.

Entering data: the INPUT *statement*
The operator can give values to variables in a program by using the INPUT statement. This consists of the word INPUT followed by the name of the variable to which that value is to be given. When the INPUT statement is executed the microcomputer will display a prompt (usually a question mark) to indicate that the data should be typed in. Thus the statement

```
100 INPUT A
```

will prompt the user to input a value for the variable A.

Displaying results: the PRINT *statement*
Data can be displayed or printed by means of the PRINT statement. This uses the keyword PRINT followed by a list of variables and/or expressions separated by commas or semicolons. If a semicolon is used to separate two items in a list the values displayed will be separated by a space, while if a comma is used they will be 'tabulated' into columns. Messages can be included in an output to make it look less cryptic by enclosing the text to be used in double quotes. For example, if R has the value 0.5, the statement

```
110 PRINT "RADIUS=";R,"CIRCUMFERENCE=";2*3.14159*R
```

will produce the output

```
RADIUS= 0.5     CIRCUMFERENCE= 3.14159
```

The results from each PRINT statement will be displayed on a separate line unless they are terminated by a semicolon.

A programming example
A program can be written using only the three types of statement encountered so far, using INPUT statements to put information into the program, LET statements to calculate the answer, and PRINT statements to display the results. A BASIC program is typed in as a list of numbered lines, although they need not be numbered consecutively nor even be entered in numerical order. When the program runs they will be executed in numerical order, starting with the lowest-numbered line.

This example requests the radius R and height H of a cylinder and then calculates its surface area and volume. The lines are numbered at intervals of 10, which leaves space for up to nine new lines to be inserted should the program need to be modified

```
10 INPUT R
20 INPUT H
```

```
30 LET P=3.14159
40 LET A=2*P*R*(H+R)
50 LET V=P*H*R^2
60 PRINT "AREA=";A
70 PRINT "VOLUME=";V
```

2.3.3 Controlling the flow of a program

Normally a program starts by executing statements in numerical order, but this sequence can be changed simply by means of special statements which allow the 'flow' of the program to be modified.

The GOTO *and* ON-GOTO *statements*
This consists of the keyword GOTO followed by a line number. For instance, if the line

```
15 GOTO 10
```

were inserted in the last example the program would be forced to return to line 10. The effect would be to make the program request a radius value over and over again, but it would never be able to proceed beyond line 15. If the GOTO had been put in as line 80

```
80 GOTO 10
```

then the whole program would be repeated indefinitely. These are examples of **loops** in which the program continues endlessly around a sequence of operations and are used quite often in control and instrumentation applications.

A variant of the GOTO statement allows control to be passed to different points in the program according to the value of an expression. This expression is evaluated by the program and then rounded to obtain an integer, whose value is then used to select a line-number from the list following the GOTO

```
50 ON X+2 GOTO 100,150,300,400
95 ON Y^2-Z GOTO 50,70
```

If the expression has the value 1 the program continues from the first line-number, if it has the value 2 it continues from the second line number, and so on. If there is no line number corresponding to the value of the expression, an error results.

The IF *statement*
The IF statement consists of the keyword IF followed by a comparison between two expressions and the word THEN followed by a line number, for example:

```
100 IF A < 0 THEN 100
```

The comparison is carried out using one of the six **relational operators** shown below:

=	equal to	<>	not equal to
>=	greater than or equal to	<=	less than or equal to
>	greater than	<	less than

If the test condition is satisfied, then execution of the program continues from the specified line-number, otherwise it continues from the line following the IF statement. For instance, if in the example which computed the cylinder area and volume, the line

```
80 IF V <> 0 THEN 10
```

had been used then the program would have repeated indefinitely until data was used which gave a volume of zero (i.e. R or H was set to 0). IF statements can be used to prefix any other kind of statement, including other IF statements, and they allow the program to modify its action according to the data which it is handling. This is obviously important in engineering applications where, for example, it might allow a computer to stop a piece of equipment if it detected a malfunction.

The IF statement can be used to make a program repeat the same set of statements with different values. For example, this program prints out the integers from 1 to 10 with their squares:

```
10 LET I = 0
20 LET I = I + 1
30 PRINT "I=";I,"I SQUARED IS";I^2
40 IF I < 10 THEN 20
```

The FOR *and* NEXT *statements*
BASIC has a special kind of statement for the type of application illustrated above, in which a loop is constructed. It starts with the keyword FOR and indicates the range of values to be used and (optionally) the amount by which the value is to be increased at each repeat. The last example could be coded using a FOR statement like this:

```
10 FOR I = 1 TO 10
20 PRINT "I=";I,"I SQUARED IS";I^2
30 NEXT I
```

Note the use of another kind of statement, NEXT which is used to 'bracket' the line or lines which are to be repeated. The next example shows how the 'cylinder' program could be modified to make a table of values for the area and volume. Here the optional step size is included using the word STEP:

```
10 LET P=3.14159
20 FOR R = 0 TO 1 STEP 0.2
30 FOR H = 0 TO 5 STEP 0.5
40 LET A=2*P*R*(H+R)
50 LET V=P*H*R^2
60 PRINT "RADIUS=";R,"HEIGHT=";H;
65 PRINT "AREA=";A,"VOLUME=";V
70 NEXT H
80 NEXT R
```

In fact this example uses two **nested** FOR **loops**, that is, one loop within another. The microcomputer will start with R equal to 0 and repeat lines 40 to 65 with values of H from 0 to 5 in increments of 0.5. Then it will set R equal to 0.2 and repeat lines 40 to 65 for the same range of values of H as before. The program will continue until R equals 1 and H equals 5.

The END *and* STOP *statements*
The END statement is used both to mark the physical end of the program and to stop a program and return control of the microcomputer to the user. The end of a program should always be marked with an END statement although in the examples given so far it is not strictly necessary. The STOP statement can be used at any point in a program and causes the program to stop executing.

Subroutines: GOSUB *and* RETURN
It frequently happens when writing a program that the same group of instructions is used more than once. In such cases the repeated group can be considered as a separate **subroutine** and included as a separate block of instructions appearing just once. This block is terminated with a RETURN statement which indicates the end of a subroutine in much the same way as END marks the end of the main program.

The statements in the subroutine are 'called to action' by means of a GOSUB (GO to SUBroutine) statement. This is much the same as a GOTO statement in that it consists of a single keyword followed by a line number. In a GOSUB statement, however, a record is kept of the point in the program where the subroutine was called so that it can return to the next instruction when the subroutine has finished.

For example, suppose that a program in a machine controller needs the operator to type in the speeds required of three shafts in a machine, making sure that each shaft does not rotate at more than 1000 r.p.m. clockwise or anticlockwise, (i.e. plus or minus 1000 r.p.m. clockwise). The same input statement and checking of speeds will be needed by each shaft, and to avoid having to repeat the same group of lines three times a subroutine could be used. This is shown in Figure 2.3. The part of the program which reads in values will then consist of a set of PRINT statements to prompt the user as to which value is to be typed in next, alternated with subroutine calls to input the value to the program and ensure that it is within the correct limits.

Main Program:

```
10 PRINT "SPEED OF SHAFT 1";
20 GOSUB 200
30 LET A=X
40 PRINT "SPEED OF SHAFT 2";
50 GOSUB 200
60 LET B=X
70 PRINT "SPEED OF SHAFT 3";
80 GOSUB 200
90 LET C=X
100 ... (remainder of program)
195 END
```

Subroutine:

```
200 INPUT X
210 IF X>+1000 THEN 240
220 IF X<-1000 THEN 260
230 RETURN
240 X=+1000
250 RETURN
260 X=-1000
270 RETURN
```

Figure 2.3 Controlling the speed of a shaft

At the end of this fragment of program the variables A, B and C will each have been loaded with values in the permitted range from the keyboard, and in the remainder of the program these values could be used to set the speed of rotation of the corresponding shafts; the problem of exactly how that is done will be postponed until later!

2.3.4 Functions

The examples described so far have used one or more of the arithmetic operations described in section 2.3.1. In addition to these operations, BASIC also provides some numeric functions, each called by a single keyword. The most common functions are supplied as part of the BASIC language and are known as **intrinsic functions**. Six of these functions are

Table 2.1 BASIC functions

Function	Returns	$x > 0$	$x = 0$	$x < 0$
ABS(x)	Absolute value of x	$+x$	0	$-x$
EXP(x)	e to the power x	e^x	1	e^x
LOG(x)	Natural log. of x	$\log_e(x)$	Error	Error
SGN(x)	Sign of x	$+1$	0	-1
SQR(x)	Square root of x	$x^{0.5}$	0	Error
INT(x)	Largest integer $\leq x$	(-1 is higher than -2)		

listed below, together with their action according to whether the variable on which they operate is positive, zero, or negative, as shown in Table 2.1.

Functions may be used in expressions in exactly the same way as numerical variables. This can be seen from the next example of Figure 2.4 which calculates the roots of a quadratic equation of the form

$$ax^2 + bx + c = 0 \qquad (2.1)$$

If A is zero, the equation is linear and the solution is simply

$$x = -\frac{c}{b} \qquad (2.2a)$$

Otherwise, the two roots of this are given by the expression

$$x = \frac{-b \pm \sqrt{b^2 - 4ac}}{2a} \qquad (2.2b)$$

If the value of $b^2 - 4ac$ is negative, its square root will be imaginary and the roots of the equation will be complex. The BASIC system will report an error if it attempts to work out the square root of a negative number, and to avoid this happening the program must calculate this value (line 40) and then check to see if it is negative. If this is the case, the program cannot proceed, and prints a message to that effect; otherwise it prints the values of the two roots.

Minimal BASIC also includes the four trigonometrical functions shown in Table 2.2.

These functions assume that all angles are measured in radians which means that numbers must be converted to and from degrees if necessary. This is illustrated in the example of Figure 2.5 which calculates the refractive index of a medium given that the angles of incidence and refraction of a ray of light are given in degrees.

Another intrinsic function, RND, returns a random number between 0 and 1. The usefulness of a random element may be obvious in electronic games programs, but the ways in which an unpredictable component could be used in an engineering or scientific program are less apparent. Nevertheless, they do exist, although it must be remembered that because any

```
10 INPUT A
20 INPUT B
30 INPUT C
40 IF A=0 THEN 130
50 LET Z = B*B - 4*A*C
60 IF Z<0 THEN 110
70 LET X = (SQR(Z)-B)/(2*A)
80 LET Y = (SQR(Z)+B)/(2*A)
90 PRINT "ROOTS ARE";X;" ";Y
100 STOP
110 PRINT "NO REAL ROOTS"
120 STOP
130 IF B=0 THEN 110
140 PRINT "ROOT IS ";-C/B
150 END
```

Figure 2.4 Calculation of roots of quadratic equations

Table 2.2 BASIC Trigonometrical functions

Function	Calculates
$SIN(x)$	The sine of x
$COS(x)$	The cosine of x
$TAN(x)$	The tangent of x
$ATN(x)$	The angle whose tangent is x

```
10 INPUT I
20 INPUT R
30 LET A = 3.14159/180
40 LET M = SIN(I*A)/SIN(R*A)
50 PRINT "REFRACTIVE INDEX IS";M
60 END
```

Figure 2.5

computer is inherently a predictable machine, the numbers only appear to be random. Some automatic control techniques depend upon a source of random numbers, and a random element is also useful in some computer communication systems such as the CSMA technique referred to in Chapter 8.

The value returned by RND is in fact the next number in a sequence of pseudorandom numbers uniformly distributed in the interval 0 to 1. This sequence always starts at the same point, which is useful when the program is being tested, but which could be troublesome in other applications. For this reason a facility for producing a further random element is provided by the RANDOMIZE statement, which generates an unpredictable starting point for the sequence of pseudorandom numbers.

In addition to the intrinsic functions described, BASIC allows the programmar to define up to 26 **extrinsic functions** using the DEF statement. These functions are named FNA, FNB, ... FNZ, and once they have been defined they can be used later in the program in the same way that an intrinsic function is used. For example,

DEF FNA = 2*3.14159

DEF FNB(Z) = SQR(Z^2-4*A*C)

Note that function FNB is defined in terms of variables A, C, and Z where Z is the argument of the function.

2.3.5 Subscripted variables and arrays of information

Sometimes it is convenient to use a set of variables which all have similar names but which have serial numbers in the same way as subscripted variables in algebra. This can be done by declaring an **array** of variables with a specified range of subscripts using a DIM (DIMension) statement. For example,

10 DIM T(50),S(8,6)

instructs the BASIC system to reserve spaces in the microcomputer memory for two arrays. The first is a one-dimensional array T which can contain up to 51 values and which can be visualised as a set of numbers arranged in a straight line. The second is a two-dimensional array S containing up to 63 values comprising a rectangular pattern of values arranged as six rows and eight columns. The DIM statements in a program should appear before any attempt is made to read the values from the array or to store values in it.

An array consists of a number of **elements** which is referred to in the program by following the array name with a number called a subscript. This is known as a **subscripted variable**. The elements of array T in the example would be referred to as T(1), T(2), etc., while those of array S would be referred to with two subscripts: S(1,1), S(1,2), and so on. Unfortunately, different implementations of BASIC interpret DIM statements differently. Minimal BASIC numbers subscripted variables from 0, while others start numbering from 1, and this can lead to problems where programs which run satisfactorily with one version of BASIC will not operate correctly with another version. If a BASIC program is to be 'portable', that is, capable of being transferred to different microcomputers, programs should be written with subscripts starting from 1.

Although 'Minimal BASIC' allows subscripts of zero, it is possible to declare that all subscripts have a lower bound of one by using the OPTION BASE statement. Thus

```
OPTION BASE 1
DIM T(20)
```

will reserve space for twenty elements of array T, T(1) to T(20). 'Minimal BASIC' also allows the DIM statement to be omitted, but it is good programming practice to declare all arrays. In cases where arrays are not declared, it is assumed that all subscripts have an upper bound of ten.

Once an array has been dimensioned, all references to subscripted variables must have the correct number of subscripts and each subscript must lie in the range which was defined in the DIM statement. It is not possible to redefine an array once it has been declared, for example if it is discovered during the operation of a program that the original array did not have enough elements.

The subscripted variables can be used just like ordinary variables, and the use of the subscript allows them to be used in a systematic manner, particularly when they are within a program loop. Arrays can be used for a wide range of purposes, for example to hold sets of readings from an instrument, or look-up tables to correct readings. The earlier example of a program for controlling shaft speeds could be written more compactly using an array to hold the speeds of the different shafts, and this is shown in Figure 2.6.

An array can also be used to hold the elements of a matrix when carrying out matrix arithmetic. The next example is for those who are familiar with matrices and vectors; it asks for the elements of a one-dimensional matrix $[I]$ and a two-dimensional matrix $[R]$ and carries out a matrix multiplication to produce a one-dimensional matrix result called $[V]$. This can be written in matrix notation as

$$[V] = [R].[I] \tag{2.3}$$

which is a mathematical 'shorthand' for

$$\begin{bmatrix} v_1 \\ v_2 \\ v_3 \end{bmatrix} = \begin{bmatrix} r_{11} & r_{12} & r_{13} \\ r_{21} & r_{22} & r_{23} \\ r_{31} & r_{32} & r_{33} \end{bmatrix} \begin{bmatrix} i_1 \\ i_2 \\ i_3 \end{bmatrix} \tag{2.3a}$$

and which in turn describes a set of equations

$$v_1 = i_1 r_{11} + i_1 r_{12} + i_1 r_{13} \tag{2.3b}$$
$$v_2 = i_2 r_{21} + i_2 r_{22} + i_2 r_{23}$$
$$v_3 = i_3 r_{31} + i_3 r_{32} + i_3 r_{33}$$

Main Program

```
10 DIM S(3)
20 FOR I = 1 TO 3
30 "SPEED OF SHAFT";I;
40 GOSUB 200
50 NEXT I
```

... (*remainder of program*)

```
195 END
```

Subroutine

```
200 INPUT S(I)
210 IF S(I)>+1000 THEN 240
220 IF S(I)<-1000 THEN 260
230 RETURN
240 LET S(I)=+1000
250 RETURN
260 LET S(I)=-1000
270 RETURN
```

Figure 2.6 Modified shaft control program

This set of equations describes the relationship between the voltages and currents in an electrical network consisting entirely of resistors. The program can be written quite compactly using subscripted variables; first the variables must be declared

```
10 DIM I(3),V(3),R(3,3)
```

Next the input values must be entered. This can be done with a loop containing a PRINT statement to indicate to the user what information is to be typed in next and an INPUT statement to accept the value entered:

```
20 FOR K=1 TO 3
30 PRINT "ENTER VALUE OF I";K;":";
40 INPUT I(K)
50 NEXT K
```

The nine resistor values must be typed in next. Here the principle is the same except that two 'nested' FOR loops are needed, one for each subscript

```
 60 FOR J=1 TO 3
 70 FOR K=1 TO 3
 80 PRINT "ENTER VALUE OF R";J;K;":";
 90 INPUT R(J,K)
100 NEXT K
110 NEXT J
```

Finally a pair of nested FOR loops are needed to allow the matrix equation to be calculated. Each voltage v_j is defined by an equation

$$v_j = i_j r_{j1} + i_j r_{j2} + i_j r_{j3} \tag{2.3c}$$

which can be rewritten as a summation

$$v_j = \sum_{k=1}^{3} i_j r_{jk} \tag{2.3d}$$

and which can be computed using K as a loop variable. J is used to allow the program to step through each of the voltages v_j, which are represented by V(J) in the program.

```
120 FOR J=1 TO 3
130 V(J)=0
140 FOR K=1 TO 3
150 LET V(J) = V(J) + I(K)*R(J,K)
160 NEXT K
170 PRINT V(J)
180 NEXT J
190 END
```

Subscripted variables are allowed in any expression where simple variables are permissible, and the subscripts can themselves be variables or even expressions. However, the subscript should be an integer because the way in which non-integer subscripts are handled depends upon the version of BASIC. If the expression used to calculate the subscript is liable to generate a fractional part, for example the line

```
100 LET X = Y + Z(A/3+1)
```

which produces non-integer subscripts unless A is divisible by three, it is safest to ensure that the subscript is an integer expression if the program is intended to be portable

```
100 LET X = Y + Z(INT(A/3+1))
```

In 'Minimal BASIC' the value of each subscript is rounded up to the nearest integer. Consequently, if the function INT is not included a value of subscript different from that intended by the programmer may be used. For example, 'Minimal BASIC' interprets the statement

100 LET X = Y + Z(A/3+1)

to give the same subscript as

100 LET X = Y + Z(INT(A/3+1+0.5))

2.3.6 Representing characters

In the examples discussed so far, all variables have been assumed to be numeric. BASIC also has a type called **string**, indicated by placing a dollar sign ($) at the end of the variable name. All versions of BASIC allow 26 string or **literal** variables, and some allow even more. There is a limit to the maximum number of characters that can be stored in a single variable, which in 'Minimal BASIC' is eighteen. String variables may be defined using LET or INPUT statements and printed using PRINT

```
10 LET A$ = "ANSWER="
20 INPUT X;Y
30 LET Z = X*Y
40 PRINT A$,Z
50 END
```

Subscripted string variables are also allowed, and arrays of textual information can be defined and manipulated in much the same way as numerical data

10 DIM B$(4), C$(6,10),Z(4)

50 LET B$(I) = C$(I,J)

70 IF B$(I) = "STOP" THEN 150

Other versions of BASIC also allow operations upon BASIC strings, but unfortunately the operators and functions differ between versions, and readers intending to use BASIC in this way are advised to refer to the programming manual of the machine which they are going to use and to bear in mind that there may be problems in transferring the programs to other machines.

2.3.7 The DATA, READ and RESTORE statements

The DATA and READ statements provide a facility for values to be assigned from a list within the program. DATA statements are used to create a sequence of data values which are assigned to variables by the READ statement. Consider the example.

```
10 DATA "JANUARY",31,"FEBRUARY",28,"MARCH",31
20 FOR I=1 to 3
30 READ M$,D
40 PRINT M$;" HAS";D;"DAYS"
50 NEXT I
60 END
```

This will print out the names of the first three months of the year together with the number of days in each month

```
JANUARY HAS 31 DAYS
FEBRUARY HAS 28 DAYS
MARCH HAS 31 DAYS
```

Any number of DATA statements may be included in the program, and data from all the statements are collected into a single data sequence. The order of the data is determined by the order in which the DATA statements appear in the line-numbered program.

The RESTORE statement is used to reset the pointer for the data sequence to the beginning of the first DATA statement. Consequently, if

```
45 RESTORE
```

were included in the above example, the first line of output would be repeated

```
JANUARY HAS 31 DAYS
JANUARY HAS 31 DAYS
JANUARY HAS 31 DAYS
```

2.3.8 Comments: the REM statement

Ideally a computer program should be completely self-explanatory to anyone reading it who understands the language in which it is written. Unfortunately this ideal is rarely achieved in practice and BASIC programs are notoriously bad in this respect. In common with other programming languages, however, BASIC allows comments to be included to make the program easier to read. Anything on a line starting with the word REM (short for 'REMark') will be ignored by the microcomputer when the program executes.

2.4 Extensions to 'Minimal BASIC'

The version of BASIC described in the ANSI Minimal BASIC standard is used by few, if any, personal computers without some extensions to improve the usefulness of the language. Unfortunately, different manufacturers have extended BASIC in different ways, so that difficulty is often encountered when transferring programs from one kind of computer to another. For this reason the programmer should use extensions to BASIC with care, as it can affect the 'portability' of programs from one machine to another. In the next few pages some of the more common extensions to BASIC will be introduced.

2.4.1 Variables and expressions

Variable names
The use of single letters or letter-number combinations for variable storage leads to programs which are very difficult to decipher, and many newer versions of BASIC allow names to be several letters long; often the name can be as long as the programmer wishes.

Logical operations
Although 'Minimal BASIC' does not provide any logical operations, some versions of BASIC have operators called AND, OR and NOT, but because they operate on what are notionally numerical variables their operation needs some explanation.

What happens is that the corresponding bits in each number are ANDed or ORed together to produce the bit pattern in the result. For example, the sequence

    50 A = 170
    60 B = A AND 3

would assign the value 2 to variable B. If each quantity were represented by an 8-bit number, the bit-wise AND operation would look like this:

quantity	binary	decimal
A	1010 1010	170
3	0000 0011	3
B	0000 0010	2

The bit pattern in B will contain a 1 only where the bit patterns for A and 3 contained a 1. Note that the number 3 has here acted as a **mask**, isolating the last two bits from the pattern in A and setting all the other ones to 0. The OR and NOT operations also affect all the bits in a pattern.

These logical operators are useful for manipulating some types of engineering and scientific data. BASIC does not provide a data type which allows logical variables to be handled directly. However, if the least significant bit of a number were used to store a 1 or 0 according to the state of a logical variable, it would be set or cleared according to the normal rules of boolean algebra.

Integer variables

The use of real variables to store integers is inefficient in terms of information storage. Some versions of BASIC allow integer variables to be used, and these are indicated by appending a percentage sign (%) to the variable name in much the same way as strings are indicated by appending $ to the variable name.

Operations on strings

Although Minimal BASIC allows the use of string variables, it does not permit the value of a string variable to be operated on in any way. Most implementations of BASIC allow strings to be joined end to end to make longer strings, a process which is known as **concatenation**. The symbol used for this operation is usually +, although & is occasionally used. The resulting string contains the component strings in the order in which they appeared in the concatenating expression, for example:

```
10 A$ = "A COMPO"
20 B$ = "SITE STRING"
30 C$ = A$ + B$
40 D$ = B$ + A$
```

will assign the value

    A COMPOSITE STRING

and

    SITE STRINGA COMPO

to variables C$ and D$ respectively.

One string may be equated to part of another by means of a set of functions LEFT$ and RIGHT$, which are used to extract a specified number of characters from the left or right of a string expression. A third function, MID$, is used to extract a specified number of characters from the middle of a string expression. For example,

```
10 A$ = "BASICALLY QUITE NICE"
20 B$ = LEFT$(A$,5)
30 C$ = RIGHT$(A$,4)
40 D$ = MID$(A$,6,4)
50 PRINT C$+" "+D$+", "+B$
```

will cause the message

    NICE ALLY, BASIC

to be displayed.

Two further functions, ASC and CHR$, return the value of the number used by the microcomputer to represent a character, and the character

2.4.1

corresponding to a given numerical code. Thus

```
100 PRINT ASC("A"),CHR$(90)
```

will produce the output

    65   Z

ASC is an abbreviation for 'ASCII' and refers to the listing of ASCII standard character set which was given in Table 1.1. Some microcomputers allow the character codes which are conventionally assigned to control characters to be used to display special characters such as accented letters and fractions.

2.4.2 Statements

Multiple statements
One common enhancement to BASIC is to allow more than one statement on a single line, with colons being used to separate them, for example

```
110 FOR I=1 TO 10: A(I)=0: NEXT I
```

will clear ten elements of an array in a single line. This improves readability, but it also offers another improvement, as shown below.

Extended IF statements
In Minimal BASIC, IF statements specify only the line number to which the program should go if the condition is true, which can make programs difficult to write and understand. Almost all implementations of BASIC allow the line number to be replaced with a statement on the same line as the IF keyword and the test condition. This statement is executed only if the condition is true, for example:

```
100 IF A < 0 THEN PRINT "NEGATIVE"
```

If several statements appear on the same line, they will be executed only if the condition is true

```
200 IF C$="CLEAR" THEN FOR I=1 TO 10: A(I)=0: NEXT I
```

Some versions of BASIC also allow an ELSE clause to be included to allow the programmer to stipulate what is to happen if the logical test in the IF statement fails.

Improved prompts in INPUT statements
The ability to place more than one statement on a single line allows sections of program which interact with the user to be laid out more neatly

```
200 PRINT "ENTER DATE:";:INPUT D
210 PRINT "ENTER MONTH:";:INPUT M
220 PRINT "ENTER YEAR:";:INPUT Y
```

However, most versions of BASIC permit the programmer to include strings as prompts in the INPUT statement itself

```
200 INPUT "ENTER DATE:", D
210 INPUT "ENTER MONTH:", M
220 INPUT "ENTER YEAR:", Y
```

Graphics
Modern microcomputers have much more sophisticated display facilities than the computers which existed when BASIC was first developed. Usually these displays enable graphical output to be produced, and the versions of BASIC with which they are used contain special graphical commands to make use of these capabilities. Unfortunately there is little standardisation between manufacturers, and these special commands will not be discussed further.

2.4.3 Accessing the microcomputer hardware

BASIC, like other high level languages, is designed to be as machine-independent as possible. However, access to the internal operation of the machine is needed for special purposes such as controlling interfaces with external equipment. Most microcomputer versions of BASIC provide a PEEK function which allows the user to 'peek' into a memory location and observe its contents. The argument of the function is the address of the location being examined, and the value returned lies in the range 0 to 255. Thus the sequence

```
10 FOR I = 0 TO 65535
20 PRINT "ADDRESS =";I;" CONTENTS =";PEEK(I)
30 NEXT I
```

will print out the contents of all the memory locations accessible on an 8-bit microcomputer ... this will prove extremely boring to most users after a while, as the equivalent of approximately 1000 pages of printout will be produced if a standard computer printer is used!

The opposite facility, that of setting the contents of a specified location to a given value, is provided by the POKE command. The keyword is followed by two values or expressions separated by commas: the first is the address of the location to be modified, while the second is the value to be placed in that location. This command must be used with care because it can affect the contents of locations used by BASIC for its internal operation, and injudicious use of POKE can cause the microcomputer to 'crash', requiring it to be restarted. If the reader is determined to crash a microcomputer, the sequence

```
10 FOR I = 0 TO 65535
20 POKE I,0
30 NEXT I
```

will oblige in virtually all cases, as it sets all memory locations to zero.

Two more keywords are often provided to allow the ambitious programmer to incorporate sections of program written in assembly language into a BASIC program. `SYS` is a command, while `USR` is a function. Each can be used to call a section of program written in machine language, but the precise manner in which they are used varies from machine to machine.

2.5 A second language: Pascal

The other programming language to be described in this chapter is rather more modern; Pascal. This was first introduced by Nicklaus Wirth in 1970 (Jensen and Wirth, 1978), and although programs written in Pascal look very different from those written in BASIC they nevertheless have many aspects in common. Pascal programs are usually written using lower case letters, and this will help us to distinguish between Pascal and BASIC examples throughout the remainder of this book. Pascal is designed to encourage a **top-down approach** to problem solving; that is, it allows the programmer to analyse a problem into a structure of successively smaller problems until a solution becomes apparent. The solution to each constituent problem will form a **program block** and the form of the overall problem will be reflected in the structure in the program as it is implemented. Each block may contain component blocks, and these blocks may also contain other blocks to as many levels as needed. For this reason Pascal provides a much wider range of program structures than does BASIC.

Programs written in Pascal also consist of sequences of statements, but they are not restricted to one statement per line as is the case in Minimal BASIC, neither do the line numbers have any significance as they do in BASIC's `GOTO` statements, for example. In fact, Pascal programs do not have line numbers. Each statement in a Pascal program is separated by a punctuation mark, and the position of the statement or part of a statement on a line is unimportant. This lets the programmer lay out the text of a program in such a way as to make it easy to follow. Pascal also allows variables and constants to be given full length names with the result that programs written in Pascal can be much less cryptic than those written in BASIC, and are easier to read and understand.

It is sometimes useful to treat a block of statements as if it were a single statement in its own right, and this Pascal does by bracketing the block with the words **begin** and **end.** Several examples of **compound statements**, as such blocks are called, will be encountered in this section.

2.5.1 Constants and variables

Assignment statements
In Pascal, as in BASIC and other programming languages, information is processed by means of assignment statements which assign values to variables. Pascal uses a special **assignment symbol** ':=', which may be thought of as meaning 'becomes equal to'. Pascal's assignment statements consist simply of a variable name, the assignment symbol, and a variable, constant, or mathematical expression, the value of which is to be assigned to the variable on the left of the assignment symbol, for example:

height := 1.90

Declaring data types
Pascal insists that the name and type of each variable is declared before it is used, by means of a special **var** statement. Similarly, constants must be given names and values in **const** statements, the types being deduced from the values given to the named constants:

const *pi* = 3.14159;

var *date, month, year* : *integer*;
temperature, pressure : *real*;

Variables can also be of type **char** which indicates that they are capable of holding a single character:

var *thisletter*: *char*

Constants of type char are identified by enclosing them in single quotes, so that the statement

thisletter := '*z*'

will assign the character '*z*' to the variable *thisletter*.

A new type may be defined by means of a **type** declaration statement in which all the possible values which that type can take are enumerated, e.g.

type *colour* = (*red, blue, green, yellow*);
direction = (*up, down, left, right*);

Subranges
Pascal allows one type to be defined as a subrange of another. For example, a date can contain an integer in the range 1 to 31 to indicate the day of the month, and a type 'date' could be defined by the statement

type *date* = 1..31;

which shows that 'date' can take a value which is a subrange of type 'integer'. Similarly, if the days of the weeks were defined by

type *day* = (*mon,tue,wed,thur,fri,sat,sun*);

subrange types could be defined for workdays and weekend days:

type *workday* = *mon..fri*;
weekend = *sat..sun*;

Arrays
Variables may be subscripted in Pascal in much the same way as in BASIC except that the subscript is enclosed in square brackets and that the range of the subscript, or **index** must be completely defined; it is not adequate to assume that it starts from some predefined value. The index may be a scalar or subrange type, but types *real* and *integer* are not allowed because they do not have a defined range. Arrays can have more than one dimension; typical declarations of arrays are

voltage,current : **array**[1..10] **of** *real*;
resistance : **array**[1..10,1..10] **of** *real*;
income : **array**[1980..2020] **of** *real*;
profit : **array**[*weekday*] **of** *real*;
string : **array**[1..80] **of** *char*;

Whenever an element of an array is referred to in an expression, it must have the appropriate number of subscripts, for example:

profit[*mon*] := 800.00;
voltage[*i*] := *voltage*[*i*] + *current*[*j*]**resistance*[*i,j*];

Records
Records provide templates for information structures whose parts may have quite distinct characteristics. For example, a variable representing a date can be defined as

date = **record**
  d: 1..31;
  m:(*jan,feb,mar,apr,may,jun,jul,aug,sep,oct,nov,dec*);
  y: *integer*
end

Once the variable *date* has been defined in this way, the various components of it, known as 'fields', can be referenced by means of the variable name appended with a full stop and the field identifier. Thus

date.d := 8; *date.m* := *may*; *date.y* := 1945

would set *date* to a historical value.

Arrays of records are useful when tables of disparate information are to be handled. For example, a program to keep track of electronic components in a store would require a record to be kept of the identifying number, stock level, and price of each type of component handled. If the identification of each part were a string of up to 10 characters, a type 'part' might might be defined like this:

type *part* = **record**
 ident: **array**[1..10] **of** *char*;
 level: *integer*;
 price: *real*
end

Now variables can be defined in terms of this type, for example:

var *thispart*: *part*;
var *stocklist*: **array**[1..999] **of** *part*

Pointers

Pointers are variables of a special type which, as their name indicates, 'point' to information stored in the memory of the microcomputer. This means that the contents of a pointer is the address where a piece of data is stored rather than data itself, the importance of which will become clearer in the next chapter. Pointers allow 'data structures' to be constructed in a Pascal program using records which are very useful in certain types of programming, such as the maintenance of lists of names which are liable to change from time to time.

If the contents of the stock list of the previous example never changed, it would be better to store it as an array of records, possibly arranged in order of the identifying numbers of the parts. Suppose, however, that a new item were added to the list; the contents of each element with a higher subscript number would need to be moved; a very time-consuming business. This type of situation arises quite commonly in some types of computing, for example personnel records. However, such restructuring of lists of data rarely occurs in engineering or scientific applications except where 'databases' of information are being updated. The writing of database management software is a somewhat specialised task, and as several good programs are available commercially, pointers will not be discussed further in this introduction to Pascal.

2.5.2 Operations

Expressions in Pascal are made up of constants, variables, and operations just as in BASIC. In Pascal, however, the type of the quantity is much more important.

Arithmetic operations
The four main arithmetic operations (addition, subtraction, multiplication and division) can be carried out on real data. However, Pascal does not allow one quantity to be raised to the power of another. Note that adding, subtracting, multiplying or dividing real numbers produces real results, that is, the result is of the same type as each of the operands.

The situation is rather different with integers because dividing one integer by another does not in general yield an integer as a result. Pascal uses a different pair of operators for the division of integers; **div** produces an integer remainder and discards the quotient, while **mod** yields the remainder and leaves the quotient. For example, seven divided by three gives two, remainder one, so the program fragment

quotient := 7 **div** 3;
remainder := 7 **mod** 2

would assign the value 2 to quotient and 1 to remainder. Note that the operators 'div' and 'mod' both operate upon integer data to yield integer results.

Logical operations
Boolean algebra provides rigorous techniques for deciding whether propositions are true or false provided only that they can be expressed in terms of other propositions that are each known to be true or false. This is important in microcomputer applications because it provides the basis for decision making and condition testing in programs. In Pascal, two operators, **or** and **and**, can be used to combine two propositions to form a result, while a third operator, **not**, can be used to change the result of a proposition from true to false, or vice versa.

Pascal has a data type called **boolean** which can take only the values *true* and *false*, and variables of type boolean can be manipulated with the three operators **and**, **or**, and **not**.

var *teawith, teawithout, tea, milk, sugar*: **boolean**;
   drink, coffee, heat: **boolean**;
   temp: **integer**;

teawith := *tea* **and** *milk* **and** *sugar*;
teawithout := **not** *sugar* **and** *tea* **and** *milk*;
drink := *coffee* **or** *tea*;
if *temp* < 100 **then** *heat* := *true* **else** *heat* := *false*;
heat := *temp* < 100

Relational operations

Pascal has the same six relational operators which were encountered in BASIC's IF statements:

$$<, \quad <=, \quad =, \quad >=, \quad >, \quad <>$$

Note that the operators '=' and '< >' may give unpredictable results when used with real quantities. This is because although in principle real numbers have an infinite number of possible values, in practice they must be represented to finite accuracy which results in **roundoff errors**. These errors accumulate during a calculation involving real quantities so that the result will rarely be exact, and if a real quantity does not have exactly the same value as another one, a test for equality will fail.

The six relational operators can each be used to compare two quantities to produce propositions that may be true or false, that is, they produce boolean results. In Pascal it is possible to construct assignment statements where the result of a comparison is assigned to a boolean variable. For example, given

var *number* : **integer**;
   propose : **boolean**;

the sequence

number := 4;
propose := (*number*=6)

will assign the value *false* to the boolean variable *propose* because the proposition *number*=6 is false.

2.5.3 Entering and displaying information

'read'

In Pascal values can be entered from the keyboard by means of a *read* statement. More than one value can be entered using the same statement by means of a list of variable names enclosed in brackets, but it does not permit messages to be printed.

'write' and 'writeln'

Similarly, values can be displayed by using a *write* or a *writeln* statement. The difference between these two kinds of output statement is simply that writeln causes the display to start a new line after it has been executed, while write does not. Messages can be included in these statements by enclosing them in single quotes. The list of variables and messages is enclosed in brackets and the items in the list are separated by commas, never by semicolons.

These statements are used to program all the interaction with the user of the computer. For example, to input the radius and height of a cylinder and calculate the volume, the following program fragment might be used:

    write('Enter the radius and height of the cylinder');
    read(radius,height);
    writeln('volume is ', pi*radius*radius*height)

2.5.4 Constructs in Pascal

Pascal offers many more methods for structuring programs than does BASIC. The starts of lines are usually indented to the right by programmers to emphasise the starts and ends of the constructs used.

'for' loops

Pascal also uses the word **for** to indicate the start of loops, and the first example of a FOR loop in the section on BASIC would be translated as:

    for *i* := 1 **to** 10 **do**
      writeln('I= ',i,'I squared is',i*i)

The control variable, the initial value, and the final value must all be of the same scalar type, excluding **real**. If more than one statement is to be included in the body of the loop, a compound statement bracketed with the words **begin** and **end** can be used; the example below does the same as the last one, but with less compact coding

    for *i* := 1 **to** 10 **do**
    begin *square* := *i*i;*
      writeln('I= ',i,'I squared is',square)
    end

In Pascal the variable can be incremented only by one, but it is possible to decrement it by one by using the word **downto** instead, so that

>   **for** *i* := 10 **downto** 1 **do**
>   *writeln('I=',i,'I squared is',i*i)*

will produce the same ten lines of output as the previous example, but in reverse order.

'while' *loops*
Pascal contains two other **constructs** for making conditional loops. The first of these is the **while ... do** loop which causes the computer to carry on executing the same statement or compound statement as long as some condition is true. For example,

>   **var** *number, answer: integer;*
>   *answer* := 1;
>   **while** *answer* > 0 **do**
>   **begin**
>   *write('type in a number');*
>   *read(number);*
>   *answer* := *number*number;*
>   *writeln('the square of that number is',answer)*
>   **end**

'repeat' *loops*
The second type of conditional loop is the **repeat ... until** loop which carries out the same sequence of operations repeatedly until the condition at the end of the loop is true. For example:

>   **repeat**
>   *write('type in a number');*
>   *read(number);*
>   *answer* := *number*number;*
>   *writeln('the square of that number is',answer)*
>   **until** *answer* = 0

The differences between the various kinds of loop are summarised in Table 2.3.

The **'if'** *statement*
Pascal allows **if** statements to be constructed as follows:

>   **if** *volume* > 1000 **then** *writeln('VOLUME IS TOO LARGE')*

The comparisons in Pascal 'if' statements use the six relational operators discussed in section 2.3.7. Compound statements can be used with 'if' state-

Table 2.3 Comparison of loop types in Pascal

for	repeat	while
Executes statements in loop at least once.	Executes statements in loop at least once.	May not execute statements in loop at all.
Executes statements until control variable is greater than (less than with **downto**) final value.	Executes statements in loop while condition is false.	Executes statements in the loop while condition is true.
Only one statement or compound statement may appear in loop.	**repeat** and **until** act as brackets with one or more statement or compound statements between them.	Only one statement or compound statement may appear in loop.

ments in just the same way as a simple statement. Here is an example which uses a compound statement to convert (integer) inches to feet and inches.

    if *inches* >= 12 **then**
    begin
     feet := *inches* **div** 12;
     inches := *inches* **mod** 12
    end

One additional feature of Pascal is that an **else** clause can be included in an 'if' statement, so that the action to be taken if the condition is not true can be specified as well. For example:

    if *volume* > 1000 **then**
     writeln(*'VOLUME IS TOO LARGE'*)
    else
     writeln(*'VOLUME IS O.K.'*)

The 'case' *statement*
Pascal's '**if** ... **then** ... **else** ... ' construct allows one of two courses of action to be taken according to the value of a boolean quantity. Frequently a program must take one of more than two possible courses of action according to the value of a non-boolean quantity. Consider a fragment of a program which examines mathematical symbols such as '+', '−', and '*', and calls

```
if symbol = '+' then answer: = var1 + var2
else begin
  if symbol = '-' then answer: = var1 - var2
  else begin
    if symbol = '*' then answer: = var1*var2
    else begin
      if ...

{'if' statements could be nested indefinitely...}

    end
  end
end
```

Figure 2.7 Multiple 'if' statements

the appropriate operation upon two variables *var1* and *var2*.

Although this could be programmed with a set of **if** statements, as shown in Figure 2.7, the program would become very difficult to follow if more than two or three **if** statements were nested one within another. Pascal allows multi-way branching within a program by means of the **case** statement. Here the value of a controlling variable, the **selector**, is compared in turn with test values, and if a match is found, the corresponding statement is executed. The multiple-if example can be programmed much more neatly using a 'case' statement, as shown in Figure 2.8.

```
case symbol of
  '+': answer: = var1 + var2;
  '-': answer: = var1 - var2;
  '*': answer: = var1*var2;

{more statements could be included here...}

end
```

Figure 2.8 Using a 'case' statement

The variable *symbol* is assumed to have been declared with a type which allows it to take the values listed. Any variable can be used provided that it is both scalar and not of type 'real'. If the same statement is to be executed for more than one possible value of the selector, the appropriate values should be listed, with commas separating them, before the colon.

2.5.5 Program blocks

Pascal programs can be divided up into subunits or **blocks**, each of which is given an identifying name.

Procedures

An improved equivalent to a subroutine in BASIC is provided by a Pascal **procedure**, although there are considerable differences as illustrated in Figure 2.9, in which the procedure *surface* works out the surface area of a rectangular solid.

---

program *solid(input,output);*

var *length,height,width,cube,rect: real;*

procedure *surface(side1,side2,side3: real;* **var** *area: real);*
var *temp: real;*
begin
  temp:= side1*side2;
  temp:= temp + side1*side3;
  temp:= temp + side2*side3;
  area := 2*rect
end;

begin
  read(length);
  surface(length,length,length,cube);
  writeln('the surface area of the cube is ',cube);
  read(length, height, width);
  surface(length,height,width,rect);
  writeln('the surface area of the solid is ',rect);
end.

Figure 2.9

The first thing to notice is that the start of a procedure is indicated by a name, which amongst other things makes the programs easier to read. The second is that variables used within a procedure are **local variables** to that procedure, which means that they cannot be accessed from outside the procedure. Values can be communicated between the program calling the procedure and the procedure itself only by means of the parameters appearing in brackets after the procedure name. Any other variables or constants used by the procedure are defined in the usual way. The procedure is called by the main program or another procedure simply by using its name with the values to be given to its parameters being appended in brackets.

Closer examination of the example will show that there are two types of parameters associated with the procedure *surface*. The first three parameters are used to pass values from the calling program to the procedure; the procedure receives these values which it can use in any way which the programmer desires, because these values cannot be returned to the calling program. Parameters of this type are called **value parameters** and can be used to transfer values of constants, variables, or expressions to the procedure.

The fourth parameter of *surface* is used to transmit the answer back to the calling program, and is called a **variable parameter**. When the procedure is called, the value of a variable parameter is taken from the corresponding variable in the list of parameters and used by the procedure if required. When execution returns from the procedure to the calling program, variable parameters are given appropriate values by the procedure.

Notice also that the overall form of a procedure in Pascal is just the same as a Pascal program. The 'parameters' of the program definition (notice that the program has a name) indicate the source of input and the destination of output; input/output will be discussed later in this book. The variables and constants used by the program are then defined, and next comes the part of the program which actually carries out the operations required. The last piece of 'punctuation' in a Pascal program is the full stop after the word 'end'.

---

function *area*(*side*1,*side*2,*side*3: *real*) :*real*;
var *temp*: *real*;
begin
  temp: = *side*1***side*2;
  temp: = *temp* + *side*1***side*3;
  temp: = *temp* + *side*2***side*3;
  area: = 2**temp
end

Figure 2.10

Functions

A **function** in Pascal is similar to a procedure except that only a single value is returned to the calling program, by means of the function's name. Some intrinsic functions such as 'sin' and 'cos' are often provided in Pascal, but others can be added by defining functions in much the same way as procedures, for example function *area* of Figure 2.10. Function 'area' is essentially the same as the example procedure described in the previous section. The sequence of operations which make it up are better expressed as a function since only a single value is returned to the calling program. Once it has been defined, this function can be used in a program as with mathematical expressions, for example:

$$surface := area(\ length, height, width\ )$$

Pascal contains a large number of **standard functions** and a complete list would be impossible here. However, Table 2.4 shows some of the more commonly used functions.

Table 2.4 Some commonly used Pascal functions

Function	Returns	$x > 0$	$x = 0$	$x < 0$
$abs(x)$	Absolute value of x	$+x$	0	$-x$
$exp(x)$	e to the power x	e^x	1	e^x
$log(x)$	natural log. of x	$\log_e(x)$	Error	Error
$sgn(x)$	Sign of x	$+1$	0	-1
$sqr(x)$	Square root of x	$x^{0.5}$	0	Error
$int(x)$	largest integer $\leq x$	(-1 is higher than -2)		

2.5.6 Comments

Pascal does not have a special type of statement for comments. Instead, comments are enclosed in curly brackets, for example:

$$triang := base*height/2 \ \{compute\ area\ of\ triangle\}$$

2.6 Design examples

In this section the programs to be used by each of the example systems will be described. Wherever possible these should be written using a high level language such as those described in this chapter, because they allow the problem to be addressed much more directly than the lower level assembly language to be introduced in the next chapter. Sometimes, however, high level languages do not allow sufficient versatility when tackling specialised applications and other approaches must be used. Some of these problems will be left until after assembly language has been introduced, in Chapter 3.

2.6.1 Controller

The function of a controller is to examine the output from a system and compare its value with a **set point**, the value which it is supposed to have. The difference between the actual value and the set point is the **error signal** which is used by the controller to modify the input to the system.

If a controller were not used, it would still be possible to vary the output of a system by varying its input, but such **open loop systems** are susceptible to errors and malfunction because they have no way of monitoring their own performance. Although it is possible to control the temperature of a saucepan of milk by varying the amount of heat applied to the bottom of the pan, it would be almost impossible to cause the milk to simmer without careful adjustment of the setting based upon the behaviour of the milk. In practice, **feedback** is applied to make sure that the pan of milk behaves itself; as the contents of the pan approach boiling, the amount of heat applied to the pan is reduced so that it eventually just reaches the correct temperature. Normally this feedback is carried out manually, but one of the major applications of microcomputers in engineering is in automatic control systems.

The simplest control systems are **on-off** or 'bang-bang' controllers. These simply switch on and off the input to the system being controlled to try and maintain the correct output value. Often, as for example in central heating control systems, on-off control is adequate, but it does not allow very precise control of the output value. If the output is too low, the input to the system is switched on, while if it is too high the input is switched off. The result is that the output oscillates above and below the set point.

If the behaviour of the system is examined in terms of the error signal and the signal applied to the input of the system it will be seen that the signal is 'on' if the error is negative and 'off' if it is positive. More precise control is possible if the input to the system is varied in a manner which is proportional to the error signal. This is **proportional control**, which allows a much more constant output to be achieved in many systems.

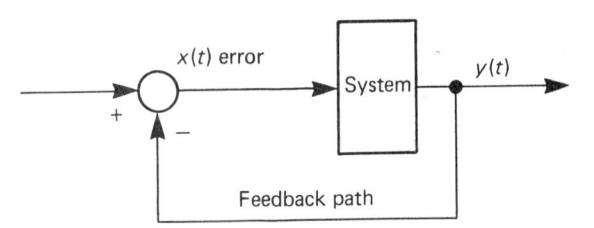

Figure 2.11 Feedback control system

The behaviour of the system being controlled varies widely from system to system. Many systems have a 'dead band' where the output does not vary even though the input signal changes. For example, a position control system must contend with friction in the drive mechanism, and if the position is controlled by varying the input to an electric motor using only proportional control, the system will tend to 'stick' before it reaches the desired position. When the motor stops, the error signal is no longer large enough to cause the motor to overcome the residual error in position. If the error signal is integrated as a function of time, the integral of the error will continue to increase even when the motor stops. If the input signal to the system is made up of a component proportional to the integral of the error signal as well as one which is proportional to the error signal itself, the controller will be able to control position much more accurately. This is known as 'two-term', or 'proportional plus integral' control.

2.6.1

Unfortunately, the effect of including an integral component in the controller tends to give rise to an 'overshoot' in the response or even instability. This can be combated by including a third term proportional to the derivative of the error signal. The result is a **three term controller** in which the signal applied to the input of the system being controlled contains three terms depending upon the error signal, its time integral, and its time derivative. $x(t)$ and $y(t)$ are the error signal and the output signal from the controller at time t, and the coefficients K_P, K_I, and K_D denote the sizes of the proportional, integral, and derivative components in the output signal.

$$y(t) = K_P x(t) + K_I \int x(t) dt + K_D \frac{dx}{dt} \tag{2.4}$$

The first thing to be noted when programming a computer to compute this equation is that computers cannot simulate continuously varying quantities, because they are digital machines which pass through discrete states at fixed time intervals. This means that equation 2.4 must be rewritten in a discrete form and special mathematical techniques based upon the z-transform should be used (Jury, 1964). A discussion of the z-transform lies outside the scope of this book, and a simpler, but an approximate approach will be used here. In practice, the distortion due to the sampling effect of the microcomputer can be kept at an acceptably low level, if the sampling rate is appreciably higher than the frequencies being processed by the system. Equation 2.4 then becomes

$$Y_N = K_P X_N + K_I \sum_{i=0}^{N} X_N \Delta T + K_D \frac{X_N - X_{N-1}}{\Delta T} \tag{2.4a}$$

Here X_N and Y_N are samples of the error signal and the controller output signal at sampling time N, and ΔT represents the time interval between samples.

A subroutine to evaluate these samples Y_N can be written quite easily in either BASIC or Pascal. The example of Figure 2.12 updates the value of the summation SM by adding the value of X to it each time that the subroutine is called.

Each time that the subroutine is called with a new sample X of the error signal, the subroutine will compute a new sample Y of the signal which should be applied to the input of the system being controlled. The values of the coefficients KP, KI, and KD, and the sampling interval DT, are assumed to be defined in the main program. The error signal X is calculated by subtracting the actual value of the system's output signal from the set point. When a computer is used, the set point can be made a function of time, so that very sophisticated control systems can be implemented. The controller might also be required to monitor the output from the system being controlled and sound an alarm or take special action if the output exceeds a value known as the **alarm point**. This can be done easily in software.

```
200 REM Compute summation SM:
210 LET SM=SM+X
220 REM Compute difference DF between this
230 REM sample X and the last one L:
240 LET DF=X-L
250 REM Update last sample:
260 LET L=X
270 LET Y=KP*X + KI*SM*DT - KD*DF/DT
280 RETURN
```

Figure 2.12

2.6.2 Waveform capturing instrument

This instrument is intended to read in a sequence of samples of a waveform at regular intervals and then to display them either directly or after processing. A number of different parameters can be calculated for a waveform once it has been read into the instrument, or 'captured'.

program *instrument(input,output)*;
{constant declarations}
{type declarations}
{variable declarations}
{procedure definitions}

begin
  {initialise system}
  while *true* **do**
    {i.e., do this for ever}
  begin
    {get a command from the console}
    case *command* **of**

      {commands and appropriate actions}

    end
  end
end.

Figure 2.13

By way of contrast to the previous example, the waveform capturing instrument will be programmed in Pascal, and because Pascal supports a 'top-down' approach we shall start by looking at the overall form of the program. Virtually all instruments are operated by means of a control panel of some sort, and in this case the system will use the keyboard and display. This control program provides the starting point for the program design: it monitors the keyboard until a command is typed in, and the command is then compared with a list of known commands and the appropriate action is taken.

Figure 2.13 shows the outline of a program called *instrument* which will be the program which 'drives' the instrument itself. This consists of a number of procedures, for example a procedure *getcommand* must be written to get a command from the console. This command will be stored as the value of a variable with the enumerated type called *commandtype*. The form of *getcommand* depends upon the form of the console, and may need to be written in assembly language, which will be introduced in the next chapter. However, the writing of a form of *getcommand* to get values from a keyboard is not difficult in Pascal, and is left as an exercise for the reader! The values which the command can take will depend upon the operations which the instrument is required to carry out. One will be *getdata* which will get a sequence of samples, or a 'record' of the input waveform from the analog input. This record can then be processed in a very large number of ways, for example to calculate the mean (DC) component, the peak value, mean power, and frequency spectrum. Each of these processes can be carried out by an appropriate procedure and there will be a corresponding command for invoking this procedure.

As an example, the **peak-to-peak range** of the waveform can be calculated by testing each sample in turn and comparing it with the maximum and minimum values read thus far, for example with procedure *peaks* shown in Figure 2.14.

The mean value of the sampled signal is calculated by adding up all the samples S_i and dividing this total by the number of samples. The result is the DC component of the signal waveform:

$$\text{mean} = \frac{1}{N} \sum_{i=1}^{N} S_i \tag{2.5}$$

Similarly, the waveform's Root Mean Square value can be calculated by computing the mean of the squares of all the samples, dividing the result by the number of samples, and taking the square root of the answer:

$$\text{rms} = \sqrt{\frac{1}{N} \sum_{i=1}^{N} (S_i)^2} \tag{2.6}$$

The writing of procedures for calculating these quantities is not difficult, and will be left as an exercise for the reader. However, other properties of the sampled signal, for example its probability density function, spectrum, power spectrum, and autocorrelation function, are not so easy to calculate. The computation of the signal's spectrum will be used as an example in the next section.

```
procedure peaks;
{local variables for this procedure}
var max,min,peak: real;
    i: integer;
{data and n are external to this procedure}
begin
  max := data[ 1 ];
  min := data[ 1 ];
  for i := 2 to n do
  begin
    if data[ i ] < min then min := data[ i ];
    if data[ i ] > max then max := data[ i ];
  end;
  peak := max − min;
  writeln( 'Range',peak,' Max',max,' Min',min )
end
```

Figure 2.14

Computing the spectrum: the Fourier transform

Any waveform which our instrument can sample can be viewed in two 'domains': in the time domain, in which we can see the way in which it varies as a function of time, or in the frequency domain, in which we can see the different components of its spectrum. A detailed explanation of the theory by which a spectrum is calculated would require too much space to be included here, but many references exist, for example (Froehlich, 1969). Essentially, the waveform can be represented as the sum of a number of sine and cosine waves of different frequencies, and with amplitudes which depend upon the form of the signal. Because the waveform is represented by N samples in the time domain, the corresponding spectrum is represented by N samples in the frequency domain, each of these representing the amplitude and phase of the corresponding frequency component. Similarly, the total number N of sines and cosines needed to represent the signal is equal to the number of samples needed to represent the waveform in the time domain. The frequencies of the sine and cosine waves are such that each wave completes an integer number of cycles within the overall duration of the waveform that has been sampled, as shown in Figure 2.15.

The conventional method of computing the amplitudes of the sine and cosine waves is to multiply each sample of the waveform by the corresponding sample of the sinusoidal wave and to calculate the mean of the products:

$$A_J = \frac{1}{N} \sum_{K=1}^{N} S_K \cdot \cos\left(\frac{2\pi JK}{N}\right) \qquad (2.7a)$$

2.6.2

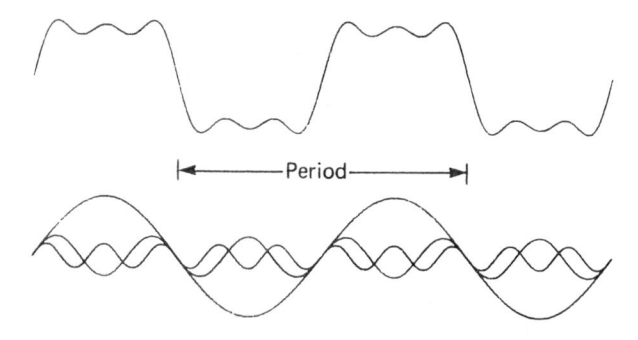

Figure 2.15 Analysing a waveform into sine waves

$$B_J = \frac{1}{N} \sum_{K=1}^{N} S_K \cdot \sin\left(\frac{2\pi JK}{N}\right) \tag{2.7b}$$

This is known as the **Discrete Fourier Transform (DFT)** after the mathematician Fourier, who first investigated the problem of transforming a time waveform into a frequency spectrum. Equations 2.7a and 2.7b can be expressed quite compactly in Pascal, as the program fragment of Figure 2.16 shows

```
a[j]:=0;b[j]:=0;
for k:=1 to n do
begin
    a[j]:=a[j]+s[k]*cos(2*pi*j*k/n);
    b[j]:=b[j]+s[k]*sin(2*pi*j*k/n)
end;
a[j]:=a[j]/n;b[j]=b[j]/n
```

Figure 2.16

The type declarations are not shown in this fragment, but the reader should be able to guess which are integer and which are real. In fact the real numbers occur as complex pairs, and Pascal can express the nature of these pairs of variables quite elegantly. The spectrum consists of the set of elements of arrays A and B arranged in order of their subscripts. The 'A' and 'B' coefficients show the amplitudes of the sine and cosine components. Pairs of sine and cosine waves of the same frequency can be 'lumped together' as sinusoids with amplitudes and phases given by:

$$\text{amplitude} = \sqrt{A_J^2 + B_J^2} \tag{2.8}$$

$$\text{phase} = \tan^{-1}\left(\frac{A_J}{B_J}\right) \tag{2.9}$$

These 'A' and 'B' coefficient pairs can be regarded as the real and imaginary parts of complex coefficients C_J such that

$$C_J = A_J + iB_J \tag{2.10}$$

where i is the 'imaginary' operator of complex arithmetic, equal to the square root of -1. In neither BASIC nor Pascal is there any built-in provision for handling complex numbers. However, the ability to define new data types in Pascal allows a data type *complex* to be introduced quite straightforwardly. This could be done as an array with two elements, but a more elegant way is to define the type complex as a record with two elements with suffices *re* and *im*. The array of data samples could be held in a variable of type *buffer*; here there would not be any point in defining type buffer to be a record because all the elements are the same. Thus the type declarations might have the form:

type *complex* = **record** *re,im* : *real* **end**;
type *buffer* = **array**[1..256] **of** *complex*

This would allow all the elements of a buffer to be copied to another one with a simple assignment statement. The previous example of Figure 2.16 can now be rewritten using an array *c* of complex numbers, as shown in Figure 2.17.

$c[j].re:=0;c[j].im:=0;$
for $k:=1$ **to** n **do**
begin
 $c[j].re:=c[j].re+s[k].re*cos(2*pi*j*k/n);$
 $c[j].im:=c[j].im+s[k].im*sin(2*pi*j*k/n)$
end;
$c[j].re:=c[j].re/n;c[j].im=c[j].im/n$

Figure 2.17

This example shows that user-defined types have a limitation: the mathematical operations add, subtract, and multiply, which are needed by various Fourier transform techniques, are not defined for the new data type complex. These can be provided as further functions before starting the main program — addition can be carried out very simply

2.6.2

```
function cpxadd( a,b:complex ): complex;
begin
  cpxadd.re := a.re + b.re;
  cpxadd.im := a.im + b.im
end
```

and subtraction can be programmed in a similar fashion. Multiplication is only marginally more complicated

```
function cpxmul( a,b:complex ): complex;
begin
  cpxmul.re := a.re*b.re - a.im*b.im;
  cpxmul.im := a.re*b.im + a.im*b.re
end;
```

Unfortunately these operations cannot be provided as functions because functions can return only scalar values, and complex numbers are in effect vectors.

Examination of equations 2.7a and 2.7b will show that the number of multiplications required to work out each coefficient is proportional to N. As there are N coefficients to be calculated, the number of multiplications to be carried out by the program is proportional to N^2. This threatens to take an excessive amount of time if N is large. The program also requires a large number of sines and cosines to be computed, which is both time-consuming and wasteful since many of these values will be recalculated several times.

Fortunately algorithms have been devised which yield the values of A_J and B_J with much less computational effort than the direct approach already outlined. Using **Fast Fourier Transform (FFT)** techniques (Nussbaumer, 1981) the computation required for the different coefficients is shared, so that instead of deriving the value of each of the A_J and B_J separately the process carries out a number of steps, each of which is a sequence of operations on all the waveform samples, until finally a set of complex numbers corresponding to the coefficients C_J is obtained.

Most of the FFT techniques which have been developed require the number of samples N to be a power of two, say, 2^n. The number of multiplications needed by these methods is much smaller than that for the DFT approach, being proportional to $n.N$ instead of $N.N$. For $N=1024$, this represents roughly a hundred-fold improvement in performance.

The penalty associated with using the FFT is that the mathematics is more difficult to follow than is the case with the DFT. A peculiarity of the FFT algorithm to be used is that it requires the input data to be sorted in a rather curious fashion called **bit reversed order**. Because there are 2^n samples,

```
procedure reverse; {Sort array into bit reversed order}
var temp: complex;
    i,j,k: integer;
begin
  j := 1;
  for i := 1 to n-1 do
  begin
    if (i < j) then
    begin        {Swap over elements}
      temp := data[j]; data[j] := data[i]; data[i] := temp
    end;
    k := n div 2;
    while (k < j) do
    begin
      j := j-k;
      k := k div 2
    end;
    j := j+k
  end;
end
```

Figure 2.18

their subscripts could be written as n bit binary numbers. When sorting into bit reversed order, the Jth sample S_J is swapped with the Kth sample S_K, where the subscript K is formed by writing the n bits of J in reverse order.

This is performed by procedure *reverse* in Figure 2.18. High level languages such as Pascal frequently have difficulty in manipulating bit patterns, and in the next chapter an assembly language method for bit-reversed sorting will be examined.

After the data has been sorted into bit reversed order it can be transformed by means of the FFT algorithm, which makes use of complex arithmetic and the result

$$\exp(ix) = \cos(x) + i\sin(x) \qquad (2.11)$$

to compute all the points in the frequency spectrum with many fewer multiplications and additions than the DFT approach. In procedure *transform* shown in Figure 2.19 the quantities *data*, *n*, *ln*, and *pi* are assumed to be declared externally.

A few points are worth noting in this listing. At the heart of the procedure is a group of operations which will also be explored in the next chapter. This group, for reasons which will become apparent, is called a 'butterfly':

2.6.2

```
    temp := cpxmul( data[k],u );
    data[k] := cpxsub( data[i],temp )
    data[i] := cpxadd( data[i],temp );
```

```
procedure transform;
var u,w,temp: complex;
    points: real;
    i,j,k,l,le,le1: integer;
begin
  points := n;
  le := 1;
  for l := 1 to ln do
  begin
    le := 2*le;
    le1 := le div 2;
    u.re := 1.0; u.im := 0.0;
    w.re := cos(pi/le1); w.im := -sin(pi/le1);
    for j := 1 to le1 do
    begin
      i := j;
      while i <= n do
      begin
        k := i + le1;
        temp := cpxmul( data[k],u );
        data[k] := cpxsub( data[i],temp )
        data[i] := cpxadd( data[i],temp );
        i := i + le
      end;
      cpxmul( u,w ); u := result
    end
  end;
  for i := 1 to n do
  begin{Divide all data points by n}
    data[i].re := data[i].re/points;
    data[i].im := data[i].im/points;
  end
end
```

Figure 2.19

The statements which make up the butterfly are repeated many times during the calculation of an FFT because they lie within three nested loops. Another point is that the values which the complex variable u takes are all

```
program instrument( input,output );

const pi = 3.141593;
      ln = 8; {defines number of samples}
      n = 256; {n is 2 to the power ln}

type complex = record re,im : real end;
     buffer  = array[1..n] of complex;
     commandtype = ( none,cmget,cmfft,cmshow );

var data: buffer;
    element,result: complex;
    command: commandtype;
    ln,n: integer;

{Define procedures cpxadd, cpxsub, cpxmul,
                getdata, showdata, etc.}

begin
  initialise;
  while true do
  begin
    getcommand;
    case command of
      cmget: getdata;
      cmfft: begin
               reverse;
               transform
             end;

      {Other operations...}

      cmshow: showdata;
    end;
  end;
end.
```

Figure 2.20

of unit magnitude. This is because it is obtained by multiplying by successive values of w which all have unit amplitude.

A skeletal form of the main program is shown in the listing of Figure 2.20. In a practical implementation of the instrument being outlined here, many more procedures would be needed to give a useful range of functions. Here lack of space precludes a detailed examination of all the algorithms which might be used to process a record of data read from an analog input.

2.7 Questions

The questions in this section are intended to provide 'food for thought' about programming techniques rather than as academic examples with a single correct answer.

1. Write a BASIC subroutine to convert fractions of a day to hours, minutes, and seconds. Write a Pascal procedure to perform the same task.
2. When listing items in sequence, 'ordinal' numbers are used. These are often abbreviated as one or more digits followed by two letters, for example, '1st', '2nd', and '3rd'. Write a Pascal program to print out the ordinal numbers from '1st' to '99th'.
3. Devise BASIC equivalents for the 'repeat' and 'while' constructs using IF.
4. Translate the modified shaft control program of section 2.3.5 into Pascal.
5. Write a BASIC subroutine to print a number in the range 0 to 255 as a two-digit hexadecimal number. (*Hint*: Use the CHR$ *function!*)
6. Modify the BASIC program to calculate complex roots as well as real ones; complex roots should be printed in the form 1.23456+J0.12345.

Chapter 3 INSIDE THE MICROPROCESSOR

The high level description of information processing used in Chapter 2 does not pay any regard to the way in which the microcomputer actually functions. One of the principal reasons for programming in a high level language is that the resulting program should work on any machine which supports that language. However, the result of making programs **machine-independent** is to hide the structure of the microcomputer from the programmer. High level languages provide means for communicating with devices such as terminals and printers without the programmer having to be concerned about the details of the interface but they are often inadequate when programming microcomputers for engineering and scientific applications.

The microcomputer based systems used in instrumentation and control almost always have special purpose interfaces for transducers and actuators, and they must interact with external changes as they happen, in **real time**. These factors mean that the designer must have a close understanding of the way in which the microcomputer and software operate. Fortunately he or she need not be concerned with the electronic details of the circuits so much as a view of the microcomputer as a logical machine whose operation can be described using a **machine language**. Although this kind of language is more 'foreign' and difficult to use than high level languages it allows the designer to express much more precisely what the microcomputer is required to do.

In this chapter the operation of a typical microcomputer will be explored from the programmer's point of view, with the instructions appropriate to each section of the microprocessor being introduced as it is described. Despite the fact microcomputers and minicomputers all appear at first sight to be very different, they all operate on the same general principles, and there is no need to concentrate on one type to the exclusion of all others. However, to ease confusion between types of microprocessor, most of the programming examples used in this chapter will be based upon the 6809, which is an 8-bit microprocessor with a particularly elegant design. Examples based upon other types will be included only when appropriate, for example where a comparison between designs is made.

3.1 Processing information

In the conventional approach to designing electronic systems each function is carried out by a separate circuit within the system. If the system were required to count pulses, a counting circuit would be used. A counter contains a register to store the current value of the count and logic to allow that number to be increased by one each time a pulse occurs. In this approach each function of a system is carried out using a different section, so that if six quantities were to be counted, six counters would be necessary.

The structure of a microcomputer based system is fundamentally different, information being stored in its memory locations but processed by its microprocessor. In our example the counting function, increasing a number by one, is separate from the storage function of the memory location. Six counters could be implemented by using six memory locations to store the counts and the microprocessor to increment the numbers stored in these locations as required.

3.1.1 Instructions

We can indicate to the microcomputer that we want it to increase the contents of a register by one using a **machine instruction** which consists of two parts: an **operation code** or **opcode** which states the operation to be carried out and an **operand** which specifies the data or register to be used. This instruction might look something like this.

        INC   1234

The word INC is short for 'increment by one' while the number '1234' defines the operand, which in this case is the address of the location to be used as a counter. In the language used by the microprocessor itself an opcode consists of a binary pattern which is understood by the microcomputer to mean 'increment by one'. In assembly language, words like INC are used simply to make the opcode easier to remember, and these words are often called 'mnemonic opcodes' or just **mnemonics**.

3.1.2 Machine code

Both instructions and data are stored within the microcomputer using memory locations. A single instruction may require one or more locations depending upon the particular instruction and the type of microprocessor being used, and the contents of each location will appear as binary patterns. For example, the instruction

        INC   1234

would require three bytes in machine code, and if the opcode were stored in location 1000 the instruction would occupy the three locations 1000, 1001, and 1002.

location address	contents hex.	contents binary	meaning of contents	
1000	7C	01111100	opcode	INC
1001	12	00010010	address	(byte 1)
1002	34	00110100	address	(byte 2)

Even when it is written using hexadecimal code rather than binary, machine code is almost incomprehensible, which is one reason why it is almost never used for programming. Another reason is that it is almost impossible to edit programs written in this form without virtually rewriting them completely, which is very wasteful of programming effort. A program written in machine code can be guaranteed to work only on the type of microcomputer for which it was written, and extensive modification may be necessary to transfer it to a different type of machine.

3.1.3 Assembly language

Despite the fact that machine language appears as a more or less unintelligible mass of binary numbers it is possible to put a more 'human' face on it by using an **assembly language**.

Assembly languages represent an attempt to overcome the near-impossibility of programming directly in machine code. Because each instruction in machine code can be represented by a corresponding instruction in an assembly language, no loss in efficiency of the final program results from the use of assembly language in preference to machine code. An assembly language program must be translated into machine code before it can be run on a microcomputer, but this task of 'assembling' the program can be carried out automatically by a program called an **assembler**. It is possible to carry out this operation manually, but it is very tedious.

In addition to instructions, assembly languages contain **assembler directives** which are instructions to the program (or unfortunate person) translating the assembly language into machine code. These directives look like instructions but do not usually produce any machine code. For example, the directive ORG is used to indicate the starting address (the 'origin') of a program in many assembly languages, including that of the 6809. The directive

ORG   256

indicates that the machine code instruction immediately following this line is to be stored in locations starting with address 256.

3.1.4 Symbolic values

Instead of numbers, names can be used to indicate operands to make a program more readable, and because they are used in the same way as numerical values, such names are called **symbolic values**. For example, since location 1234 was used as a counter in the example, it would make the meaning of the instruction more obvious if this location were given a name like COUNT. The assembly language instruction

    INC   COUNT

provides a concise description of what it does: it causes a memory location called COUNT to count up by one.

The actual value corresponding to each symbolic value must, of course, be defined somewhere in the program, and there are various ways of doing this. The most direct method is to directly assign a value to a symbol with an 'equate' or EQU directive such as

    COUNT   EQU   1234

which assigns the value 1234 to the symbol COUNT. Note that this is not the same as programming an assignment statement such as

    count := 1234

because no machine code is produced by this instruction. It corresponds instead to the Pascal declaration

    const *count* = 1234

Each time that the constant 'count' is used in a Pascal program after a declaration of this type, the effect is precisely the same as if the number 1234 had been used directly. The special symbol '*' is used to indicate 'this point in the program'. For example, the statement

    HERE   EQU   *
              INC   COUNT

gives to the symbol HERE the value of the address of the location where the opcode INC is stored. The symbol HERE is referred to as a **label** because of the way in which it is used to 'label' a point in a program.

A more straightforward way to do this is to use a label at the left-hand side of a line containing an assembly language instruction. For example:

    HERE   INC   COUNT

will also cause the label HERE to be given a value equal to the address of the memory location containing the first part of the instruction INC COUNT. Any reference to the label HERE in the program will be treated as if it were a reference to this address. Assembly languages often use capital letters throughout, although many newer systems do not insist upon this.

3.1.5 Declaring variables

When a memory location is used to store the value of a variable the programmer must indicate to the assembler that one or more memory locations are to be used to store the value of that variable. In the assembly language of the 6809 a directive called RMB (Reserve Memory Byte) is used. Thus the statement

```
COUNT   RMB     1
```

means 'reserve one byte of memory and call this address COUNT'.

If more than one memory location is to be reserved, RMB should be used to reserve any number of bytes with the appropriate 'operand'. A symbolic value can be used for this purpose if desired, so that the sequence

```
NBYTES EQU     100
ARRAY  RMB     NBYTES
```

will reserve 100 bytes starting with the address labelled ARRAY.

3.1.6 Declaring constants

Sometimes constants must be declared which are to be stored in memory, for example tables of values. These can be declared by means of two special directives FCB (Form Constant Byte), which stores a constant byte, and FDB (Form Double constant Byte) which stores a constant as a pair of bytes. Each of these directives can be used with more than one 'operand', so that the sequence

```
FRED   EQU     10
BERT   EQU     11
       FCB     1,2,FRED
       FDB     FRED,BERT
```

will cause seven bytes to be stored in memory. The first three (1, 2, and 10) are produced by the FCB directive, while the last four (0, 10, 0, and 11) are produced by the FDB directive.

3.1.7 Assembly-time arithmetic

One useful feature offered by many assembly languages is the ability to perform simple arithmetic when working out values. Thus

```
VALUE  EQU     5
RESULT EQU     2*VALUE+1
```

will give the value 11 to the label RESULT. Again, this calculation is carried out at the time at which the program is translated from assembly language to machine code, known as **assembly time**. The programming of assignment statements which will be executed when the program runs, that is at **run time**, will be discussed in the next section.

3.2 Assignment statements

Most microcomputers allow a wide range of operations to be carried out on a single register. Two further examples are `DEC` which decrements the contents of a register by one, and `CLR` which resets all the bits in a register to zero. Together with `INC` these instructions allow the programmer to use a memory location as an up/down counter which can be reset to zero.

The operations carried out by the three assembly language instructions

```
CLR     COUNT
INC     COUNT
DEC     COUNT
```

could be expressed quite easily in terms of either of the languages described in the last chapter, for example in Pascal:

count := 0;
count := *count* + 1;
count := *count* − 1;

Assignment statements in high level languages are actually carried out by means of data manipulating instructions such as `INC` and `CLR` which take information from a register, operate upon it in some way, and store the result in a register. Often a sequence of such instructions is needed to perform the arithmetic operations specified by a single statement in a high level language.

Each of the three instructions `CLR`, `INC`, and `DEC` operates on only a single piece of data: the contents of a memory location. Other operations, such as addition, subtraction, and logical AND require two operands. The result must be stored somewhere, and this will not necessarily be in one of the operand registers.

In general, therefore, an instruction would appear to need at least three addresses in order to allow statements such as

total := *amount* + *tax*

to be executed. However, the use of instructions that reference more than one register does not imply that microprocessors need to have more than one address bus. In practice, virtually all microcomputers specify the operands one after another using the same address bus.

3.2.1 The accumulator

Almost all 8-bit microcomputers contain one or more registers called **accumulators**. During a calculation this register 'accumulates' intermediate results in the same way that the total accumulates in a calculating cash register, or intermediate results appear in the display of a pocket calculator. The accumulator is involved in the majority of instructions that a microprocessor can perform, because most instructions involve the manipulation of data.

Any operation on two registers in an **accumulator-based microcomputer** will involve the accumulator and another register, usually a memory location. When the microprocessor carries out any operation involving two items of data, the result is normally placed in the accumulator which was used to hold one of the operands, referred to as the **destination register**; the other operand is held in the **source register**. There are, however, a few types of single-chip microcomputer in which the result can be sent directly to a memory location or an output register. In accumulator-based microcomputers the source register is usually a memory location, but we shall see later in section 3.5 that other microcomputer designs allow both source and destination registers to be either memory locations or registers within the microprocessor. Most microcomputers also allow single operand instructions to be carried out on either the accumulator or on memory locations.

In order to program the equation

$$p = l + m - n$$

the microcomputer is required to add together the contents of two memory locations L and M and subtract the contents of a third location, N. The result must then be stored in a fourth location P, so the operation must be programmed in four steps:

> Load the accumulator with the number in L,
> Add to the accumulator the number in M,
> Add to the accumulator the number in N,
> Store the answer in the accumulator in P,

Each operation in this sequence must be programmed as a separate operation in assembly language, which is equivalent to having to program

$accum := l;$
$accum := accum + m;$
$accum := accum - n;$
$p := accum$

instead of simply writing

$p := l + m - n$

Each of the four steps requires a different type of instruction, as this short fragment of assembly language shows:

3.2.1

LDA	L	Load accum. with number in L
ADDA	M	Add number in M to accum.,
SUBA	N	Subtract number in N from accum.,
STA	P	Store answer in P.

This is written in the assembly language of the 6809, which is a relatively sophisticated 8-bit microprocessor design with two accumulators called 'A' and 'B'. The 6809 will be used as the main example of a microprocessor throughout this chapter. In this example each of the operation codes ends with the letter 'A', which is used to indicate that it is the 'A', as opposed to the 'B', accumulator which is being used. The basic opcodes used are thus:

LD	LoaD register
ADD	ADD to accumulator
SUB	SUBtract from accumulator
ST	STore contents of register

Table 3.1 Arithmetic operations

Opcode	Operation
ADD	Add without carry
SUB	Subtract without carry
ADC	Add with carry
SBC	Subtract with carry
NEG	Negate
CLR	Set to zero
INC	Increment by one
DEC	Decrement by one

3.2.2 Arithmetic operations

Four arithmetic operations have now been described: addition and subtraction, and incrementing and decrementing by one. From the mathematical point of view, the latter pair are simply special cases of the former, but they are normally used quite differently. A list of the arithmetic instructions which can be carried out by the 6809 is given in Table 3.1. Those instructions which have not already been encountered will be introduced later in this chapter.

The minus sign is used in two ways in mathematics. In the statement

$total := value1 - value2$

it means that one quantity is to be subtracted from another, while in the statement

$value := -value$

it indicates that the sign of a value is to be reversed. The former example can be translated into assembly language as

```
LDA    VALUE1
SUBA   VALUE2
STA    TOTAL
```

In the latter case the 6809 instruction NEG can be used

```
NEG    VALUE
```

3.2.3 Shift and rotate operations

Although all microprocessors can add and subtract, they do not usually have instructions for multiplication and division. The programming of these operations will be described later, in section 3.10. If these operations are not available as instructions they can be carried out by programs using the shift and rotate instructions, the effect of which is to cause the contents of a register to be shifted left or right one bit position at a time. A list of the shift and rotate operations which can be performed by the 6809 is shown in Table 3.2.

Table 3.2 Shift and rotate operations

Opcode	Operation
ASR	Signed divide by two (Arithmetic shift right)
LSR	Unsigned divide by two (Logical shift right)
ASL	Multiply by two (Arithmetic shift left)
LSL	Logical shift left (same operation as ASL)
ROL	Rotate left
ROR	Rotate right

Shifting all the bits left one place in a binary number multiplies it by two in the same way that a decimal number is multiplied by ten if it is shifted left one place. A zero is inserted at the right hand side to fill the vacated position. When this is done in a microcomputer the bit which 'falls off' the left hand side of the register is kept in the carry register after the instruction is completed. The 'rotate left' instruction is similar except that the 'gap' is filled by the previous contents of the carry register. Some microprocessors also have another instruction which fills the 'gap' at one end of the register with the bit from the other end without affecting the carry register. In the 6809 the **arithmetic shift left** instruction is given the mnemonic code ASL, while **rotate left** is represented by ROL. The action of these instructions upon the contents of a register is shown in Figure 3.1.

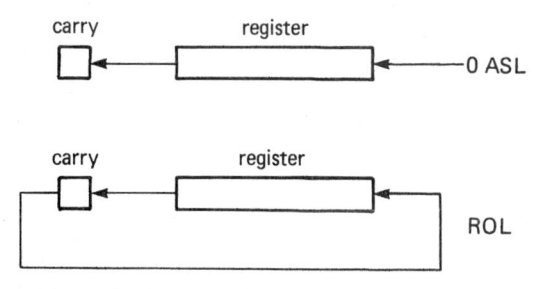

Figure 3.1 Rotate left and shift left operations

Both signed and unsigned numbers are multiplied by two by shifting left and inserting a zero at the right hand side. A binary number can similarly be divided by two by shifting it right, but in this case signed and unsigned numbers are handled differently and the programmer must take care to choose the correct instruction. The left-most bit of a two's complement number contains the sign, which must not be affected when shifting right, but the left-most bit of an unsigned number simply contains the most significant bit. The rule for shifting right with the 6809 is as follows: when the number is unsigned, use the LSR instruction which inserts a zero at the left; when it is a two's complement number use the ASR instruction which preserves the sign at the left, as Figure 3.2 illustrates.

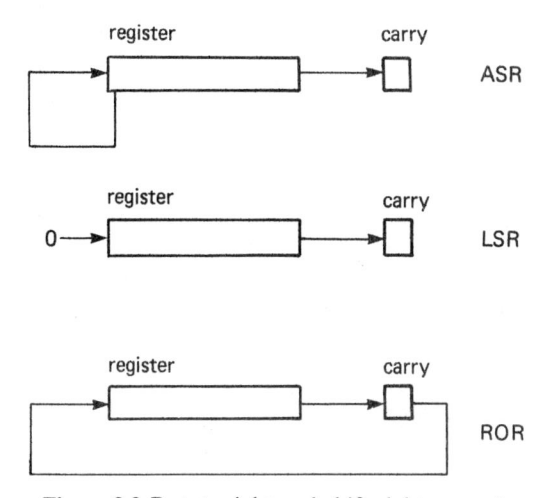

Figure 3.2 Rotate right and shift right operations

3.2.4 Logical operations

The logical operations AND, OR, Exclusive OR, and NOT can also be carried out using single instructions in most types of microprocessor. In the 6809 these are given the mnemonic codes AND, OR, EOR, and COM (for one's COMplement) respectively, as shown in Table 3.3.

Table 3.3 Logical operations

Opcode	Operation
AND	Logical 'AND'
OR	Inclusive 'OR'
EOR	Exclusive 'OR'
COM	Logical 'NOT'

Boolean expressions can be programmed quite easily using these few operations together with the load and store operations; for example

$$L = M + N.P \tag{3.1}$$

is carried out by the instructions

LDA	N	Accumulator A contains N
ANDA	P	Accumulator A contains N.P
ORA	M	Accumulator A contains $M + N.P$
STA	L	Store result in L

These instructions operate on corresponding bits in each word so that bit i of the result depends only upon bit i of each of the operands.

3.2.5 Relational operations

The six relational operations $<$, $\leq$, $=$, $\geq$, $>$ and $\neq$ are available in the 6809, but they are used to implement programming constructs such as 'if' and 'repeat', and will be discussed later in the section 3.4.3.

3.3 Addressing modes

The way in which the operand is specified within an instruction is known as the **addressing mode** of that instruction. Microprocessors offer a wide range of addressing modes, but this section will concentrate upon those modes which are common to most types of microprocessor and which allow programs to handle constants, variables, and arrays.

3.3.1 Variables: direct addressing

The mode which has been used in examples so far in this chapter, in which the operand is the contents of a memory location, is called **direct addressing**. The operand of the instruction indicates the location where the value of a **variable** is stored. The value of a variable can be changed by storing a new value into it using instructions such as ST or by directly modifying it using instructions such as INC.

Direct addressing is used where a scalar variable occurs within an assembly language program.

3.3.2 Constants: immediate addressing

Another widely used addressing mode allows the programmer to specify the actual value to be used rather than the address where it is to be found. This is called **immediate addressing** and is used where constants are to be introduced into a program. In the assembly language of the 6809 immediate addressing is indicated by prefixing the operand with a 'hash' sign, '#'. This can be read as meaning 'the number' or 'the value of', so that the instruction

        LDA   #8

can be read as meaning 'load accumulator A with the number 8'. Immediate addressing can be used with any 6809 instruction where it makes sense to do so, For example, the Pascal statement

total := *value* + 55

adds 55 to the value of *value* and stores the result as *total*. This can be expressed in assembly language as

        LDA   VALUE   Variable operand
        ADDA   #55   Constant operand
        STA   TOTAL   Variable result.

which adds 55 to the value of VALUE and stores the result in location TOTAL. However

        INC   #99

is not allowed, since it attempts to make the number 99 equal to 99+1!

3.3.3 Arrays: indirect addressing

In **indirect addressing** the operand specifies a register or location which contains the address of another location which holds the actual data to be operated upon. Although this might seem at first sight a bit obscure, it is a most important facility because by changing the contents of the first **pointer** register it allows variable effective addresses to be used, for example in accessing the elements of **subscripted arrays**.

Suppose that an array of 100 elements is stored in successive locations in memory, as shown in Figure 3.3. If the address of the 'zeroth' element, the one with subscript zero, is ARRAY, that of the element with subscript one will be ARRAY+1, and that of the *i*th element, the element with

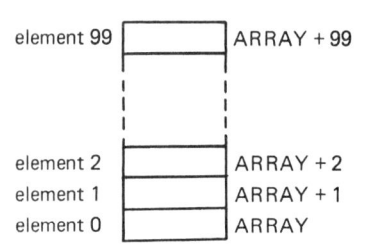

```
                element 99  ┌──────┐  ARRAY + 99
                            │      │
                            :      :
                            :      :
                 element 2  ├──────┤  ARRAY + 2
                 element 1  ├──────┤  ARRAY + 1
                 element 0  │      │  ARRAY
                            └──────┘
```

Figure 3.3 Array of 100 words

subscript i, will be ARRAY + i. This address can be calculated quite easily in assembly language; first the address of the start of the array is loaded into the accumulator, and then an amount equal to the subscript i is added to produce the address of the corresponding element. This address is stored in address INDRCT.

```
    LDA     #ARRAY      Get address of first element
    ADDA    I           add subscript and store address
    STA     INDRCT      of ith element in location INDRCT.
```

In fact, this will not work as written with the 6809, because the accumulators of the 6809 handle 8-bit data, while addresses are 16 bits long. There is a simple solution to this which will be discussed more fully in section 3.4, but for now it is sufficient to note that the sequence

```
    LDD     #ARRAY      Get address of first element
    ADDD    I           add subscript and store address
    STD     INDRCT      of ith element in location INDRCT.
```

will cause the 6809 to compute a 16-bit result suitable for use as an address.

The use of the indirect addressing mode is indicated in 6809 assembly language by means of square brackets. If the contents of the ith element of the array were to be loaded into accumulator A, the instruction

```
    LDA     [INDRCT]
```

would be needed after this program fragment.

3.3.4 Indexed addressing

Most 8-bit microprocessor designs have **index registers** which allow variable addresses to be used in a program. The function of an index register is to contain an address which 'points' to locations in memory like an imaginary 'index finger'. In this respect it closely resembles the pointers used in Pascal. The index register allows an addressing mode similar to indirect addressing to be used, and the last example could be reprogrammed to use the index register:

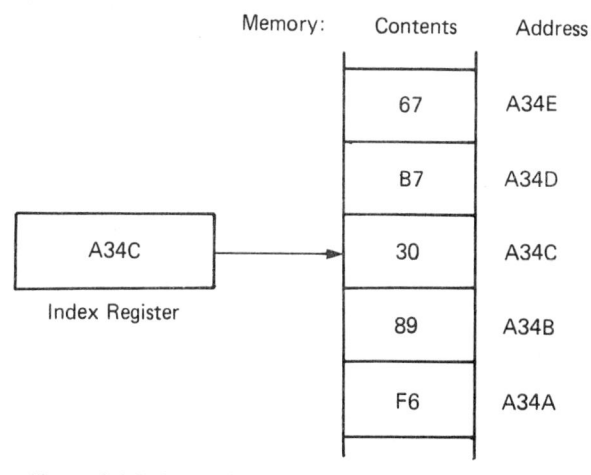

```
Memory:     Contents    Address

              67        A34E
              B7        A34D
A34C  ──▶    30        A34C
Index Register
              89        A34B
              F6        A34A
```

Figure 3.4 Index register 'points' to a memory location

```
LDD     #ARRAY      Get address of first element
ADDD    I           add subscript and transfer address
TFR     D,X         thus generated to index register X.
```

Here the instruction TFR serves to TransFeR the contents of accumulator D to the index register X; this instruction will be discussed further in section 3.5.1. Now accumulator A can be loaded from the ith element of the array simply by using the index register as a pointer:

```
LDA     ,X
```

Thus far, indexed addressing has been shown to closely resemble indirect addressing. However, the advantage of indexed addressing stems from its ability to add an **offset** to the number in such a register to create an **effective address**. For example, the 6809 instruction

```
LDA     8,X
```

means 'take the number currently in the index register and add eight to it. The result is the address of a memory register; load the contents of that register into accumulator A.' If the index register contains the hexadecimal number 4321, the effective address will be

$$4321 + 8 = 4329$$

Most 8-bit microprocessors have index registers which can be used with offsets, but the 6809 offers many variations in the way in which the index register can be used. For example, either accumulator A or B can be used to provide the offset information, or the concatenated accumulators A and

B can be used as a single D accumulator to provide the offset. The sequence

```
LDX     #ARRAY      Point to start of array
LDD     I           Pick up offset
LDA     D,X         Load accumulator A with element i
```

loads accumulator A with the contents of element i of an array in the same way as the previous examples; the operand D,X indicates that the effective address is the sum of the contents of index register X and accumulator D.

Instructions are available for manipulating the contents of the index register, for example to load it with a new value or to store the current value in memory. Thus the sequence

```
LDX     #2000
STX     1234
```

loads the number 2000 into the index register and then stores that value at address 1234. Note that the same two codes ST and LD are used with the letter 'X' indicating that it is index register X which is to be loaded and stored.

3.3.5 A comparison of addressing modes

The way in which these addressing modes can be used to access the information used by an operation code are summarised in Figure 3.5. This is by no means an exhaustive list of the modes used by microcomputers, and many more will be introduced before the end of this chapter.

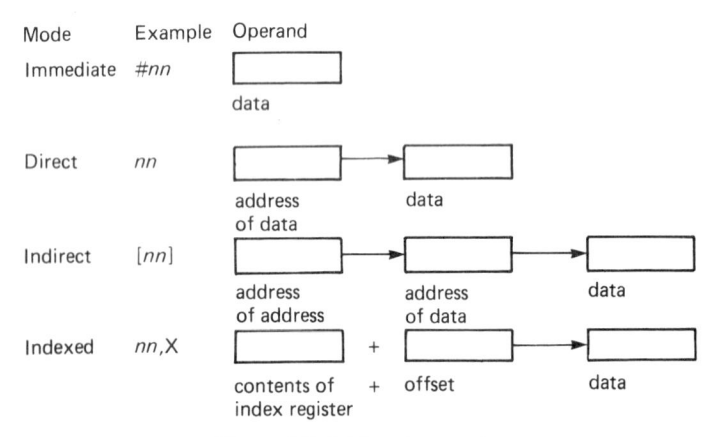

Mode	Example	Operand		
Immediate	#nn	data		
Direct	nn	address of data	→ data	
Indirect	[nn]	address of address	→ address of data	→ data
Indexed	nn,X	contents of index register	+ offset	→ data

Figure 3.5 Addressing modes

3.3.6 Representing constants

The examples which have been used during this discussion of addressing modes have each assumed that all the constants used were decimal numbers. However, the assembly language programs which run on real microcomputers often use constants which are not decimal numbers. These assembly languages usually permit the programmer to insert special types of constant into a program. In 6809 assembly language the 'percentage' symbol (%) is used as a prefix to indicate binary numbers and the 'dollar' symbol ($) can be used in the same way to indicate hexadecimal numbers. For example, the three instructions

```
        LDA     #%01010101
        LDA     #$55
        LDA     #85
```

each load accumulator A with the same bit pattern 01010101, which can be written in hexadecimal form as 55 and which corresponds to the decimal value 85. The choice of representation is at the discretion of the programmer, and depends upon the particular program. Each of these instructions will translate into the same binary machine code instruction. Yet another prefix character, the single quote character or apostrophe (') is used to indicate that the ASCII code of the following character is to be used. Thus the instruction

```
        LDA     #'U
```

will produce the same machine code as the previous examples, because the ASCII code for the capital letter 'U' is 01010101.

As an example of the use of these different number representations, consider the problem of converting the ASCII code for a decimal digit into a single byte containing a binary number in the range 00000000 to 00001001. The symbols '0' to '9' are represented in ASCII code by the hexadecimal codes from 30 to 39. When characters are entered from a keyboard they are in ASCII form, but they can be changed to their numerical equivalents either by 'masking' them with the binary pattern 00001111 or by subtracting 00110000. The program fragment below shows how an ASCII code in register 'CODE' can be converted to a number in register 'VALUE'.

```
        LDA     CODE            Get ASCII code
        ANDA    #%00001111      Strip off leading '3'
        STA     VALUE           Save result
```

This example could be rewritten more neatly as

```
LDA     CODE        Get ASCII code
ANDA    #$0F        Strip off leading '3'
STA     VALUE       Save result
```

The machine code produced in either case will be identical, since hexadecimal 0F is equivalent to binary 00001111. The same result could have been obtained by subtracting hexadecimal 30 from the code to leave the value:

```
LDA     CODE        Get ASCII code
SUBA    #$30        Subtract ASCII code for '0'
STA     VALUE       Save result
```

This can be expressed more readably by using an ASCII code:

```
LDA     CODE        Get ASCII code
SUBA    #'0         Subtract ASCII code for '0'
STA     VALUE       Save result
```

Again, the machine code produced by these two examples will be the same because the ASCII code for the symbol '0' has the hexadecimal value 30.

3.4 Data types

Most of the examples which have been used in the first part of this chapter have not indicated the form of the data being operated upon. In fact the 6809 is an 8-bit microprocessor; each memory location contains exactly one byte of information and the accumulators A and B also contain one byte. We have seen that the 6809 offers a solution to the problems posed by this restricted wordlength by enabling its two accumulators to be used end-to-end as a single 16-bit accumulator called D. Unfortunately, few microcomputer designs offer a double-length accumulator, of this sort, and other techniques must be used.

3.4.1 Multiple-word arithmetic

It might be tempting to assume that the maximum size of number which can be used in an arithmetic expression is limited by the wordlength of the microprocessor used. However, this is not the case, because another register is involved with arithmetic operations: the single bit **carry register**. This is set to 1 if the result of an addition or subtraction falls outside the range of numbers which can be held in the accumulator; otherwise it is reset to 0. If the carry register is set it implies that the composite calculation needed to carry to, or borrow from the next column. To show this more clearly, consider the effect of addition in an 8-bit machine; eight bits allow unsigned numbers in the range 0 to 255 to be represented.

$$100 + 100 = 200 \tag{3.2}$$

The result is within the range which can be held in the accumulator, so the accumulator contents will be 200 and the carry register will contain 0. Suppose that the addition had been

$$200 + 200 = 400 \tag{3.3}$$

The answer now exceeds the permitted range of values which the microcomputer indicates by placing a 1 in the carry register. The accumulator will contain 144 ($= 400 - 256$).

A 16-bit addition can be programmed as two consecutive 8-bit additions. First the two less significant bytes of the numbers are added using the ADD instruction to give the less significant byte of the result:

```
    LDA     LLESS       The LSB of number L
    ADD     MLESS       The LSB of number M
    STA     NLESS       The LSB of the result
```

Next the two more significant bytes must be added; this time the contents of the carry bit must be included in the addition and a different **add with carry** instruction is used

```
    LDA     LMORE       The MSB of number L
    ADC     MMORE       The MSB of number M
    STA     NMORE       The MSB of the result.
```

In general the ADD instruction is used with the least significant word and ADC is used with all subsequent words when a multiple-word addition is carried out using a 6809. Similarly, SUB is used with the least significant word of a subtraction and SBC with the other words. In this way numbers of any desired length can be added using the carry bit to link together the successive phases of the operation.

The address calculation example can be rewritten as a pair of

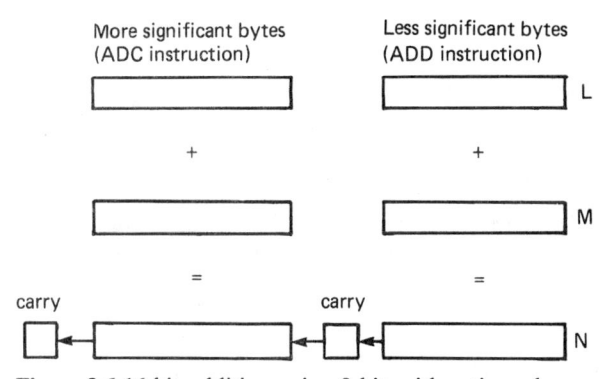

Figure 3.6 16-bit addition using 8-bit arithmetic and carry

single-byte additions producing a two-byte address in a pair of memory locations:

```
LDA   #<ARRAY        Get less significant byte of
ADDA  I+1            address ARRAY, add less significant
STA   INDRCT+1       byte of I and store as LSB of INDRCT.


LDA   #>ARRAY        Get more significant byte of
ADDA  I              address ARRAY, add more significant
STA   INDRCT         byte of I and store as MSB of INDRCT.
```

The subscript is assumed to be in a pair of memory locations I and I+1 and the result is a two-byte address in locations INDRCT and INDRCT+1 (the next location in memory). Note the use of the operators < and >. These are used in 6809 assembly language to indicate the less significant and more significant bytes respectively of a two-byte quantity. Note that

```
<ARRAY = ARRAY mod 256
>ARRAY = ARRAY div 256
```

Some assembly languages allow the programmer to use the operations 'mod' and 'div' explicitly.

3.4.2 Multiple-byte shift and rotate operations

The carry bit is used to link together successive shift and rotate operations when numbers consisting of more than one word are to be shifted. A 16-bit number stored in locations LMORE and LLESS can be shifted right in two steps by 8-bit microprocessors such as the 6809:

```
ASR   LMORE          More significant byte
ROR   LLESS          Less significant byte
```

Figure 3.7 shows how the bit lost from the register at the left is kept in the carry register after the shift instruction, and passed to the next register in the rotate instruction that follows. Multiple shifts of any number of bytes and in either direction can be programmed by using a shift instruction followed by the appropriate number of rotate instructions. When shifting right the process should start with the left-most word and then rotate successive words to the right. Left shifts should start with the right and work towards the left. The two examples which follow treat a set of three registers called MOST, MIDDLE and LEAST as a composite 24-bit register and shift it left and right.

```
ASL   LEAST          Composite shift left...
ROL   MIDDLE
ROL   MOST
```

3.4.2

```
LSR    MOST         Composite shift right...
ROR    MIDDLE
ROR    LEAST
```

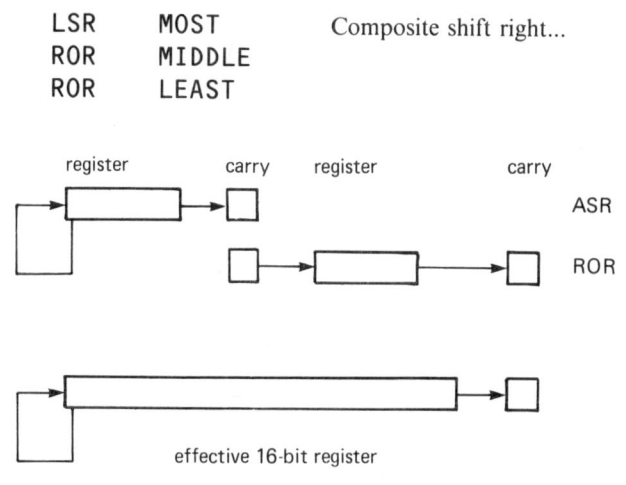

Figure 3.7 Programming a 16-bit arithmetic shift right

3.4.3 Binary coded decimal numbers

All the arithmetic examples in this section have assumed that data is represented as binary numbers which are one or more bytes in length. Sometimes **Binary Coded Decimal (BCD)** representation, which packs two decimal digits each represented as 4-bit binary numbers into a single byte, is more attractive. Curiously, BCD constants must be represented in assembly language with the prefix $ used to represent hexadecimal numbers. Thus, while the instruction

```
LDA    #55
```

loads accumulator A with the binary representation of the decimal number 55, the instruction

```
LDA    #$55
```

will load it with the BCD representation of 55.

In order to program BCD addition, the 'decimal adjust accumulator' instruction DAA must be used immediately following an ADDA or ADCA instruction, that is, an addition which references accumulator A. This corrects the result of the addition to produce one which corresponds to BCD, as opposed to binary, addition. For example,

```
        LDA     BCDNUM          Get BCD number,
        ADDA    #$12            add 12 in BCD format,
        DAA                     convert result to BCD form,
        STA     BCDNUM          and store sum.
```

adds the number 12 to the BCD number BCDNUM. Multiple-byte BCD additions can be carried out just like multiple-byte binary addition.

```
        LDA     BCDLO           Get less significant byte,
        ADDA    #$34            add BCD 34,
        DAA                     correct sum,
        STA     BCDLO           and store.
        LDA     BCDHI           Get more significant byte,
        ADCA    #$12            add BCD 12,
        DAA                     correct sum,
        STA     BCDHI           and store.
```

Figure 3.8 Four-digit BCD addition of 1234

The example of Figure 3.8 adds the four-digit BCD number 1234 to the contents of the two-byte number stored in locations BCDHI and BCDLO. Unfortunately the instruction DAA works only for addition and for accumulator A, and it does not work after subtraction.

In order to program the BCD subtraction

$bcd1 := bcd2 - bcd3$

the second operand 'bcd3' must be negated and then added to 'bcd2'. The 'trick' is to subtract the BCD number from 99 and add one to the result. The result is then added to the other operand; an example of this rather tortuous procedure is shown in Figure 3.9. Note that this subtracts 'bcd3' from the BCD constant 9999, and then adds 1.

Left and right shifts of decimal numbers involve four successive shifts of BCD numbers, since four bits are used to represent each decimal digit. This can be programmed with BCD variables of any number of digits simply by carrying out the appropriate multiple word shift sequence four times.

3.4.4 Combining shifts with logical operations

Bit patterns can be moved to the left or right using the shift and rotate instructions, and this ability to shift bit patterns is especially important when used in conjunction with the AND and OR instructions. In the example of Figure 3.10, the ASCII codes stored in registers DIGIT1 and DIGIT2 are

Programming the BCD subtraction $bcd1 := bcd2 - bcd3$

First compute $temp := 9999 - bcd3$:

BCDSUB	LDA	#$99	Get BCD value 99
	SUBA	BCD3L	subtract l.s. byte of BCD3
	STA	TEMPL	and put into temporary storage.
	LDA	#$99	Get BCD value 99
	SUBA	BCD3M	subtract m.s. byte of BCD3
	STA	TEMPM	and put into temporary storage.

(*Repeat second part for as many bytes as necessary.*)

Next compute $bcd1 := bcd2 + temp + 1$:

LDA	BCD2L	Get l.s. byte of BCD2
SEC		set C=1, and add this and
ADCA	TEMPL	l.s. byte of 'temp'; correct
DAA		l.s. part of sum 'bcd2+temp+1',
STA	BCD1L	and store it.
LDA	BCD2M	Get m.s. byte of BCD2
ADCA	TEMPM	add m.s. byte of 'temp'
DAA		correct result
STA	BCD1M	and store it.

(*Repeat second part for as many bytes as necessary.*)

Figure 3.9

---

converted first to a pair of decimal digits and then into a pair of binary coded decimal digits packed into one byte.

3.5 Data handling in other microcomputers

There is a fundamental difference between the accumulator and index registers. Accumulators are used to handle data during the course of a calculation, while index registers are used to handle addresses. Larger microprocessor designs may have several data registers and several address registers, and in some cases the same registers may be used to handle both data and addresses.

```
ASCBCD  LDA   DIGIT1      Strip hexadecimal '30' off
        ANDA  #$0F        each digit;
        STA   DIGIT1
        LDA   DIGIT2
        ANDA  #$0F
        STA   DIGIT2
        LDA   DIGIT1      then get first digit and
        ASLA              shift it four places left;
        ASLA
        ASLA
        ASLA
        ORA   DIGIT2      combine with second digit
        STA   BCD         and put result into register BCD.
```

Figure 3.10 ASCII to BCD conversion

3.5.1 Register-to-register architectures

The availability of several data and address registers within a microprocessor allows a more flexible approach to the problem of addressing the operands of an instruction. Most larger microcomputers and minicomputers are of the **register-to-register** type, which means that where necessary an instruction can specify two memory addresses. This is easier to do in a machine with a wordlength of 16 or 32 bits than is the case with an 8-bit microcomputer.

One example of such a microprocessor is the 68000, the circuit of which contains eight accumulators known as **data registers**, numbered D0 to D7. It also has eight registers which can be used like index registers to 'point' to memory locations. These **address registers** are numbered A0 to A7. Programming the 68000 is somewhat more complicated than programming the 6809, but only because each instruction offers the programmer a range of addressing modes for each operand. For example, the instructions

```
ADD   #10,D0
SUB   D1,D2
```

add a constant 10 to data register D0 and subtract the contents of data register D1 from those of D2. The result of the operation is stored at the address indicated in the second operand, which for this reason is known as the **destination operand** while the first operand is known as the **source operand**.

In general the opcodes used by the 68000 are the same as those of the 6809, the main difference being the general purpose operation MOVE which can be used to load a register from a memory location and to store the contents of a register in memory:

3.5.1

```
MOVE    #10,D0
MOVE    SNOOPY,D0
MOVE    D0,SNOOPY
```

Indirect addressing is specified by enclosing the operand in brackets; thus

```
MOVE    (A0),D1
```

means 'move to D1 the number in the location whose address is in A0'.

The data and address registers within the 68000 are each 32 bits long and opcodes can be used with a suffix to indicate the number of bytes to be used in an instruction. The three possible suffixes are .B (for 'Byte') for single byte operands, .W (for 'Word') for two byte operands, and .L (for 'Long') when four byte operands are used:

```
MOVE.B  MYWORD,D0    Move a byte from MYWORD to D0
MOVE.W  MYWORD,D0    Move two bytes from MYWORD to D0
MOVE.L  MYWORD,D0    Move four bytes from MYWORD to D0
```

Despite having 32-bit registers within the microprocessor and a 16-bit data bus, the 68000 addresses memory as if it were organised as 8-bit bytes. When an instruction moves several bytes at a time, the bytes will be moved to or from a block of locations starting at the address specified in the instruction's operand.

The LSI11 microprocessor is another microprocessor with a register-to-register design. It has eight registers, each 16 bits long, which can each be used as data or address registers. The instructions used to manipulate and move data within the microcomputer are nevertheless similar to those of the 68000. For example the LSI11 instruction

```
MOV     R0,R1
```

means 'copy the number in register R0 to register R1'. Thus R0 is the source register and R1 is the destination register. The LSI11 microprocessor has the same instruction set as the PDP11 series of minicomputers.

The example above used **register direct addressing**; the symbols R0 and R1 directly specified the registers whose contents were to be used. If the operands had been bracketed, **register indirect addressing** would have been implied:

```
MOV     (R0),(R1)
```

This copies the number in the location whose address is in register R0 to the location whose address is in register R1. In the assembly language of the LSI11 there is another way of indicating indirect addressing, which consists of placing an 'at' symbol before the operand concerned:

```
        MOV      @R0,@R1
```

Note that registers R0 and R1 are here used to hold addresses, while in the case of direct addressing they were use to hold data.

Although some microprocessor designs such as that of the LSI11 allow almost any register to carry out almost any function, there must always be some degree of specialisation. This arises partly from the limitations imposed by integrated circuit technology, although these are no longer as restrictive as they were during the 1970s. More important is the fact that if two complete addresses had to be specified with every operation, instructions would become too long to be accommodated in a single word. In virtually all practical microcomputer designs some registers, normally those of the microprocessor itself, are referenced in some 'shorthand' way rather than using addresses.

3.5.2 Accumulator-based architectures

The three types of microprocessor discussed so far provide useful examples because they have particularly 'regular' instruction sets with few 'gaps'; instructions exist to allow the programmer to express a wide range of arithmetic and logical operations directly.

One of the main differences between the instruction sets of different microprocessors lies in the way in which the programmer can instruct the microprocessor to move information between registers of the microprocessor and to and from memory locations. Some microcomputers have a range of special instructions indicating moves between registers within the microprocessor itself. For example, in the 6800 microprocessor, the instruction TAB copies the contents of accumulator A to accumulator B, while in the 6502, the instruction TXS copies the contents of an index register called X to another register called S.

The 6809 uses a more generalised instruction called TFR, which is followed by a pair of single letter operands indicating the two registers involved. Thus the two instructions

```
        TFR      A,B
        TFR      X,S
```

copy the contents of accumulator A to accumulator B and the contents of the X register to the S register respectively. In this respect TFR resembles the 'move' instructions of the 68000 and LSI11, except that they may be used only with register direct addressing.

Some 8-bit microprocessors such as the 8085 do have a general purpose 'move' instruction. Thus the 8085 instruction

```
        MOV      B,C
```

causes the contents of register C to be copied to register B. Unfortunately the order of the source and destination operands is reversed in the assembly language of this microprocessor, which can provide a trap for the unwary. It would also be tempting to assume that if the constant 50 were to be loaded into register B, the use of immediate mode could be indicated by a '#' symbol as with the microprocessors already discussed:

> MOV   B,#50   WRONG!!

but the assembly language for this microprocessor uses a special 'move immediate' opcode for this purpose, and the instruction should be coded as

> MVI   B,50   CORRECT

3.5.3 Binary coded decimal arithmetic

Because 8-bit microcomputers are frequently called upon to manipulate binary coded decimal (BCD) numbers, most types have special instructions to facilitate the handling of them. Many types, including the Z80, 8085, and 6800, have the **decimal adjust accumulator** instruction which was described in section 3.4.3. Instead of using a 'decimal adjust' instruction, the 6502 can be switched between **decimal** and **binary arithmetic modes** by special instructions.

Left and right shifts of decimal numbers involve four successive shifts of BCD numbers, since four bits are used to represent each decimal digit. A few types of 8-bit microcomputer have special four-bit shift operations to simplify the programming of this sort of operation.

A more detailed comparison of five commonly used types of microprocessor is given in Appendix I.

3.6 The processor status register

The **carry flag C** is one of several single bit registers which contain information concerning the current status of the processor. These individual bits are usually grouped together to form a composite register known variously as the **processor status register**, the **condition code register**, or the **flag register**, This is because when the contents of the register are set to logic '1', it is to draw attention to the fact that something has happened in the system, rather like raising a flag at a football match or athletics meeting to indicate that some transgression of the rules of the game has occurred! Whatever it is called, this composite register contains a summary of the result of the last operation: the sign of the result, whether it was zero, and so on.

3.6.1 The carry flag

As we have seen in the preceding section, the function of the **carry flag C** or **link bit** is to link together multiple operations such as additions, subtractions, and shifts. It can also be interpreted as indicating whether the result of an arithmetic operation lies outside the range of unsigned numbers which can be represented within the wordlength being used: for an 8-bit microcomputer this range is 0 to 255. The two functions are sometimes performed by different flags, for example the 68000 uses a flag designated 'X' to link shift operations.

3.6.2 The overflow flag

In addition to C, most microcomputers have an **overflow flag**, or **V bit** which is set to 1 if the result of an arithmetic operation caused the answer to fall outside the two's complement range of the accumulator. For an 8-bit microcomputer this range is from -128 to $+127$; if we carried out the addition

$$100 + 100 = 200 \tag{3.4}$$

using two's complement notation, the result would fall outside this range and the V bit would be set to 1. On the other hand, after the addition

$$50 + 50 = 100 \tag{3.5}$$

the V bit will be cleared to zero because the result lies within the range -128 to $+127$. In effect, the overflow bit raises a 'flag' indicating that something has happened in the operation of the program; later in this chapter we shall see how this flag information is used to modify the flow of a program.

Note that it is the overflow flag V which tests for overflow of signed two's complement numbers while it is the C flag which indicates whether unsigned results fall outside the permitted range. In some microcomputers the V bit is also set if a shift or rotate left causes the sign bit to change, indicating that two's complement overflow has taken place.

3.6.3 The negative flag

The sign of a two's complement number is contained in the left hand bit of the number representation. If this is a '1' the number is negative and the **N bit** or **Negative flag** (some microcomputers use the letter S for 'sign') is set, otherwise it is cleared to '0'.

3.6.4 The zero flag

The function of the **Zero flag** or **Z bit** is to indicate whether the result of the last operation resulted in zero or not. If the result was zero, the Z bit is set to '1', otherwise it is reset to '0'.

3.6.5 Other flags

The flags described above are found in most microcomputers, but some have additional flags to indicate further conditions such as parity (whether the number of 1's in the binary result is odd or even). One flag, the **half-carry**

3.6.5

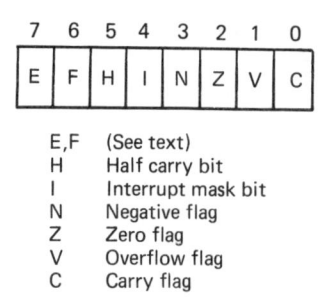

```
 7  6  5  4  3  2  1  0
┌──┬──┬──┬──┬──┬──┬──┬──┐
│E │F │H │I │N │Z │V │C │
└──┴──┴──┴──┴──┴──┴──┴──┘
```

E,F (See text)
H   Half carry bit
I   Interrupt mask bit
N   Negative flag
Z   Zero flag
V   Overflow flag
C   Carry flag

Figure 3.11 Processor status register of 6809

flag or **H bit**, is used by the 'Decimal Adjust Accumulator' instruction. It indicates whether there was a carry from the right-hand block of four bits to the left-hand block. The processor status register also contains information as to whether normal execution of the program can be interrupted by external events. This important topic of interrupts will be discussed later in Chapter 4. The E, F, and I bits of the 6809's processor status register, shown in Figure 3.11, are associated with the handling of interrupts.

3.6.6 Setting and clearing the flags

Although the principal function of the flag bits is to indicate the outcome of an arithmetic or logical operation, special instructions are available to directly set and clear these bits under program control. In most microprocessor designs these instructions control individual bits, so that SEC and CLV would set the C bit and clear the V flag respectively. The 6809 recognises these operation codes, but a more general pair of operations is available:

ANDCC   performs a logic 'AND' between an immediate byte and the contents of the processor status register and leaves the result in the PSR, and is used to clear selected bits in the PSR.

ORCC   performs a logic 'OR' between an immediate byte and the contents of the processor status register and leaves the result in the PSR, and is used to set selected bits in the PSR.

3.7 Controlling program flow

The instructions which have been encountered up to this point in the chapter have all been concerned with the manipulation of data. In this section a new group of instructions will be considered. These instructions affect the flow of execution of a program and they will be used to implement the programming constructs which were discussed in Chapter 2.

3.7.1 The program counter

The microcomputer uses a special purpose register called the **program counter (PC)** to keep track of where it is in the program. The program counter register always contains the address of the next instruction to be carried out, and as the program executes the address which the PC contains it is automatically incremented to point to successive instructions.

Microcomputer instructions are carried out in two phases called **fetch** and **execute**. During the first phase the program counter provides addresses for the binary opcode and operand information to be fetched from memory and in the second phase the microcomputer acts upon this information and executes these instructions. The program counter is so called because its contents are incremented by one each time a word is fetched from memory, and most of the time it continues in this mode with the microcomputer fetching the next instruction when it has finished executing the current one.

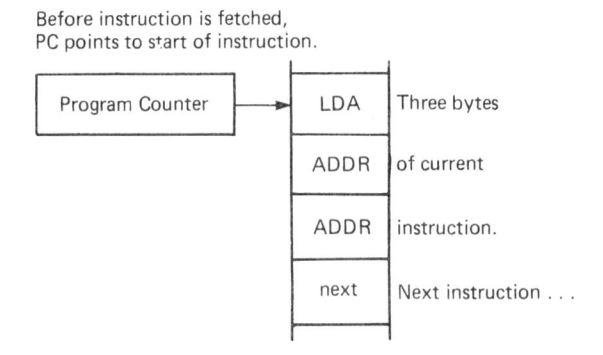

Before instruction is fetched,
PC points to start of instruction.

Program Counter →	LDA	Three bytes
	ADDR	of current
	ADDR	instruction.
	next	Next instruction . . .

After instruction has been fetched,
PC points to start of next instruction.

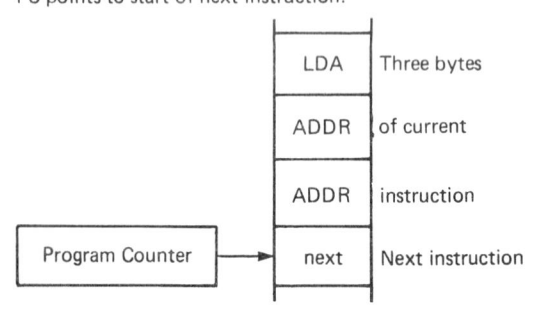

	LDA	Three bytes
	ADDR	of current
	ADDR	instruction
Program Counter →	next	Next instruction

Figure 3.12 Behaviour of the program counter

Note that because the program counter has been used to fetch all the bytes of the current instruction into the microprocessor by the time that it starts to execute the instruction, the program counter will point to the start of the next instruction while it is executing the current one. This is illustrated in

Figure 3.12. The program counter is used whenever instructions are fetched from memory and ensures that they are executed in numerical sequence. If this were all that the program counter were capable of, however, its usefulness, and the usefulness of microcomputers in general, would be severely limited.

3.7.2 The jump instruction

Practical programs contain **jumps** where execution of the program continues at a different part of the program instead of at the next line, like GOTO in BASIC. These operations must be programmed using a different type of instruction, one which affects the flow of the program rather than processing data.

Transferring control to a different part of the program is achieved quite simply by loading the program counter with the address of the next instruction to be carried out. In the 6809 this is done using the instruction JMP. The 'nonsense' program below is a loop which runs indefinitely incrementing the contents of a register each time the program restarts:

```
BEGIN   INC   COUNT      Add 1 to the number in COUNT
        JMP   BEGIN      Go back and do it again...
```

Note the label BEGIN in this example. The use of a label on any line is optional, but labels are often used to identify a line so that it can be referenced elsewhere in the program as the destination of jump instructions.

3.7.3 Branch instructions

The limitation of the jump instruction is that it is unconditionally obeyed. Another type of instruction is needed to allow the equivalent of IF statements to be programmed, which is where the flag bits become useful. **Branch** instructions are essentially the same as jumps with the exception that they are obeyed only if the flag bits are set to the correct state for the appropriate instruction to be effective. If the flag bits are not in that state the branch instruction is ignored and the program goes on to the next instruction instead.

The range of branch instructions available to the programmer varies considerably between microcomputer types. The instructions described here will be those for the 6809 which has a fairly large range; the 6800 and 6502 both use a subset of this range. The simplest branch instructions test individual bits of the flag register to determine whether or not to branch. Thus BCS will cause the program to branch if the carry bit is set to 1 while BCC will cause it to branch if C is 0. The BVS and BVC instructions act similarly with the overflow bit V. The instructions which test the sign bit N are BPL (Branch on PLus, when $N=0$) and BMI (Branch on MInus, when $N=1$). The Z bit is often tested after subtracting one quantity from another to compare them. BEQ (Branch on EQual) will cause the branch to be taken if they were the same so that $Z=1$ and BNE (Branch if Not Equal) will cause the branch to be taken if $Z=0$.

The 6809 thus has eight instructions for testing the status of the four bits N,Z,V and C as shown in Table 3.4.

Table 3.4 Branch on flag bits

Branch if:	Bit=0	Bit=1
N (Negative)	BPL	BMI
Z (Zero)	BNE	BEQ
V (Overflow)	BVC	BVS
C (Carry)	BCC	BCS

It is also useful to have branch instructions which directly test combinations of the flag bits for conditions such as 'greater than', 'less than or equal to', and so on, because branch instructions are used to implement high level constructs such as IF. The rules for determining these conditions are different for signed and unsigned numbers, as may be seen in Table 3.5, which shows two sets of 'arithmetic' branch instructions. On some designs of microcomputers the appropriate combination of branch instructions from Table 3.4 must be used, but in the case of the 6809 all of the 'arithmetic' branch instructions shown in Table 3.5 are available.

Table 3.5 Flag bits and arithmetic tests

Branch if	Two's Comp. Signed		Unsigned	
	Opcode	Test on Flags	Opcode	Test on Flags
$\neq$	BNE	(Z=0)	BNE	(Z=0)
>	BGT	(Z=0) and (N=V)	BHI	(Z=0) and (C=0)
$\geq$	BGE	(N=V)	BHS	(C=0)
=	BEQ	(Z=1)	BEQ	(Z=1)
$\leq$	BLE	(Z=1) or (N$\neq$V)	BLS	(Z=1) or (C=1)
<	BLT	(N$\neq$V)	BCS	(C=1)

In addition to the conditional branches shown in Tables 3.4 and 3.5, the 6809, in common with some other microprocessor designs, offers two instructions which are described as branches but which do not cause a 'branching' of program flow into one of two possible directions. The first of these is BRA (branch always) which always causes a jump to the indicated address, and BRN (branch never) which never causes a jump. The former is useful because although it has the same effect as a jump instruction it is coded more compactly in machine code. The latter is almost an 'accident' of the way in which the microprocessor functions, but it is sometimes used in testing programs; substituting a BRA or a BRN opcode for a the opcode of a conditional branch can force a section of program to select one or other path when it is executed.

3.7.4 Data test instructions

Although two quantities can be compared by subtracting one from the other and testing the result, it is useful to be able simply to compare the contents of a register with a test value without losing the data currently in that register. In many microprocessors this can be done by using a **compare** instruction which sets the flag bits in the same way as subtract but which discards the answer instead of leaving it in the accumulator. This instruction, called CMP in the 6809, is an example of a data test instruction. Other such data test instructions available in many microprocessor designs are the **Bit test** instruction BIT which performs an AND operation but discards the answer and the **Test** instruction TST which allows the contents of the accumulator or a memory register to be compared with zero. Note that the compare and test instructions treat the contents of a register as a numerical value, while the bit test instruction tests bit patterns in the accumulator.

Programming **goto** *in assembly language*
The GOTO statement of BASIC is programmed by means of a JMP instruction with direct addressing. In fact this is equivalent to a 'load program counter' instruction with immediate addressing, except that such an instruction does not exist! The example of section 3.7.2 might have been programmed in BASIC as

```
10 LET COUNT = COUNT + 1
20 GOTO 10
```

Programming **on ... goto**
The equivalent of the ON ... GOTO statement of Minimal BASIC can be programmed by using a JMP instruction with a variable address. There are two main ways to do this: indirect addressing and indexed addressing. Indirect addressing, for example

        JMP   [INDRCT]

will cause the program to jump to the address stored in locations INDRCT and INDRCT+1. Note that this corresponds to 'load program counter direct', but that it is an *address* rather than *data* which is being loaded. Indexed addressing, for example

        JMP   X

will cause the program to continue from the address currently held in index register X. This copies the contents of the index register into the program counter. An offset can be used if desired.

    In order to program the equivalent of ON ... GOTO an expression must first be evaluated using a sequence of arithmetic and/or logical instructions. Next, the resulting value i in the accumulator must be used to

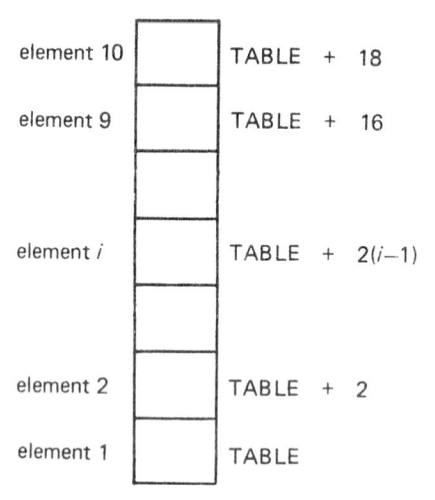

```
                element 10  ┌─────┐  TABLE  +  18
                            │     │
                element 9   │     │  TABLE  +  16
                            ├─────┤
                            │     │
                            │     │
                element i   │     │  TABLE  +  2(i−1)
                            │     │
                            │     │
                            ├─────┤
                element 2   │     │  TABLE  +  2
                            │     │
                element 1   │     │  TABLE
                            └─────┘
```

Figure 3.13 Table of 10 jump addresses

reference a table of addresses, such as that illustrated in Figure 3.13, to pick up the jump address corresponding to the value obtained. This will require an extra level of indirection; first the address of the entry in the table must be calculated, then this address must be loaded into a pointer register (either a pair of memory locations or an index register), and then the address must be transferred to the program counter to effect the jump instruction.

Each entry in the table requires two bytes, and so the subscript must be multiplied by two before being added to the address of the base of the table. The result will be the address of the *i*th entry in the table, assuming the entries to be numbered from zero. In BASIC the table of line numbers used by an ON ... GOTO statement begins with the line number corresponding to a value 1.

If indirect addressing is used, the program will take several lines to produce the desired address, as shown in the listing of ELEMNT in Figure 3.14.

If indexed addressing is used, the 6809 program can be made shorter. First the subscript must be multiplied by two to produce the required offset from the base address of the table.

```
        ASLB                    and multiply by 2
        LDX     #TABLE-2        Point to start of table
```

```
              { Work out address of element number i in array
                TABLE.    Assume that 'i' is in accumulator B }
ELEMNT  CLRA                       Clear accumulator A
        DECB                       Decrement B by 1 (so that a value
                                   of 1 selects first entry in table)
        ASLB                       Multiply number in A and B by two
        ROLA                       because table has two bytes/entry.
        ADDD    #TABLE             Add start address of table
        STD     INDRCT             and store 2-byte address;
        LDD     [INDRCT]           get address from table
        STD     INDRCT             and put that in INDRCT instead
        JMP     [INDRCT]           then jump to that address
```

Figure 3.14

This offset must then be added to the base address to point to the appropriate entry. Here the X register has been loaded with an address two less than that of the table so that the subscript can be offset by one:

TABLE + 2(i − 1) = (TABLE − 2) + 2i

If the subscript is held as a single byte in accumulator B (which implies that i lies in the range 0 to 127), the address of the ith entry in the table can be calculated by adding the contents of accumulator B to the contents of index register X. The 6809 has a special instruction ABX which does precisely this. Thus the next stage could be programmed as

```
        ABX                Now X points to element i
        LDX     0,X        Load element i into X register
        JMP     0,X        Jump to this address
```

However, the flexible addressing modes of the 6809 allow an even more compact method of programming this. Firstly, the contents of one of the accumulators can be used to provide the offset, so that if the indexed addressing mode B,X is used, the base and offset can be added while the effective address is being computed. Secondly, the extra LDX 0,X can be avoided by including an extra stage of indirection. Thus the single instruction

JMP   [B,X]

replaces all three of the instructions described above. If the offset is larger than 127, the contents of accumulator D the two accumulators A and B placed end to end, can be used in a similar fashion:

JMP   [D,X]

Program 'decisions'
As their name implies, the data test instructions are used to test data and affect the bits of the flag register before a branch instruction without affecting any other register. They can be combined with branch instructions to produce many different programming constructs. As a matter of principle, the programmer should try to make use of the constructs described in Chapter 2, because this usually speeds up both the writing and testing of assembly language programs.

TST	ITEM1	'Branch if item1 <> 0 '
BNE	...	
LDA	ITEM1	'Branch if item1 > 12 '
CMPA	#12	
BGT	...	
LDA	ITEM1	'Branch if item1 = item2 '
CMPA	ITEM2	
BEQ	...	
LDA	ITEM1	'Branch if item1 AND item2 = 0 '
BITA	ITEM2	
BEQ	...	

Figure 3.15

The diamond-shaped 'decision box' used in flow charting can be programmed as a combination of a sequence of data handling and/or data test instructions combined with a conditional branch. A program fragment which illustrates this is shown in Figure 3.15.

Programming **if ... then**
In an **if ... then** construct, a section of program is carried out only if the logical expression bracketed by the words **if** and **then** is true. The minimal BASIC statement

{*line number*} IF {*logical expression*} THEN {*line number*}

can be represented using a 'template' of the form

```
IF      EQU     *
{Evaluate logical expression}
        Bxx     LINENO
```

Here `LINENO` is a label used to mark the address of the section of code to which the program is to jump if the logical expression proves to be true and `Bxx` indicates the appropriate branch opcode. For example, the BASIC statement

100 IF L+M>5 THEN 200

can be translated using this template to give the assembly language sequence shown in Figure 3.16.

```
IF      EQU     *               '100 IF'
        LDA     L               'L'
        ADDA    M               '+M'
        CMPA    #5              '>5'
        BGT     LINENO          'THEN 200'
```

Figure 3.16

A line containing `EQU *` can always be removed from programs by amalgamating it with the line which follows, and distinctive labels should be used because a program is likely to contain several `IF` statements. In the shortened version of this example shown in Figure 3.17 the label `LAB100` is needed only if the line is referenced in another statement.

```
LAB100  LDA     L               '100 IF L'
        ADDA    M               '+M'
        CMPA    #5              '>5'
        BGT     LAB200          'THEN 200'
```

Figure 3.17

The **if ... then** construct used by Pascal (and most versions of BASIC other than ANSI minimal BASIC) has the form

```
IF      EQU     *
{evaluate expression}
THEN    Bxx     ENDIF
{Section of program executed if expression is true.}
ENDIF   EQU     *
```

Thus the Pascal statement

if $l+m>5$ **then** $n:=5$;

can be translated using this template to give the assembly language sequence shown in Figure 3.18.

```
IF      EQU     *               'if'
        LDA     L               'l'
        ADDA    M               '+m'
        CMPA    #5              '>5'
THEN    BLE     ENDIF           'then'
        LDA     #5              'n:=5'
        STA     N
ENDIF   EQU     *
```

Figure 3.18

Note that the branch instruction causes the program segment bracketed by THEN and ENDIF to be omitted if the test proved false, and for this reason the branch instruction always tests for the opposite condition. In this case BLE ('branch if less than or equal') is used although the condition being tested is 'greater than'.

Programming **if ... then ... else**
The assembly language implementation of the **if ... then** construct is easily extended to include an **else** clause.

```
IF      EQU     *
{evaluate expression}
THEN    Bxx     ELSE
{Section of program executed if expression is true.}
        JMP     ENDIF
ELSE    EQU     *
{Section of program executed if expression is false.}
ENDIF   EQU     *
```

Again, the template approach allows the Pascal statement

if $l+m>5$ **then** $n:=5$ **else** $p:=8;$

to be translated into the assembly language routine of Figure 3.19 without difficulty. As before, the lines containing label definitions of the form

LABEL   EQU   *

can be eliminated and the labels defined more directly once the whole program has been written, although no loss of efficiency in the final program will result from leaving them in place.

'**if** $l+m>5$ **then** $n:=5$ **else** $p:=8;$'

IF	EQU	*	'if'
	LDA	L	'l'
	ADDA	M	'$+m$'
	CMPA	#5	'>5'
THEN	BLE	ELSE	'then'
	LDA	#5	'$n:=5$'
	STA	N	
	JMP	ENDIF	
ELSE	EQU	*	'else'
	LDA	#8	'$p:=8$'
	STA	P	
ENDIF	EQU	*	

Figure 3.19

Programming **case** *statements*

The **case statement** represents an extension of the **if ... then ... else** construct to allow multiple branching, and may be programmed as a sequence of comparisons and conditional branches as shown in the example of Figure 3.20. Here the label SWITCH marks the start of the start of the statement, corresponding to the position of the word **switch** in Pascal, the labels starting with the word CASE mark the separate cases, and ESAC marks the end of the case statement.

Programming **repeat ... until**

The **repeat ... until** loop can be represented in terms of a sequence of instructions forming the 'body' of the loop followed by an **if** statement which causes the program to jump back to the beginning of the sequence if the test expression proved false.

Thus the loop will have the general form

Programming a **case** statement with
a series of branch instructions.

```
SWITCH EQU     *
```
{Evaluate expression and put value in accumulator A.}
```
CASE1  CMPA    #VALUE1
       BNE     CASE2
```
{Section of program executed if value in A = VALUE1.}
```
       JMP     ESAC

CASE2  CMPA    #VALUE2
       BNE     CASE3
```
{Section of program executed if value in A = VALUE2.}
```
       JMP     ESAC

CASE3  CMPA    #VALUE3
       BNE     CASE4
```
{Section of program executed if value in A = VALUE3.}
```
       JMP     ESAC

CASE4  CMPA    #VALUE4
```

{Continue for as many cases as necessary...}

```
ESAC   EQU     *
```

Figure 3.20

```
REPEAT EQU     *
```
{Body of loop comes here}
```
UNTIL  EQU     *
```
{Conditional expression}
```
       Bxx     REPEAT
```

The rather trivial example of Figure 3.21 shows how a sequence of Pascal statements can be translated directly into assembly language using this template.

Removing the EQU directives simplifies Figure 3.21 to:

```
         LDA    #-10         i := -10;
REPEAT   EQU    *            repeat
         INCA                i := i+1
UNTIL    EQU    *            until
         CMPA   #10          i > 10;
         BGT    REPEAT
```

Figure 3.21

```
         LDA    #-10         i := -10;
REPEAT   INCA                repeat i := i+1
UNTIL    CMPA   #10          until i > 10;
         BGT    REPEAT
```

Programming **while ... do**

The **while ... do** construct differs from the previous one only in the fact that the conditional test precedes the body of the loop rather than following it. Thus the template has the form

```
WHILE    EQU    *
{Conditional expression}
         Bxx    ENDW
DO       EQU    *
{Body of loop comes here}
         JMP    WHILE
ENDW     EQU    *
```

Again, translation of **while ... do** constructs from Pascal into assembly language is relatively straightforward using this template approach, and the Pascal fragment

$i := -10;$
while $i <= 10$ **do** $i := i+1$

translates directly as shown in Figure 3.22.

The use of lines containing EQU * is inefficient insofar as it requires the programmer to put extra lines into the program, and the example of Figure 3.22 could be rewritten as

```
          LDA    #-10        i := -10;
WHILE     CMPA   #10         while i <= 10
          BGT    ENDW
DO        INCA               do i := i+1;
          JMP    WHILE
```

This would produce exactly the same machine code when translated.

```
          LDA    #-10        i := -10;
WHILE     EQU    *           while
          CMPA   #10         i <= 10
          BGT    ENDW
DO        EQU    *           do
          INCA               i := i+1;
          JMP    WHILE
```

Figure 3.22

Programming **for ... next**

The **for ... next** loop differs from other types of loop in explicitly requiring a variable to be increased or decreased by a specified amount each time the loop is traversed. In each type of loop something must change or the program would never leave the loop; sometimes this is what the programmer intends, and the skeletal Pascal constructs

    while *true* **do**   **repeat**
       {*loop body*};   {*loop body*}
                           until *false*;

will each repeat indefinitely. However, in the other types of loop the change may be due to any cause, for example a particular bit changing in a register. Again, a template of a **for ... next** construct can be used. Assume initially that the loop variable is a single byte. The Pascal fragment

    for $i := first$ **to** *last* **do**
    {*loop body*}

could be translated as shown in Figure 3.23.

Note that there is no need to store and recover the loop variable in ITEMP if the contents of the accumulator A are not affected by the instructions contained in the loop body.

BASIC allows the step size to be other than one, and in this case the INC instruction must be replaced with an addition; the BASIC fragment

```
FOR       LDA      FIRST
TO        CMPA     LAST
          BGT      ENDFOR
          STA      ITEMP
          {loop body}
          LDA      ITEMP
          INCA
          BRA      TO
ENDFOR    EQU      *
```

Figure 3.23

```
FOR I = FIRST TO LAST STEP INCR
{loop body}
NEXT I
```

might be translated as shown in Figure 3.24.

```
FOR       LDA      FIRST
TO        CMPA     LAST
          BGT      ENDFOR
          STA      ITEMP
          {loop body}
NEXT      EQU      *
          LDA      ITEMP
          ADDA     INCR
          BRA      TO
ENDFOR    EQU      *
```

Figure 3.24

3.7.5 More about the index register

Loops are frequently used to examine successive locations in memory, for example to search for the occurrence of certain data. In such cases it becomes attractive to use an index register as a loop counter. This allows a pointer to be changed automatically each time that the instructions in the loop are executed. In those microprocessor designs which have 16-bit index registers it also offers the possibility of two-byte loop variables; the examples which were used in the last section all used a loop variable which could be represented as a single byte in the accumulator.

Most microcomputers have instructions for incrementing and decrementing the contents of the index register by one. These allow the index register to be used to 'point' to successive addresses in memory to enable the microprocessor to read or write successive entries from or to an array or to search it for a particular item of data.

The example of Figure 3.25 considers a program which searches an array called ARRAY for a location containing the number 55. The listing shows how a BASIC version could be translated into an assembly language equivalent.

BASIC Program:

```
10 FOR I = 1 TO 100
20 IF ARRAY(I)=55 THEN 120
30 NEXT I
40 ...
   ...
120 ...
```

Assembly language version:

```
FOR     LDX     #1              ' 10 FOR I = 1 ...'
TO      CMPX    #100            '... TO 100'
        BGT     ENDFOR
        LDA     ARRAY,X         ' 20 IF ARRAY(I) ...'
        CMPA    #55             '... = 55 ...'
        BEQ     FOUND           '... THEN 120'
NEXT    INX                     ' 30 NEXT I'
        BRA     TO
ENDFOR  EQU     *               ' 40 ...'
        ...
FOUND   EQU     *               '120 ...'
```

Figure 3.25

Note the use of the 'compare index register' instruction CMPX in the example. This is a data test instruction similar to the 'compare accumulator' instruction already encountered.

The INX instruction causes the contents of index register X to be incremented by one so that X points to the next address. The program 'loops' until a match is found or the whole array has been searched. There is a similar DEX instruction which decrements the contents of X by one. Most microprocessors allow their index registers to be incremented or decremented by values equal to one, which fixes the step size in a **for ... next** loop at one, as is the case in Pascal. If larger steps are required, more flexible instructions are needed. The ABX instruction, which is available in some microprocessor designs in addition to the 6809, allows the contents of accumulator B to be added to those of index register X. If the loop variable is held in X and the step size in B, X can be incremented by the appropriate amount each time.

The 6809 has a more generalised instruction called LEA ('load effective address') that works out the address specified in the opcode field and loads it into the index register. The instruction

LEAX   1,X

adds one to the value currently in X using the normal indexed addressing mode, but then stores the result back in X, which is precisely how INX operates.

LEAX   -1,X

is equivalent to DEX in the same way. The power of LEA becomes more apparent when it is realised that LEAX   10,X has the effect of ten successive INX instructions. This instruction would allow a step size of +10 to be used. The ability of the 6809 to use an accumulator in computing effective addresses means that a variable step size can be used, for example LEAX D,X could be used to add a step size held in accumulator D to register X. Note that LEAY   D,X adds together the contents of the D accumulator and the X register and stores the result in the Y register.

Effect of instruction 'LDA X+'

Contents of Memory

Location	Before		After
A34D	BB		BB
A34C	CC	Index →	CC
A34B	89 (Index →)		89
A34A	12		12

Index Register: A34B → A34C

Accumulator A: A3 → 89

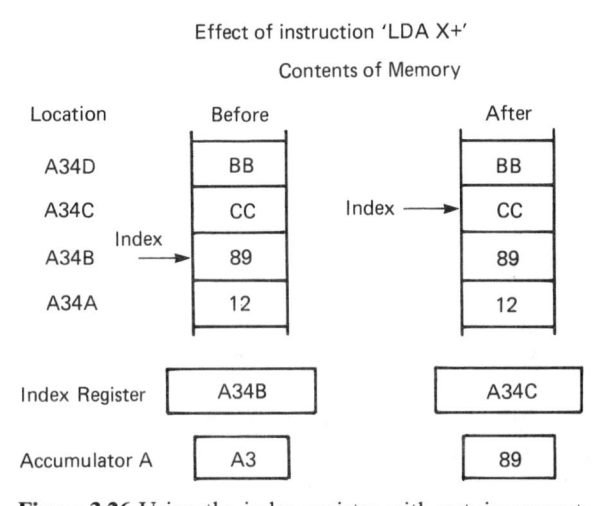

Figure 3.26 Using the index register with autoincrement

Although LEAX 1,X is not quite as 'neat' as INX, the latter is rarely necessary with the 6809 which has the ability to automatically increment or decrement an index register while it is being used in addressing. This feature is found on many of the more sophisticated microprocessors such as the LSI11 and 68000. For example the sequence

```
LDA    0,X
INX
```

can be replaced in the 6809 by the single instruction

```
LDA    0,X+
```

The contents of the index register X are incremented by one after it has been used to generate the address used by the instruction. If this instruction had been used when the index register was pointing to address A34B, as shown in Figure 3.26, the index register would afterwards point to address A34C and the contents of accumulator A would be 89. Similarly,

```
LDA    0,X++
```

causes the index register to be incremented by two after it has been used to generate the address. Addressing modes such as this are called **autoincrement addressing modes**, and corresponding **autodecrement addressing modes** also exist; the autodecrement instructions of the 6809 perform precisely the inverse functions of the autoincrement examples:

```
STA    -X
STA    --X
```

Although these automatically decrement the contents of the index register, they usually do so before the index register is used to generate the address used by the instruction. The reason for this will be explained shortly, in section 3.8.1, which describes the stack pointer register.

3.7.6 More on programming case statements

Unfortunately, the method of programming 'case' statements which was described earlier in this chapter can use a lot of storage if there are many different cases to be programmed. When there is a large number of cases to be tested it becomes more attractive to use a loop in which the value of an expression (the selector) is compared with successive values in a table until a match is found. Then the program counter is loaded with the address corresponding to this value in much the same way as in the ON ... GOTO construct. This address marks the beginning of the sequence of operations to be carried out when the selector has the corresponding value.

In the next example it is assumed that accumulator A contains the variable to be tested and that index register X points to the start of a table containing test values for the selector and their corresponding jump addresses. Each entry in the table will consist of three bytes:

Value i	Address i	
Byte 1	Byte 2	Byte 3

The first byte is the value to be tested, while the next two bytes contain the starting address of the corresponding piece of program. Each such piece of program must terminate with a jump to the end of the case statement, which is marked with the label ESAC, while each entry in the table can be programmed with a pair of lines of the form:

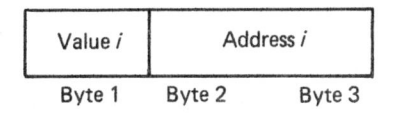

```
FCB    VALUEI    i th test value for selector
FDB    CASEI     Address of corresponding code.
```

The table contains three bytes for each value which the control variable might take: the first byte is the value to be tested, while the next two bytes contain the starting address of the corresponding piece of program. FCB and FDB are in fact **assembler directives**, instructions to the software which translates the assembly language program to machine code rather than operation codes for the machine itself. They stand for 'Form Constant Byte' and 'Form Double Byte', and allow single or double byte constants to be inserted into the program. Different microcomputer manufacturers use different conventions for these operations.

The program fragment CASE, of Figure 3.27, compares the number in accumulator A with each test value in turn until a match is found. The address of the corresponding set of operations is then loaded into the index register and the program jumps to that address. At the end of the sequence the program will continue from the point labelled ESAC.

Here a potential danger lies in the possibility of CASE being used with a variable in the accumulator which does not have a corresponding entry in the table. In Pascal, such statements are normally called with variables of enumerated type, and provided that all eventualities are catered for, this problem cannot arise. Other languages such as 'C' (Kernighan, 1978) allow for problems when case statements are used with non-enumerated types by including a 'default' in the case statement which specifies the action to be carried out if no match is found. The assembly language example could be extended to allow an equivalent operation. If the table began with an entry indicating how many entries there were in the table, this could be used by a modified version of CASE to determine when the end of the table was reached.

3.7.7 Position independent code

The operands used by jump and branch instructions are addresses of points in the program rather than data as in the case with the data processing and test instructions. Sometimes it is desirable to be able to change the range of

CASE	LDX	#TABLE	X points to start of jump table.
LOOP	CMPA	0,X	Does acc-A match this value?
	BEQ	FOUND	yes...
	LEAX	3,X	no, so increment X by 3
	BRA	LOOP	and try the next value...
FOUND	JMP	[1,X]	Get address of corresponding subroutine and jump to its start.
TABLE	FCB	VALUE1	First test value for selector
	FDB	CASE1	Address of corresponding code.
	FCB	VALUE2	Second test value for selector
	FDB	CASE2	Address of corresponding code.
	FCB	VALUE3	Third test value for selector
	FDB	CASE3	Address of corresponding code.
	FCB	VALUE4	Fourth test value for selector
	FDB	CASE4	Address of corresponding code.

{Continue for as many cases as necessary}

```
CASE1    EQU     *
{Instructions to be carried out if selector = VALUE1}
         JMP     ESAC

CASE2    EQU     *
{Repeat with instructions for each 'case' ...}

ESAC     EQU     *
{End of 'case' statement}
```

Figure 3.27

addresses used to store a program without having to retranslate it into machine code, which means working out the actual numerical addresses used in the system. The solution is to code the program in such a way that the operand of each jump and branch instruction is stored not as the actual address to be used but as the difference between that address and the address currently in the program counter. Such programs will operate in the same way no matter where they are stored in memory, and the machine code is said to be **Position Independent Code (PIC)**.

For example, if the program contained an instruction at address 2000 that caused a jump to address 2010, the operand might be coded as

$$2010 - 2000 = 10 \tag{3.6}$$

This value 10, the difference between the branch address and the current one, is called an **offset**. This type of addressing is provided by the 'program counter relative mode', or **relative addressing mode** for short. Many microcomputers allow this mode to be used with jump and branch instructions. The example of a jump instruction using relative addressing can be interpreted as meaning 'jump forward 10 bytes before continuing', rather than 'jump to address 2010' which is the meaning of the corresponding jump instruction with absolute addressing. Many types of microprocessor use relative addressing with conditional branches; often the opcode BRA (BRanch Always) is used for this operation. When the program runs, the microprocessor simply adds the offset to the current address to find the destination of the jump instruction:

$$2000 + 10 = 2010 \tag{3.7}$$

If the program was moved so that the instruction was at address 3000, the target of the jump address would be at

$$3000 + 10 = 3010 \tag{3.8}$$

The 6809 allows the program counter to be used as a pointer in addressing modes, so that

```
        JMP     10,PC
```

would cause the program counter to be loaded with an address ten greater than the current address. The assembly language allows addresses to be coded using references to absolute addresses which are automatically translated into offsets during conversion to machine code. For example,

```
        LEAX    TABLE,PCR
```

causes the index register to be loaded with the address TABLE when the program is executed. The difference between this and the instruction

```
        LDX     #TABLE
```

lies in the manner in which the instruction is represented in machine code. In the former example the address is coded as an offset from the current point in the program which is added to the contents of the program counter to yield the address TABLE once more. In the latter example the address is represented as an absolute quantity. Both instructions will have the same effect provided that the program is not relocated in memory, but only the first one will operate correctly if the program is relocated.

The use of this position independent addressing mode allows position independent code to be written easily. One important application of position independent code is in **silicon software**, programs in read only memory that can be used as components, and mapped anywhere in the address space of the system in which they are incorporated.

3.8 Stacks

The addressing schemes encountered so far allow information to be accessed at known places in memory. Sometimes, however, it is convenient simply to put information to one side with the intention of recovering it later. In such cases the actual memory address used is irrelevant provided that information saved can be recovered when required.

If words are written sequentially into memory they can be read later in reverse sequence without the actual addresses used for storage being of importance to the program. The effect is similar to that obtained when documents are stacked up in a tray instead of filing them. Items can be found systematically in a stack of papers if they are removed one at a time in the reverse order to that in which they were placed on the stack. This idea of 'stacking' information can be contrasted with that of 'filing' it away at known memory addresses.

When data is stored on a **stack** it is customary to refer to it as being **pushed** on to the stack, and **pulled** or **popped** from the stack. A simple stack can be programmed using either indexed or indirect addressing and incrementing and decrementing the pointer (index or memory register) each time it is used, which is especially easy when autoincrement and autodecrement instructions are available. No matter how they are implemented, the key point about stacks is that they allow information to be stored in memory according to a **Last In, First Out (LIFO)** rule.

For example, a stack can be programmed in the 6809 using

```
STA     0,X+
```

to store the contents of the accumulator at the address pointed to by the index register and move up to the next address, and

```
LDA     -X
```

to move the index register back to the previous location and load the accumulator with the contents of that register. An example is shown diagrammatically in Figure 3.28. Initially the index register points to the unused location A34B (the contents of the memory locations are arbitrary). If the accumulator contains the hexadecimal number 30 and the instruction

```
STA     X+
```

is executed, the contents of the index register and memory will be modified as shown in Figure 3.28. If this is followed by the instruction

```
LDA     -X
```

3.8   INSIDE THE MICROPROCESSOR   127

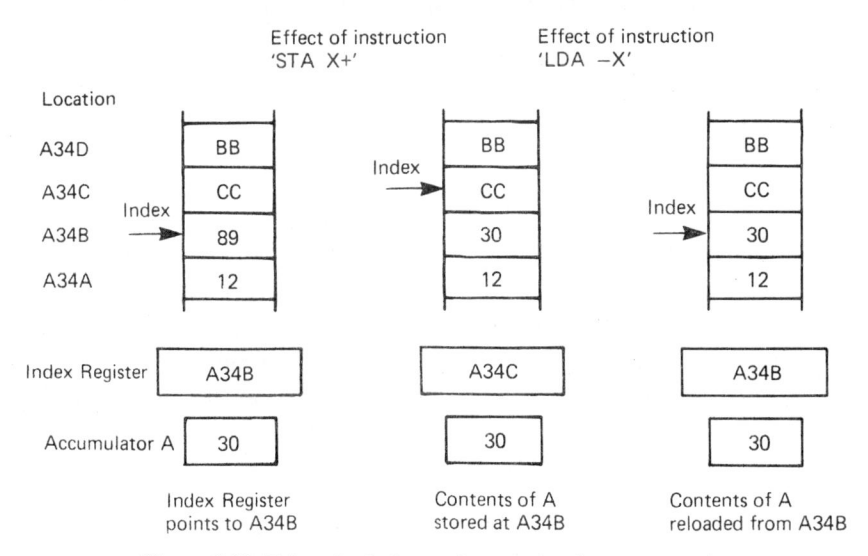

Figure 3.28 Using the index register to implement a stack

the contents of the index register will again be decremented by one and the accumulator will be loaded with the value 30 which was previously stored using the index register.

This example assumed that the index register is incremented after storing in memory and is decremented before loading from memory, so that the contents of the stack pile up in memory like an imaginary stalagmite. In practice, however, it is more common to decrement the index register after storing and increment it before loading from memory, so that the stack 'hangs' downwards from the top of the available memory space, like a stalactite. This means that the 'top' of the stack is the location with the lowest address, but despite being somewhat more difficult to visualise the principle is exactly the same. Thus it would be more common to use the instructions

         STA   −X
         LDA   X+

to push and pull the contents of accumulator A, using index register X to point to the 'top' of the stack.

The same techniques can be used in register-to-register microprocessor designs to implement stacks. For example, in the 68000 the instructions

```
MOVE    D0,-(A0)
MOVE    (A0)+,D0
```

can be used to push and pull the contents of register D0 using register A0 as a pointer.

3.8.1 The stack pointer register

Virtually all microprocessors have a special register called the **system stack pointer** (or just **stack pointer**) which has special 'push' and 'pull' instructions associated with it. The mnemonic codes used for pushing and pulling using this register vary widely between microprocessors as do the range of registers which can be directly pushed or pulled. For example only the two accumulators of the 6800 can be pushed or pulled while the 8085 allows almost any register to be handled thus.

The stack pointer must be initialised at the beginning of a program to make sure that the addresses used for the stack correspond to RAM locations. In some microprocessors this must be done by the programmer, while in others, such as the 6502, the stack pointer is initialised automatically when the microprocessor is reset. The system itself needs a stack for supervising operations such as subroutine linkage, which will be discussed shortly. However, stacks are useful in a wide range of types of software and programs often contain other stacks for various purposes.

The 6809 has two stack pointer registers: the **system stack pointer**, S, and a **user stack pointer**, U, This microprocessor is particularly flexible in that it allows any register or combination of registers to be pushed on to or pulled from any stack (except the pointer for that stack) in a single instruction. The mnemonics for **push** and **pull** using the system stack are PSHS and PULS.

Not only does the use of a special register for pointing to the memory registers used in a stack simplify programming compared with using indirect addressing or an index register, but it also enables the microprocessor itself to 'save' in memory certain information which needs to be 'put to one side' for later use. This need arises most commonly when a subroutine is called.

3.9 Subroutines

Subroutines are called in microcomputers by special **jump to subroutine** instructions which are essentially the same as the jump instructions already encountered, except in one important respect. The processor must keep a note of the point which had been reached in the calling program when the subroutine was called, so that when the subroutine has finished it can continue from the instruction following the jump to subroutine. In this respect they resemble the GOSUB instruction of BASIC.

There are several ways of storing this 'return address' but in most microcomputers it is done by pushing a copy of the return address on to the system stack before starting the subroutine. The program counter always contains the address of the next instruction after it has finished fetching the current one, so the action of a 'jump to subroutine' instruction consists of pushing the current contents of the program counter on to the stack and then loading the program counter with the effective address of the instruction. The mnemonic used for this operation varies with each microcomputer; two favourites are CALL, the word used in many high level languages, and JSR, which stands for 'Jump to SubRoutine'. Some microprocessors such as the 8085 offer conditional 'jump to subroutine' instructions, similar to branch instructions, where the subroutine call is effective only when the flags are set in the appropriate manner.

When execution of the subroutine has finished the microcomputer returns to the calling program by pulling the return address off the system stack and loading it back into the program counter. Thus the next instruction to be executed will be the one immediately following the 'jump to subroutine' instruction which caused that return address to be pushed on to the stack. This is done by means of a **return from subroutine** (RTS) instruction which is the assembly language equivalent of the 'return' statement used in high level languages. The only limitation upon the number of times that subroutines can call further subroutines without returning (this is called the **nesting of subroutines**) is provided by the finite amount of address space available for the stack. Saving the return address on the stack also allows a subroutine to call itself subject to certain conditions, a trick known as **recursion**.

Figure 3.29 illustrates the effect upon the program counter and stack pointer registers of calling a subroutine by means of the instruction

JSR   1632

stored at address 1200. The program counter will contain the address of this instruction immediately before executing it, and the stack pointer will point to the current 'top' of the stack. The contents of unused memory locations are shown starred for clarity, since their contents are irrelevant.

First the processor fetches the three bytes of the instruction using the program counter, which will then contain the number 1203, the address of the following instruction. This address is the 'return address' to which the processor will return after the subroutine has been finished, and it is pushed on to the stack. The program counter is then loaded with the address at which the subroutine starts. When the microprocessor fetches the next instruction using the program counter it will commence with the first instruction of the subroutine.

Note that in pushing the two bytes of program counter contents on to the stack, the contents of the system stack pointer register are here decremented by two, unlike the previous examples, and the stack pointer always points to the last entry to be pushed on to the stack. The last item is always referred to as being on the 'top' of the stack, despite the fact that it has a

Effect of instruction 'JSR 1632'

Location	Before	After	
01FF	89	89	
01FE 'top'→	10	10	return address on stack
01FD	**	03	
01FC	** 'top'→	12	
01FB	**	**	

Stack Pointer: 01FE → 01FC

Program Counter: 1200 → 1632

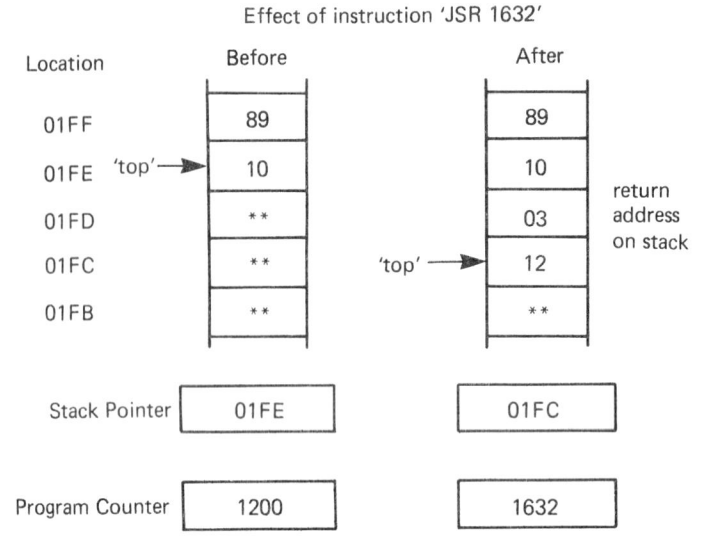

Figure 3.29 Subroutine linkage using the stack

lower address than other items which are referred to as being 'lower down' the stack. Despite the apparently confusing terminology, confusion does not seem to occur in practice!

At the end of the subroutine the return address, 1203 in the example, will be pulled again from the stack and loaded into the program counter. The stack pointer will then point once more to address 01FE as it did before the subroutine was called.

3.9.1 Passing parameters

Almost all subroutines require information to be given to them by the calling program or return information to the calling program on completion. The business of identifying the information which is to be communicated between the calling program and the subroutine is called **passing parameters**.

There is a large number of ways in which this can be carried out. The choice depending upon the microprocessor and programming language being used as well as the amount of information being handled. In every case one or more registers or memory locations are used, but the registers may be in the microprocessor itself, or a fixed set of memory registers or the stack may be used. Sometimes the values of the parameters being used are passed, while in other systems it is the addresses of the memory registers where these values are stored that are communicated. These two approaches are called **calling by value** and **calling by address**. Variables which are 'local' to a particular procedure or section of program are usually stored on the stack or in one of the registers of the processor, while 'global' variables are stored at fixed memory locations so that they may be accessed from any section of the program.

The most concise way to pass parameters when programming most 8-bit microcomputers in assembly language is to use the registers of the microprocessor itself. The accumulator(s), and index register(s) can be used to carry a small amount of data, but only enough for simple applications. The flag registers are also useful in this application because they can be tested upon return to the calling program using branch instructions. The example of Figure 3.30 shows a subroutine which tests the ASCII code in accumulator A at entry and sets the 'V' bit if it is numeric.

TEST	CMPA	#'0	Test to
	BLE	NOT	see if it is
	CMPA	#'9	numeric
	BGT	NO	
	SEV		it is, so
	RTS		V = 1
NOT	CLV		it isn't,
	RTS		so V = 0

Figure 3.30

The trouble with using a processor register in this way is that although it provides some very neat methods of calling subroutines it is difficult to use with high level languages. Different microprocessor designs have different sets of registers available, and using the registers of the microprocessor itself tends to introduce an element of machine dependency into the software.

On the other hand, all microcomputers have a large enough number of memory locations available to allow parameter passing conventions to be used that are essentially independent of the type of processor. Thus a set of registers can be assigned to hold values or pointers to values to be used by a subroutine. These memory locations must be set to contain the appropriate information before the subroutine is called; after execution of the subroutine has finished the calling program can access parameters being returned from the subroutine by picking up their values or addresses from the appropriate memory locations.

3.10 Multiplication and division

The use of the various registers within the microcomputer can be demonstrated in the programming of the multiply and divide operations. These are generally available on 16-bit microcomputers, but only rarely so on their 8-bit counterparts.

Multiplication

Although the 6809 and a few of the other 8-bit microprocessors have multiply instructions, the majority do not. This is partly because multiplication is more complicated and time-consuming than the operations met so far, and partly because of the problem of storing the product, whose representation can have twice as many bits as the multiplicand and multiplier. However, multiplication is possible with other microcomputers by using a mixture of shifts and additions.

The principle used is the familiar one of 'long multiplication' used to multiply decimal numbers manually. In the case of decimal numbers the multiplicand is multiplied by each digit of the multiplier in turn, shifted by the appropriate amount, and added to the total. At the end of the operation the total represents the product.

The principle is the same when multiplying numbers of any length or to any base, and it works just as well with binary numbers as with decimal ones. Figure 3.31 illustrates the long multiplication of 26 by 23 in both decimal and binary arithmetic.

```
        BINARY              DECIMAL

         11010                 26
       × 10111               × 23
       -------               -----
         11010                 52
         00000                 78
         11010               -----
         11010                598
         11010
       --------
       1001010110
```

Figure 3.31 Long multiplication

With binary numbers there are more digits to be handled, but each operation is easier because the multiplicand need only be multiplied by one or by zero at each step. Putting it another way, either the multiplicand is added or not at each step. This provides the basis for a multiplying program. A total is initially set to zero, and then both total and multiplier are shifted left one bit at a time. If the bit appearing in the carry bit after the multiplier has been shifted is a logic one, the multiplicand is added to the total. After all the bits in the multiplier have been tested, the process is finished and the total represents the product. The way in which this is programmed depends upon the type of microcomputer being used and the way in which the parameters are passed to the multiplying subroutine. In the example of Figure 3.32, when subroutine MULTPY is called the multiplicand is assumed to be in register `MPCAND` and the multiplier in register `MLTPLR`. At the end of the subroutine the data is in registers `HPRDCT` and `LPRDCT`.

```
MULTPY  LDA   #8          Count eight bits
        STA   COUNT       while going round loop...
        CLRA              Clear accumulator
LOOP    ASLA              Shift product one
        ROL   HPRDCT      place left.
        ASL   MLTPLR      Get a bit from the multiplier
        BCC   ZERO        and test it;
        ADDA  MPCAND      add to total if it is a one.
ZERO    DEC   COUNT       Decrement bit count
        BNE   LOOP        and repeat until done.
        STA   LPRDCT      Store LS byte of product
        RTS
```

Figure 3.32 Multiplication subroutine (6809)

Various programming tricks have been devised to speed up the multiplication process such as Booth's algorithm (Booth, 1951), but programs based upon the principle already described will be fast enough for almost all applications. The subroutine used as an example is designed to handle unsigned numbers, and if signed numbers are to be used a few extra lines of program are needed to work out the sign of the product and convert the multiplicand and multiplier to their absolute values. Then the subroutine can be called, and afterwards the quotient can be multiplied by minus one if appropriate.

Division
Division can also be programmed by returning to the arithmetic technique of long division and adapting it for the microcomputer. Here the technique consists of subtracting the largest single-digit product of the divisor which does not leave a negative remainder, as shown in Figure 3.33. Again, the process is easier using binary arithmetic because the intermediate products are all formed by multiplying by either one or zero. In the case of division there are two results: the quotient and the remainder, and a division subroutine must be able to handle each as appropriate. Subroutine `DIVIDE`, shown in Figure 3.34, takes an 8-bit unsigned dividend in a register called `DIVDND` and divides it by the 8-bit unsigned divisor in register `DIVSOR` to produce an 8-bit unsigned quotient called `QUOT` and an 8-bit remainder called `REMAIN`. Again, a small amount of extra code can be used to handle signed numbers.

There is an extra complication with division. If an attempt is made to divide by zero, special action must be taken because the answer is infinite and cannot be handled easily. Two courses of action are open. The first is to

```
              BINARY                        DECIMAL

              10111   (quotient)              23    (quotient)
      _____                           _____
11010 )1001011000                      26 ) 600
     -  11010                            -  52
        _____                               ___
        10111                                80
     -  00000                             -  72
        _____                               ___
        101110                               2    (remainder)
     -   11010
        _____
         101000
      -   11010
         _____
          11100
       -  11010
          _____
             10   (remainder)
```

Figure 3.33 Long division

halt the program and print an error message, which is the normal course of action. This is an example of a program **exception**; the normal flow of execution cannot continue when an exception occurs and special steps must be taken to handle such conditions. However, if the program is controlling a piece of equipment, it may be more satisfactory to take the largest number which can be handled with the number representation currently being used and assign that value to the answer.

3.11 Input/output instructions

Assembly languages do not offer any direct equivalents of Pascal's *read* and *write* statements, because these require specific input/output devices such as keyboards and displays. Additionally, many applications use special devices such as analog-to-digital converters, for which the Pascal input/output statements would not be appropriate. In assembly language, all input/output must be programmed by the user, which requires a closer knowledge of the machine than would be necessary with a high level language.

At the simplest level, input/output (I/O) devices appear to the microprocessor as a set of registers, like memory. Many types of microprocessor (including the 6809) do not distinguish between memory and I/O registers in any way. In fact this can be done with all types of microprocessor provided that the I/O circuits are connected in the appropriate way, a technique

```
DIVIDE  LDA   #8          Count eight bits
        STA   COUNT       of divisor.

LOOP    LDA   DIVDND      Compare dividend,
        SUBA  DIVSOR      and if it is larger,
        BCC   DIV1        restore intermediate
        ADDA  DIVSOR      result and put a 0
        CLC               in the carry bit;
        BRA   DIV2
DIV1    SEC               if smaller, put a 1.

DIV2    ASL   QUOT        Pass the bit to the quotient.

        DEC   COUNT       Decrement the count,
        BNE   LOOP        and carry on until done...

        STA   REMAIN      Then store the remainder
        RTS               and return.
```

Figure 3.34 Division subroutine (6809)

which is known as **memory mapped input/output**. The advantage of memory mapped I/O is that it permits the programmer to use a range of memory reference instructions with the registers in I/O devices.

Many microprocessor families have special instructions for handling input and output. Single chip microcomputers, because they spend a relatively large proportion of their time reading in and writing out information, often have special instructions for manipulating the bit patterns in input and output registers. The 6805 has instructions which allow a specified bit of a register to be tested and a branch to a new address to be made if the bit is set (or cleared, according to the instruction). Several single chip microcomputers, including the 6805 and 8048, allow an output register to be loaded directly with a bit pattern or logically ANDed or ORed with a bit pattern to set or clear some of the bits in that register without having to use the accumulator.

At the other end of the range of sizes, the problems associated with I/O are different; in some cases the task of input and output are delegated to another microprocessor in order not to waste the time of the main microprocessor. Even so, many 16-bit microprocessors are used with direct I/O. For example the 68000 has a 16-bit data bus which makes the use of 8-bit I/O devices difficult. This problem is overcome with a **move to peripheral device** instruction which formats data on to eight of the sixteen bits of the data bus. However, the most useful I/O instructions appear in microprocessors which handle input and output separately from memory references.

A well-known example of this type of approach is the widely-used Z80 which because it uses separate control bus signals for memory and I/O transfers has two separate address spaces. For example, to load the accumulator from memory location 50 and to store the contents of the accumulator at address 60 the following instructions would be used

```
LD      A,(50)
LD      (60),A
```

The corresponding I/O operations, to load the accumulator from I/O address 50 and to store the accumulator contents at I/O address 60, would be programmed using the instructions IN and OUT

```
IN      A,(50)
OUT     (60),A
```

The real advantages of having a separate I/O addressing scheme arise when more complicated instructions are used. The Z80 has a set of instructions for moving blocks of up to 256 bytes between memory and I/O devices. These instructions use the registers illustrated in Figure 3.35: an 8-bit register (B) to count the bytes as they are transferred, another 8-bit register (C) to hold the I/O address to be used and a 16-bit register (HL) to point to the address in memory being used. The contents of HL, and hence the memory location to be used, are automatically incremented or decremented (according to which instruction is being used) as the data is transferred.

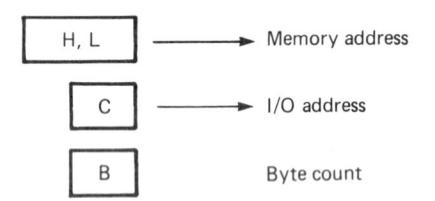

```
[ H, L ]  ──────▶  Memory address

[  C  ]   ──────▶  I/O address

[  B  ]            Byte count
```

Figure 3.35 Registers used by Z80 block I/O instructions

A further example is provided by the 9900 and related microprocessors. These use an interesting I/O scheme in which the input/output space is configured as 2048 locations, each capable of holding only a single bit of data. This not only allows any bit to be addressed separately, but also allows transfers of up to 16 bits to be made to or from any consecutive block of bits in the I/O space.

Despite the usefulness of I/O instructions they will not be discussed in the remainder of this book because they tend to be specific to particular machines. Chapter 4 will concentrate upon input/output devices and their programming.

3.12 Linking programs in different languages

Thus far, assembly language and high level language (HLL) programs have been discussed separately, although it is becoming clear that each type of language has its respective advantages in different aspects of a given problem. It is possible to write programs which are 'hybrids' (a 'chimera' might be a more appropriate analogy) of high and low level program sections. The problem lies in joining the sections in such a way that information can be passed from one part to another.

Usually it is assembly language subroutines which are called by an HLL program rather than the other way round. Most microcomputer languages provide ways to do this as extensions to the standard language definition, and unfortunately the manner of doing this varies from system to system. To some extent this does not matter, because the inclusion in an HLL program of a section that is written in assembly language limits its 'portability' from machine to machine in any case. In some versions of BASIC, for example, a program in machine code (translated from assembly language) can be called as a command by using the keyword SYS, so that the line

SYS   1024

would cause the microcomputer to start running a program that was assumed to start at address 1024. The programmer must make sure that there is indeed a program at address 1024 (corresponding to the hexadecimal address 0400) and that putting a program at this address will not cause another program to be corrupted by any other program. Similarly, a user function called USR may be defined, but the way of doing so varies from machine to machine. Assembly language procedures may be used with some versions of Pascal, for example by declaring them as **external** to the main Pascal program but specifying the address at which the program is stored and the way in which parameters are passed.

3.13 Design examples

In Chapter 2, the functions to be performed by the two examples — the controller and the instrument — were described using high level languages. Most high level languages are designed principally to allow the function of the program to be expressed as concisely and clearly as possible, and they do not usually offer the same detailed control of the manipulation of data within the microprocessor as assembly language. Despite the fact that programs in assembly language are almost always much more long-winded than their high level equivalents, assembly language still offers advantages in many circumstances, for example where speed is important, where patterns of bits must be manipulated in unusual ways, and input/output devices are being programmed.

3.13.1 Controller

The controller uses a relatively simple program which could be coded entirely in assembly language without undue difficulty. As with almost all microcomputer-based equipment, the main controlling program will need to monitor an input for commands, and when these are received, carry out appropriate action. This is programmable by means of a 'case' statement using a table of commands and a corresponding table of operations to be carried out, which can be implemented as subroutines.

Another aspect of the controller is that it must continue to control the system even while the operator is entering new information. Thus it appears that at least two programs must be running at the same time, although in fact the microcomputer switches between the programs at a rate fast enough to deceive the human operator and yet not affect the operation of the controlled system.

Case statement with jump table

The function of a case statement can be programmed in many ways using assembly language, but each uses the same basic approach: the value of the control variable is compared with a test value and if it matches, the corresponding sequence of operations to that value is carried out.

```
CASE    CMPA    0,X         Does acc-A match this value?
        BEQ     FOUND       yes...
        LEAX    3,X         no, so increment X by 3
        BRA     CASE
FOUND   LDX     1,X         get address of corresponding
        JSR     0,X         subroutine and jump to its start.
        RTS
```

Figure 3.36 Programming a case statement

The listing in Figure 3.36 shows one method in which the case statement is called as a subroutine CASE. When the subroutine is called, accumulator A contains the variable to be tested and index register X points to the start of the table to be used. If there were to be only one case statement within a particular program then the value of X could be assigned within the subroutine. In our case the subroutine might be called with a sequence such as

```
        LDA     SWITCH      Get control variable
        LDX     #SWTABL     and pointer to table
        JSR     CASE        call case statement.
```

The subroutine compares the number in the accumulator with each test value in turn until a match is found. The address of a subroutine containing the corresponding set of operations is then loaded into the index register and the program jumps to the start address of that subroutine. The sequence

```
FOUND   LDX    1,X
        JSR    0,X
        RTS
```

can be replaced with the more efficient but less obvious

```
FOUND   JMP    [1,X]
```

In either case, control will return to the instruction immediately following the call to CASE when the case statement has been executed. The table contains three bytes for each value which the control variable might take: the first byte is the value to be tested, while the next two bytes contain the address of the corresponding subroutine. In this example the case statement might be used to carry out actions corresponding to push buttons on a keypad on the front panel of the controller. The functions required would probably be as follows:

Setpoint Set the value to which the system's output quantity should be controlled.

High Set the 'output high' level at which an alarm will be sounded. If the output quantity exceeds this value, the alarm will be switched on.

Low Set the 'output low' level at which an alarm will be sounded. If the output quantity falls below this value, the alarm will be switched on.

There are four quantities which are used in the control equations: the three coefficients K_P, K_I, and K_D, and the sampling time interval T. These might also be placed on the front panel if the controller were to be used for teaching purposes. For the purposes of this example, the values of these variables will be assumed to be set in a similar fashion to the other quantities already described.

Thus the table for the case statement might take the form shown in Figure 3.37.

Number ranges
In the BASIC example in Chapter 2 real variables were used for the control calculation. However, the input to the controller would in practice come from a transducer providing an analog signal, and analog-to-digital converters do not provide the accuracy provided by real variables in most versions of BASIC. Carrying out the calculations to the full precision afforded by real variables may be too slow if high speed control is required, while integers are unsatisfactory. Real-time control calculations may use special number representations or special purpose hardware (this topic will be dis-

```
FCB     'S              Letter 'S' => 'set SETPOINT'
FDB     SETPNT          Address of 'set SETPOINT' subroutine
FCB     'H              Letter 'I' => 'set HIGH'
FDB     STHIGH          Address of 'set HIGH' subroutine
FCB     'L              Letter 'L' => 'set LOW'
FDB     SETLOW          Address of 'set LOW' subroutine
FCB     'P              Letter 'P' => 'set KP'
FDB     PROPOR          Address of 'set KP' subroutine
FCB     'I              Letter 'I' => 'set KI'
FDB     INTEGL          Address of 'set KI' subroutine
FCB     'D              Letter 'D' => 'set KD'
FDB     DERIVE          Address of 'set KD' subroutine
FCB     'T              Letter 'T' => 'set DT'
FDB     INTRVL          Address of 'set DT' subroutine
```

Figure 3.37 Table used by **case** example

cussed again shortly), but in this example we shall assume that the standard 'real' type allows calculations to be carried out fast enough.

3.13.2 Instrument

The calculations to be performed by the instrument are much more complicated than those of the controller. For this reason it might seem that assembly language programming would be unsuitable for this task, but some examples will be presented here which will show how sections written in assembly language could substantially increase the execution speed of some of the operations required.

```
BRSORT  LDA     #1          'for SUBSCR := 1 to 255'
        CLR     SUBSCR      Start at element number 0
BRSRT1  BSR     SWAP        and swap elements if appropriate...
        INC     SUBSCR      Then increment to next element
        BNE     BRSRT1      and continue until all 256 have been
        RTS                 done.
```

Figure 3.38 Sorting an array into bit-reversed order

Bit reversed sorting

The algorithm used in Chapter 2 to sort an array of data into bit reversed order demonstrates that high level languages are not always effective at manipulating bits within a word even though logical operations such as

AND and OR are available. In assembly language all the facilities of the processor are at the programmer's disposal, and sometimes considerable increases in efficiency can be achieved.

The sorting subroutine must operate upon each subscript of the array in turn, converting it into bit reversed order. This may be programmed as a simple loop, and if there are no more than 256 elements in the array each subscript can be represented as a single byte. The listing of subroutine BRSORT in Figure 3.38 shows a bit reversed sorting subroutine for a 6809. This uses a single byte SUBSCR to contain the subscript of the element currently being sorted. The use of an 8-bit byte to hold the subscript limits the size of the array to a maximum of 256 elements. However, it would not be difficult to increase the size of the subscript to two bytes if larger arrays were required. Each subscript of the array must be tested in turn, using a loop which counts through all the possible subscripts. During each pass through the loop, subroutine SWAP, shown in Figure 3.39, is called to exchange a pair of elements if appropriate.

```
SWAP   LDA    SUBSCR      Get subscript
       BSR    REVERS      and reverse it
       CMPA   SUBSCR      Is the result
       BCC    SWAP2       smaller?

       LDB    #4          It is; multiply subscript
       MUL                by element size,
       LDX    #DATA       get pointer to start of
       LEAX   D,X         array and add in offset
       LDA    SUBSCR      Similarly calculate address
       LDB    #4          of other element:
       MUL
       LDY    #DATA
       LEAY   D,Y         Now pointers X and Y are loaded
       LDA    #4          Count number of bytes
       STA    COUNT       to be swapped

SWAP1  LDA    0,X         Pick up bytes
       LDB    0,Y
       STA    0,Y+        and swap them, incrementing
       STB    0,X+        pointers...
       DEC    COUNT
       BNE    SWAP1       repeat for each byte of elements

SWAP2  RTS                and return
```

Figure 3.39 Swapping over two 4-byte words in an array

This subroutine assumes that each element of the array occupies four bytes and that the array to be sorted starts at address DATA. This number of bytes per element could easily be changed if the elements of the array were to contain a different number of bytes simply by replacing each occurrence of the number '4' with the appropriate value. Note that the form of the data in each entry does not matter since it is only being moved, and not processed in any way.

SWAP calls another subroutine REVERS, shown in Figure 3.40, to reverse the order of bits within the accumulator. This can be carried out quite easily by rotating the bits right from one register to the carry bit and then rotating that bit left into another register. The process must be done eight times to transfer all eight bits of a byte. Subroutine REVERS shows how much more easily bit patterns can be manipulated when the architecture of the particular microprocessor is known.

Once the bit reversed pattern has been produced, the corresponding elements in the array can be swapped over. This must be done once only for each pair of elements to be swapped, otherwise the array would end up in the same state as before. Here the swapping is carried out only when the bit reversed pattern corresponds to a smaller number than the original. Sixteen of the patterns are symmetrical and are the same when reversed, and of the 240 remaining patterns, 120 will produce a smaller number when reversed. First the bit reversed pattern is compared with the original, and if it is smaller, the subscripts corresponding to the two elements are calculated.

```
REVERS  STA   LOC1        This is the byte to be reversed.
        LDA   #8          Use the accumulator to count bits.
REV1    ROR   LOC1        Shift one bit right to carry bit
        ROL   LOC2        and left into another location
        DECA              Decrement counter,
        BNE   REV1        and repeat eight times.
        LDA   LOC2        Pick up reversed pattern
        RTS               and return with it in accumulator.
```

Figure 3.40 Reversing the order of bits in a byte

Rewriting the subroutine for another type of microcomputer would involve modifying the sections concerned with the calculation of the element addresses and the actual swapping of the data. The 6809 does this more concisely than most other 8-bit microcomputers, but 16-bit designs often have instruction sets which are even more effective in calculating addresses, and the larger number of registers within the microcomputer and their greater length allow operations such as this to be coded very compactly.

Each subscript of the array must be tested in turn, and the corresponding elements swapped if appropriate. This is done by a subroutine consisting of a loop which counts through all the possible subscripts, as shown in the listing.

FFT butterfly

The 'butterfly' discussed at the end of Chapter 2 derives its name from the way in which the new values of a pair of elements of the array are derived from the previous values. New values of elements d_i and d_k are calculated by

$$d_i = d_i + u*d_k$$
$$d_k = d_i - u*d_k$$

where the asterisk is used to denote complex multiplication. These operations are shown diagrammatically in Figure 3.41, with lines showing the flow of information and coefficients indicating scaling factors. A diagram of this sort is known as a **signal flow graph** and in this example the shape formed by these crossing lines is reminiscent of the wings of a butterfly.

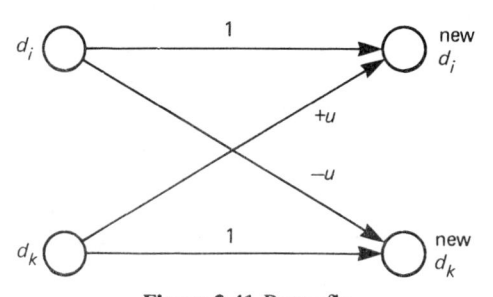

Figure 3.41 Butterfly

The sequence used in the Pascal example of the last chapter was

```
temp := cpxmul( data[ k ],u );
data[ k ] := cpxsub( data[ i ],temp );
data[ i ] := cpxadd( data[ i ],temp );
```

which although expressed more compactly than the corresponding assembly language program is inefficient in the use of the microcomputer's time. The microcomputer must calculate the addresses of the elements d_i and d_k, that is, *data[i]* and *data[k]* several times during each 'butterfly' sequence. Each complex operation uses both the real and imaginary parts of both operands, and it would be more efficient to compute a pointer to each of the elements before carrying out the calculations. Calling and returning from subroutines carries an overhead in microcomputer time, and the program would run faster if the actual operations required were written without recourse to subroutine calls.

The address of an element of an array may be calculated quite compactly using a 6809, as was seen in the listing, and the two index registers X and Y can be used as pointers to the two elements d_i and d_k. These pointers can be used each time that the contents of these elements are changed during the 'butterfly' calculation. The 6809 can perform 16-bit additions and subtractions directly using single instructions, but 16-bit multiplications must be programmed as a sequence of instructions.

This point about programming multiplications and other arithmetic operations underlines another aspect of the use of assembly language. Assembly language allows the programmer to choose the number representation to be used in the calculations with much more freedom than when a high level language is being used. Both BASIC and Pascal have real number representations which are capable of calculating to an accuracy of many decimal digits. In the case of the instrument being described here, however, reduced accuracy may be preferred if it offers more speed. Special number representations are often used in FFT calculations in order to speed up calculations.

The very large amount of computation required for an FFT requires more processing power than most other applications of microcomputers. 16-bit microcomputers based upon devices such as the 68000 have sufficient data registers to hold all the temporary results needed during the butterfly calculation and the two pointers required can be stored in address registers. This microprocessor can multiply 16-bit signed numbers directly much faster than the sequence used in the 6809. Special types of microprocessor are available for signal processing applications with instruction sets which are optimised for this task. The 32010, for example, can carry out a multiplication in well under a microsecond. This enables it to carry out FFTs at very high speed: a 256-point transform requires less than 16 ms, and a 64-point transform less than 1 ms.

The additions and multiplications required in this calculation are all complex, requiring several steps. High speed designs have been produced in which more than one processor is used to carry out these calculations in parallel. For example, the real and imaginary parts of a sum or product could be computed in parallel. If large numbers of processors are used, the time taken to compute an FFT spectrum can be reduced even further, and it is estimated that 64 32010s operating in parallel could compute a 1024-point FFT in less than 0.5 ms. (Magar *et al.*, 1982).

3.14 Questions

In attempting the following questions the reader should use the assembly language of any type of microprocessor with which he or she may be familiar, but the 6809 is recommended.

1. Write a program to read 8-bit values of variables L and M from memory and store the larger of these two values in a location N. Repeat this question for both signed and unsigned numbers.

2. Write a program to locate the first occurrence of the letter 'E' (stored in ASCII code) between locations 0 and 255.

3.14 INSIDE THE MICROPROCESSOR 145

3. Modify the program of Question 2 so that it counts the number of occurrences of the letter 'E' (again, stored in ASCII code) between locations 0 and 255.

4. Modify the program of Question 3 to allow the number of occurrences of each of the letters 'A' to 'Z' (again, stored in ASCII code) between locations 0 and 255.

5. The following example shows a recursive subroutine. Show that it returns in accumulator B the factorial of the number which was in accumulator A when called.

```
SUBR    LDB     #1
        DECA
        BEQ     EXIT
        BSR     SUBR
        INCA
        PSHS    A
        MUL
        PULS    A
EXIT    RTS
```

6. Write a program to convert a BCD number in accumulator A into its binary equivalent.

7. Write a subroutine which tests the bit pattern in accumulator A and sets the flag bits as follows:

$C=0, V=0$: character is neither a letter nor a number.
$C=1$: character is a letter.
$V=1$: character is a number.

8. Using the subroutine of Question 6, write a subroutine which searches successive memory locations starting at address 1000 until a decimal number is found, and which returns with the address of the first digit in the index register.

9. Write a subroutine to convert up to three decimal digits (occupying successive memory locations pointed by the index register X) to a binary number in accumulator A. The overflow (V) bit should be set on return if the number exceeds 255.

Chapter 4 INPUT/OUTPUT TECHNIQUES

4.1 The structure of an interface

A microcomputer interface converts information between two forms. Outside the microcomputer the information handled by an electronic system exists as a physical signal, but within the program it is represented numerically. The function of any interface can be broken down into a number of operations which modify the data in some way, so that the process of conversion between the external and internal forms is carried out in a number of steps.

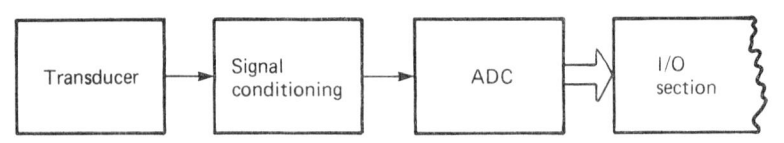

Figure 4.1 Input interface

This can be illustrated best by means of an example such as that of Figure 4.1, which shows an interface between a microcomputer and a transducer producing a continuously variable analog signal. Transducers often produce very small output signals requiring amplification, or they may generate signals in a form which needs to be converted again before being handled by the rest of the system. For example, many transducers have variable resistances which must be converted to a voltage by a special circuit. This process of converting the transducer output into a voltage signal which can be connected to the rest of the system is called **signal conditioning**. In the example of Figure 4.1, the signal conditioning section translates the range of voltage or current signals from the transducer to one which can be converted to digital form by an analog-to-digital converter.

An analog-to-digital converter (ADC) is used to convert a continuously variable signal to a corresponding digital form which can take any one of a fixed number of possible binary values. If the output of the transducer does not vary continuously, no ADC is necessary. In this case the signal conditioning section must convert the incoming signal to a form which can be connected directly to the next part of the interface, the **input/output section** of the microcomputer itself.

The I/O section converts digital 'on/off' voltage signals to a form which can be presented to the processor via the system buses. Here the state of each input line, whether it is 'on' or 'off', is indicated by a corresponding '1' or '0'. In the case of analog inputs which have been converted to digital form, the patterns of ones and zeros in the internal representation will form binary numbers corresponding to the quantity being converted.

The 'raw' numbers from the interface are limited by the design of the interface circuitry and they often require linearisation and scaling to produce values suitable for use in the main program. For example, the interface might be used to convert temperatures in the range -20 to $+50$ degrees, but the numbers produced by an 8-bit converter will lie in the range 0 to 255. Obviously it is easier from the programmer's point of view to deal directly with temperatures rather than having to work out the equivalent of any given temperature in terms of the numbers produced by the ADC. Every time the interface is used to read a transducer, the same operations must be carried out to convert the input number into a more convenient form. Additionally, the operation of some interfaces requires control signals to be passed between the microcomputer and the components of the interface (these will be discussed later in this chapter). For these reasons it is normal to use a subroutine to look after the detailed operation of the interface and carry out any scaling and/or linearisation which might be needed.

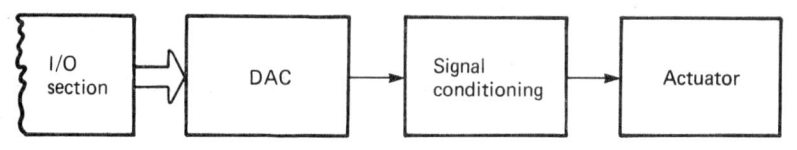

Figure 4.2 Output interface

Output interfaces take a similar form, the obvious difference being that here the flow of information is in the opposite direction; it is passed from the program to the outside world. In this case the program may call an output subroutine which supervises the operation of the interface and performs any scaling of numbers which may be needed for a digital-to-analog converter (DAC). This subroutine passes information in turn to an output device which produces a corresponding electrical signal, which could be converted into analog form using a DAC. Finally the signal is conditioned (usually amplified) to a form suitable for operating an actuator.

At the heart of any interface is the input/output device which converts information between the numerical form within the program to an electrical form outside the microcomputer. The techniques which can be used to produce subroutines for operating I/O devices have already been discussed in Chapters 2 and 3, while the problems associated with signal conditioning for various kinds of transducers and actuators will be discussed in Chapters 5 and 6. The remainder of this chapter will be devoted to an examination of the commoner types of input/output device and the ways in which they are used.

4.1.1 Input/output registers

The basic element from which all input/output devices are built is the register, which is typically selected by means of an address generated by the microprocessor. Unlike those used in the remainder of the microcomputer, these registers are accessible from outside the system by means of electrical connections.

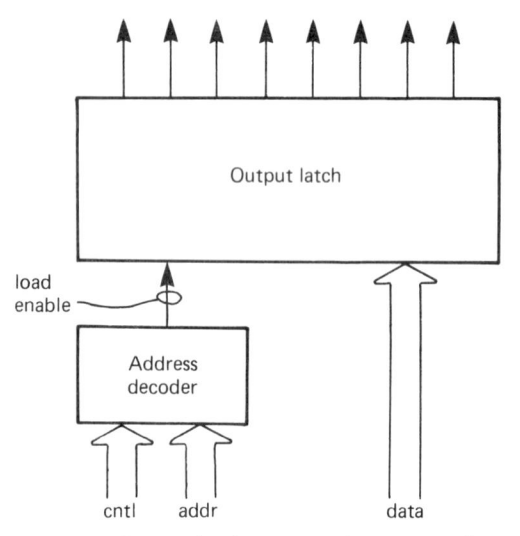

Figure 4.3 Output latch connected to system buses

Output devices consist of registers with their data inputs connected to the data bus and their outputs taken to external connections which can be used for interfacing; these are known as **output registers**. The register is operated by means of a **load enable** signal which is decoded from information on the address and control buses, as shown in Figure 4.3. The logic block which generates this enable signal from the control and address bus signals is called an **address decoder** and consists of an AND gate and inverters. Those bus lines which should be at logic '1' when the register is to be enabled are connected directly to the input of the AND gate, while those which should be at logic '0' are connected via inverters. In order to assign a unique address to a register, each of the address lines must be connected to the address decoder, which requires 16 connections in a typical 8-bit microcomputer with a 16-bit address bus.

The address is often not fully decoded in simple designs. If only the eight most significant address lines were decoded, for example, the output register would respond to any of 256 consecutive addresses which contained the appropriate binary pattern in their eight most significant bits. This partial decoding of addresses allows simpler logic circuits to be used, although the designer must ensure that no two registers can respond to the same address. If an AND gate is used in the circuit to generate the enable signal, the active level will be logic '1'. In practice, however, most interface devices have an enable input which is active at logic '0', and a NAND gate must be used instead.

When the microcomputer generates an appropriate address signal and a **write signal**, the load enable signal becomes 'active' and allows the register to be loaded with the data which is on the data bus at that instant. This data, which comes from the microprocessor, then appears on the register outputs. When the enable signal ceases to be active the data in the register, and hence the voltages on the output connections, will remain the same until the next time that the enable signal becomes active.

Input devices also use registers, but in this case it is their inputs which are brought out to external connections and their outputs which are connected to the data bus. Not surprisingly, these registers are known as **input registers**. The address decoding logic used for input registers produces a signal which connects the register outputs to the data bus when 'active' and disconnects them when not active. When the microprocessor selects the register for input, by generating the appropriate address and a **read signal**, information from the external connections flows on to the data bus and is read by the microprocessor. Usually the register inputs are 'enabled' continuously so that the information read by the microprocessor is the data on the inputs at that instant, but some designs make the input register's **load enable** signal available externally. This type of operation is useful when the input information is present only momentarily, because it allows the data to be loaded into the register while it is available and held there whilst waiting for the program to read it.

Both the input and output functions can be carried out using the same kind of integrated circuit: a register with a 'load enable' input which allows it to be loaded, and an 'output enable' input which allows the register outputs to be connected to the output pins. This ability to be connected and disconnected internally allows output connections of this type to have three possible states: 'high', 'low', and 'disconnected', and they are commonly referred to as being **three-state outputs**.

4.1.2 Parallel input/output devices

The simple latch circuit described in the last section is very useful, allowing registers of any length to be constructed. However, it is considerably simpler than the microprocessor and memory circuits which it complements. The same technology which is employed to make these more complicated devices can also be applied to input/output circuitry to produce more sophisticated, but more flexible, I/O circuits. Although the simple latch circuit can be used both for inputting and outputting information, the data in an output register cannot be read by the microprocessor when it is being used as an output register. This is not an insuperable problem, since all that is necessary when using such devices is to make a copy in memory of the information being written to the latch, but it is nevertheless inconvenient. Suppose, for example that the programmer wishes to switch one line of an output port to a different state without affecting the state of the others, for example to switch on or off one of a set of lamps on a control panel. The corresponding problem of changing the state of a bit in a memory location is quite straightforward, requiring just one line of high level language or three lines of assembly language, e.g.

LDA	MEMORY	Load Accumulator from memory
ORA	#8	Set bit 3 to logic 1
STA	MEMORY	and store result back in memory

It is common to connect the I/O registers into the microcomputer as if they were memory locations. This **memory-mapped connection** is possible with all microcomputers, and it allows memory-referencing instructions such as 'ORA' and 'SUB' to be used with input and output registers in just the same way as with memory locations. If it is possible to read back the contents of an output register then the register can be treated just like a memory device. However, if similar code were to be used with a write-only register, such as the latch used in the output register, it would not function correctly because the result of reading from a write-only register is undefined. In fact the solution is quite simple: a memory location is used to contain a copy of the contents of the output register at any time. Each time that the output state is to be changed, the copy in memory is changed first and then the new contents of this memory register are written into the output register, e.g.

LDA	COPY	Load Accumulator from memory
ORA	#8	Set bit 3 to logic 1
STA	COPY	Store the result back in memory
STA	OUTPRT	and copy into Output Register

The need for this extra line of program can be avoided by including **extra** circuitry on the I/O chip to allow the outputs of an 'output register' to be connected back to the data bus when the microprocessor wishes to read its contents. This does not represent a significant problem in the design of large-scale integrated circuits, because the registers in memory and microprocessor circuits already include circuitry of this type.

Many microcomputers use a separate signal to select input/output devices so that memory and I/O are controlled separately and occupy distinct address spaces, called the **memory address space** and the **I/O address space**. This approach has the advantage of allowing special instructions to be used with input/output devices, for example to send a sequence of characters to a display. However, it is not usually possible to use ordinary memory reference instructions in such cases, and either a copy must be kept in memory or special bit-manipulating instructions must be used if they are available.

4.1.3 Controlling data direction

An input/output circuit can be made even more flexible by designing it so that the same component can be programmed to behave as an input or as an output. This is normally done in such a way that the direction (input or output) of each input/output connection can be programmed separately. A circuit of this type may be used in a wide range of applications provided that the total number of inputs and outputs does not exceed the total number of I/O connections on the device. Using this approach there is no longer any need to discriminate between input and output circuits because one type of circuit can be programmed to provide whatever mixture of input and output connections is required. Of course this increased flexibility is achieved at the cost of greater complexity, but this is more than outweighed by the cost savings which result from larger volumes of manufacture and sales of a programmable component. Ports which can be programmed to act as either inputs or outputs are called **bidirectional input/output ports**.

These more flexible I/O devices require extra registers within them to store information about the way in which each line is programmed, that is whether it is an input or an output. This information is held in a **Data Direction Register (DDR)** which contains one bit for each I/O connection on the circuit. When any connection is to be used as an output, the corresponding bit in the DDR is programmed to the 'output' state (usually a logic '1', although there is no fundamental reason why it has to be this way round), while if the line is to be an input the bit is programmed to the other state.

Programmable input/output devices

Most manufacturers produce a kind of programmable input/output device embodying the ideas described in the last few sections. The input and output lines of these circuits can be programmed independently to act as inputs or outputs, or they can be grouped as input or output 'ports' which will allow the input or output of information in the form of words, usually eight bits wide. Unfortunately each manufacturer tends to use a different acronym for its product, although the commonest names tend to be **Parallel Input/Output (PIO)** and **Peripheral Interface Adapter (PIA)**. The number of input/output lines on one of these circuits is determined to a large extent by the size and economics of the integrated circuit packages used. Because any I/O circuit must have access to the data bus and requires connections to the address and control buses, at least a dozen of the pins of the integrated circuit are used up simply in connecting it to the rest of the system. This means that in order to get a useful number of external connections a 40-pin package must be used, which is enough to furnish two 8-bit ports with their control lines in addition to the connections to the buses and power supply.

The 6522, or **Versatile Interface Adapter (VIA)**, is one such general purpose interface circuit which provides a wide range of interface functions, and it will be used as an example throughout the remainder of this chapter. It has two 8-bit parallel ports, each with a data direction register. The two ports are given the names 'ORA' and 'ORB' (the letters 'OR' standing for 'output register'), while their data direction registers are called 'DDRA' and 'DDRB'. A logic '1' in any bit of DDRA or DDRB will configure the corresponding bit of ORA or ORB as an output, while a logic '0' will configure it as an input. Thus to program ORA to have four output lines and four input lines a sequence such as this might be used

```
LDA     #%11110000    4 outputs, 4 inputs
STA     DDRA          Configure port A
```

When port A is read with the instruction

```
LDA     PORTA
```

the bits read into accumulator A will reflect the logic levels on the input/output lines. The four most significant bits will be due to those lines configured as outputs, the levels of which are controlled by ORA. The four least significant bits read will be input levels controlled by whatever is connected to those lines.

4.2 Single-bit input/output

One use of an I/O device has already been touched upon: the input and output of individual bits. This sort of I/O is used very often in engineering applications, for example to read switches or to control on/off actuators. Because microcomputers handle data in the form of words rather than as individual bits, provision must be made for manipulating the bits within a word when the programmer wishes to input or output independent bits of data.

This can be done in the microprocessor by making use of the logical instructions provided. A bit or a group of bits can be set to logic '1' by using the 'OR' operation with the pattern of bits which are to be set. If the 'Exclusive OR' operation is used in the same way selected output bits can be 'toggled', or switched to the opposite state. Bits can be cleared selectively by using logic 'AND', but in this case the pattern to be used should contain zeroes in positions where the corresponding output line is to be switched to logic '0' and ones elsewhere. This example switches on bit 0, switches off bits 1 and 2, and switches bit 5 to the opposite of its current state

```
LDA     OPORT          Get current output state
ORA     #%00000001     Switch on bit 0
ANDA    #%11111001     Switch off bits 1 and 2
EORA    #%00100000     Toggle bit 5
STA     OPORT          and update the output state...
```

If it had not been possible to read the output port, then it would have been necessary to use a memory register instead and copy the result of any change to the output register. The same is true of microcomputers which use a separate address space for input/output devices such as the Z80. In cases such as this it is often more convenient to work with a memory register and communicate the result to the corresponding output register with an OUT instruction.

The normal way to isolate the state of a particular bit is to 'AND' it with a bit pattern which contains a single '1' corresponding to that line and to follow the AND instruction with a branch instruction which tests to see whether or not the result of that operation was zero, e.g.:

```
LDA     IPORT           Read Input Port
ANDA    #%00001000      Test bit 3
BNE     ACTION          and branch if it is set...
```

This ability to act upon individual input bits of information is important in applications such as switch panels where the state of a single line must be separated from those of the remaining lines on that port.

In the case of microcomputers intended for single-chip control applications the ability to control individual input and output lines as efficiently as possible is particularly important, and many single-chip microcomputers have special instructions to make the manipulation of individual bits easier. The 6301, for example, is a single-chip microcomputer with a basic instruction set that is very similar to those of the 6800 and 6809. However, it includes some extra instructions which allow logical operations to be carried out between 'immediate' data in the instruction and the contents of an output register and which store the result in the register instead of in the accumulator.

Similarly, the 6805 has a set of branch instructions which test the state of a specified bit in a specified register and cause the program to branch if that bit is set (or clear, according to the instruction used). A single test and branch instruction BRSET replaces the three instructions of the last example and requires only three bytes of storage instead of seven

```
BRSET3    IPORT,ACTION
```

This allows programs to be shorter which is an important consideration when the storage space is small, which is often the case with single-chip microcomputers.

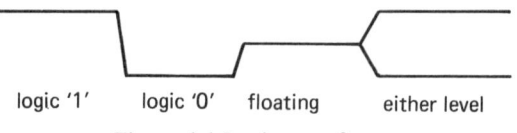

    logic '1'   logic '0'   floating   either level

Figure 4.4 Logic waveforms

4.2.1 Waveforms

The signals which appear on the external connections to the microcomputer consist of voltage levels. The actual voltage levels are not important at this stage, although they will be discussed in the next chapter. However, it is obvious that any such binary signal will contain voltages with two possible levels corresponding to the logical values '0' and '1'. The voltage signals can be drawn as graphs of voltage against time which represent the **waveforms** which would be seen if an oscilloscope — a test instrument — were connected to that point in the circuit. In fact the waveforms are usually drawn in a rather stylised form with sloping changes from one level to another. The lower of the two levels is used to represent logic '0' and the upper one to represent logic '1', as shown in Figure 4.4.

It is also possible for a connection to be 'floating', with no signal connected, which is shown in waveform diagrams such as that of Figure 4.4, by means of a line midway between the logic '1' and '0' levels. This is not what would usually be seen in practice, but it provides a clear indication that the signal must be assumed to be unknown (it could be at either logic level) when looking at waveform diagrams.

Figure 4.5 Waveform with uncertainties in timing

Sometimes a waveform can be at either logic level at certain times, which is shown by means of two parallel lines, one at level '0' and the other at level '1'. If the signal can change from one level to another at any point during a period of time, that part of the waveform is shown with several sloping 'transitions' throughout the time in question, as illustrated in Figure 4.5. Waveform diagrams are widely used to show the relationship between signals at different points in microcomputer systems. When a change between levels in one signal causes a change in another signal, this can be shown by means of a curved arrow with its tail looped around the transition in the waveform which causes the change and its head pointing at the change in the other waveform. This provides a diagrammatic method of showing the interaction between one part of a system and another. The time taken for a signal to respond to a change in another signal can be shown with a double-ended straight arrow indicating the difference in time between the two events. Figure 4.6 shows the relationship between the enable signal of a latch and the signal on its output.

4.3 Handshaking

The examples which have been discussed so far were concerned simply with reading the state of input lines, and of changing the state of output lines. Sometimes this information consists of a sequence of items of information

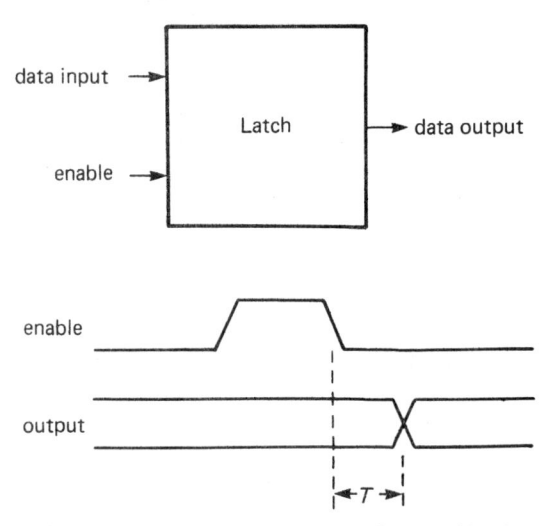

Figure 4.6 Input and output waveforms of latch

being passed from one machine to another, and in such cases it is necessary for the two machines to act in a co-ordinated manner to make sure that none of the information is lost. It is not sufficient for the receiving machine to assume that a new item of information arrives only when the input signals change because two consecutive items may be the same; extra information must be provided to co-ordinate the transfer from source to destination.

Consider a pair of machines A and B in which machine A sends information to machine B. Each time A outputs a new item of information it must also inform B that the information is available by means of a **Data Available (DAV)** signal, and when B reads that information it must tell A that it has done so by means of a **Data Acknowledge (DAK)** signal. These two signals DAV and DAK are called **handshake signals** and appear in various forms in a very wide range of input/output applications. Each I/O device that uses a handshake needs two extra connections for the handshake signals. When A outputs new data it switches its 'data available' output to the state which indicates the presence of the data and then waits for B to switch its 'data acknowledge' output to the state which indicates that it has accepted the data. Machine A then switches its DAV signal back to the original state indicating that the data is no longer available and waits for machine B to switch its DAK output in reply.

4.3.1 Automatic handshaking

The technique for managing DAV and DAK discussed above uses the microcomputer to monitor the *levels* of the control lines whereas it is *changes* in level which signal the availability and acceptance of each item of data. However most kinds of input/output device have special control lines associated with each input and output port which can be programmed to

```
┌─────────────┐   Data AVailable   ┌─────────────┐
│             │ ─────────────────► │             │
│             │                    │             │
│             │   Data AcKnowledge │             │
│  Machine A  │ ◄───────────────── │  Machine B  │
│             │                    │             │
│             │        Data        │             │
│             │ ══════════════════►│             │
└─────────────┘                    └─────────────┘
```

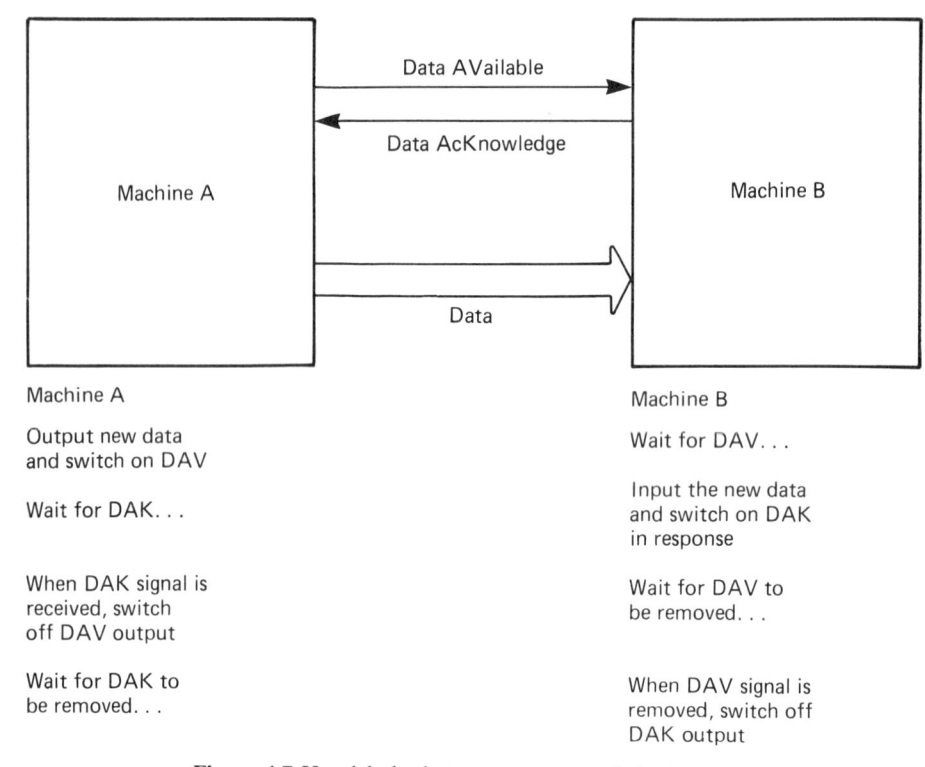

Machine A
Output new data
and switch on DAV

Wait for DAK...

When DAK signal is
received, switch
off DAV output

Wait for DAK to
be removed...

Machine B
Wait for DAV...

Input the new data
and switch on DAK
in response

Wait for DAV to
be removed...

When DAV signal is
removed, switch off
DAK output

Figure 4.7 Handshake between source and destination

supervise the passing of data automatically, without the microprocessor having to monitor the lines itself.

These lines are controlled quite differently from the ordinary I/O lines which are used for data. Usually they have two special registers associated with them; a control register which is used to control the way in which they are to respond to external signals and a status register which is used to report what is going on. The status register contains a number of 'flag bits' just like the processor status register within the microprocessor itself. Each of these flag bits is set to logic '1' when a particular event takes place on the interface between the two machines, which can be illustrated best by means of an example.

Consider an I/O device being used to input words of data from another machine with a pair of handshake signals DAV and DAK being used to supervise the transfer of information. The status register of the I/O circuit will contain a bit which is set to logic '1' when the DAV input changes level to indicate, or 'flag', the arrival of the next piece of data. This bit will remain at logic '1' even when the DAV input changes back to its original state again because it indicates the fact that data has been made available to the input; however, DAV would not normally change until DAK had been asserted. The only way to reset the bit to logic '0' again is to read the input register. When this happens the input circuit automatically changes the logic level on the DAK line to indicate that the microprocessor has 'accepted' the data.

Using the special handshake lines in this way means that the microprocessor does not need to monitor the state of the handshake lines continuously; all it needs to do is to check by reading the status register that new data has arrived before reading the input data itself. The control register is used to program the details of the handshake operation. For example, some systems use a DAV signal which goes to logic '1' when new data arrives, while others operate with DAV at logic '0' when data is available. These different modes of operation can be accommodated by programming the handshake input line to detect either changes from logic '0' to '1' or vice versa. Note that in each case it is the *change* in level which is important rather than the actual level itself. The handshake output line, which is DAK in the case of a data input, can be programmed into any one of a number of modes. It can be made to switch to logic '1' or '0' to indicate that data has been accepted by the input, and it can be programmed to remain in that state until a change in logic level on the other input is detected, or simply to produce a pulse when the microprocessor reads the input. If a handshake is not needed, it is possible with some I/O circuits to program the control line to remain at a fixed logic level under program control as an extra output and in some cases it can even be programmed to act as a spare input.

The operation of an output interface is very similar to that of an input except that the roles of the signals are reversed. Here the input line is the **data acknowledge** line and the output line is the **data available** line, and this output line must indicate when new data is written into the output register rather than read from an input register.

Handshaking with a programmable I/O device
Each port of a programmable I/O device usually has a pair of **control lines** associated with it which can be used to implement a 'handshake' if one is needed. Each of the parallel ports of the VIA has such a pair of control lines associated with it; in the case of port A these are called CA1 and CA2, while in the case of port B they are called CB1 and CB2. These lines may be used to implement a handshake when the port is used. One of the control lines of each port (CA1 and CB1) is an input which is sensitive to changes in logic level; when a change in level occurs on this input, a bit is set in a status register called the **Interrupt Flag Register (IFR)**. The IFR contains several flag bits which indicate events that have occurred in the interface, rather like the way in which the processor status register of a microprocessor indicates special events, such as a zero or negative result, in the microprocessor. Either changes from '0' to '1' or changes from '1' to '0' can be detected on these input control lines. The changes in level to which the lines are sensitive can be programmed by means of yet another register, the **Peripheral Control Register (PCR)**. Bit 0 of the PCR controls CA1 and bit 4 controls

CB1; a '0' causes the corresponding input to be sensitive to a change from '1' to '0', while a '1' causes the input to be sensitive to changes from '0' to '1'.

The second control line of each port (CA2 and CB2) is more versatile. CA2 and CB2 can be programmed to act as inputs rather like CA1 and CB1, or as outputs. When programmed as outputs the level can be set to logic '1' or to logic '0' in the same way as any of the output lines of ORA and ORB. More commonly, however, they are programmed to carry out a read or write handshake function. During a read handshake CA2 remains at logic '1' until the corresponding port ORA is read, when it switches to logic '0' to indicate that data has been read from ORA. It then remains at '0' until an 'active' transition occurs on CA1, which indicates that new data has been presented, whereupon it returns to logic '1' until the data is read by the microprocessor. A write handshake is similar; CA2 remains at logic '1' until data is written to ORA, when it switches to logic '0' to indicate the availability of new data. When an active transition is detected on CA1, which indicates that the outputs of ORA have been read, CA2 is returned to logic '0' until new data is written to ORA. The behaviour of CB2 is similar to that of CA2 except that it interacts with ORB and its input line CB1.

There are yet more modes of operation which can be programmed for control lines CA2 and CB2. Each of these lines requires three bits of the PCR to specify its mode of operation; bits 1, 2, and 3 are used to specify the behaviour of CA2, while bits 5, 6, and 7 control CB2.

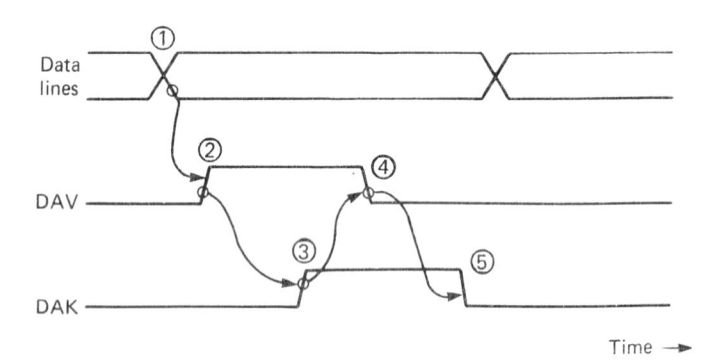

Figure 4.8 Handshake waveforms

4.3.2 Using an automatic handshake

The existence of a flag bit allows the programmer to test for the presence of data before attempting to read an input port or to make sure that the previous word of data has been taken before writing to an output port. In each case this is usually tested in a subroutine which contains a 'waiting loop'. An instruction in the waiting loop tests the flag continually until the port is ready before the program proceeds to use the I/O circuit. This process is known as **polling** and the subroutine is said to 'poll' (test) the DAV flag. In the VIA, all the flags are in the IFR; bit 1 of the IFR is set to '1' by an active transition on CA1 and bit 4 by an active transition on CB1. Thus, if port A of the VIA were used as an input, CA1 would be used as the DAV line and bit 1 of the PCR would be the DAV flag. The other control line belonging to port A, CA2, would then act as the DAK line, and if the VIA were programmed correctly the handshake would be performed automatically by the VIA.

At the beginning of the program ORA must be programmed as an input port and the PCR must be programmed to give the correct handshake behaviour:

```
CFIGIP  LDA   #%00000000   Set ORA to inputs by
        STA   DDRA         setting all DDR bits to '0'
        LDA   #%00001000   Set PCR bits 0 to 3 for handshake
        STA   PCR          and downward transition on CA1
```

Each time that port A is to be read, the flag register IFR must be polled and if necessary the microprocessor must wait for the DAV flag to be set before continuing:

```
WAITIP  LDA   IFR          Read Status Register IFR.
        BITA  #%00000010   Mask off all but DAV flag
        BEQ   WAITIP       and repeat until it is at logic 1.

        LDA   ORA          Read the data on Input Port ORA
        RTS                and return...
```

A similar subroutine can be used in the microcomputer which is providing the data. Here the port to be used must be programmed to provide output lines and the control lines must be programmed to perform an automatic handshake as before. In this example port B will be used:

```
CFIGOP  LDA   #%11111111   Set ORB to outputs by
        STA   DDRB         setting all DDR bits to '1'
        LDA   #%10000000   Set PCR bits 4 to 7 for handshake
        STA   PCR          and downward transition on CB1
```

Each time that a byte of data is to be output via ORB, the data is output first and then the subroutine waits for an acknowledgement that the data has been accepted before continuing.

```
WAITOP  STA   ORB              Send the data to ORB

WAIT01  LDA   IFR              Read Status Register IFR.
        BITA  #%00010000       Mask off all but DAK bit
        BEQ   WAIT01           and repeat until it is at logic 1.

        RTS                    then return...
```

The two subroutines running in the two machines interact to give a handshake as shown in Figures 4.8 and 4.9.

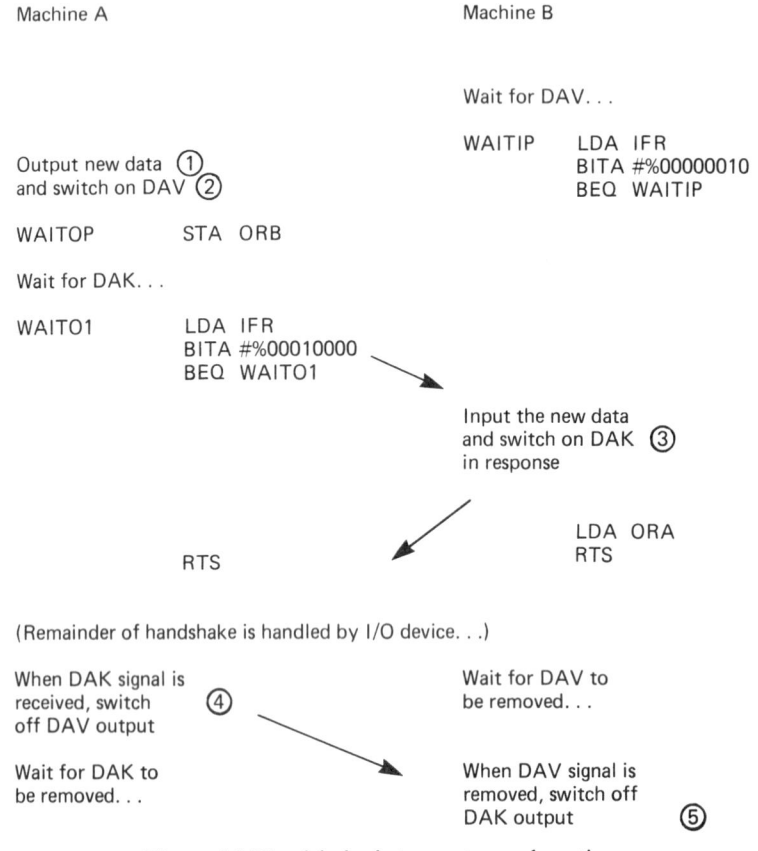

```
Machine A                                         Machine B

                                                  Wait for DAV...

                                         WAITIP   LDA  IFR
Output new data   (1)                             BITA #%00000010
and switch on DAV (2)                             BEQ  WAITIP

WAITOP            STA  ORB

Wait for DAK...

WAIT01            LDA  IFR
                  BITA #%00010000
                  BEQ  WAIT01
                                                  Input the new data
                                                  and switch on DAK  (3)
                                                  in response

                                                  LDA  ORA
                  RTS                             RTS

(Remainder of handshake is handled by I/O device...)

When DAK signal is                                Wait for DAV to
received, switch       (4)                        be removed...
off DAV output

Wait for DAK to                                   When DAV signal is
be removed...                                     removed, switch off
                                                  DAK output          (5)
```

Figure 4.9 Handshake between two subroutines

The inputs and outputs used for the DAV and DAK signals must be programmed for the particular I/O device being used, and exactly what must be programmed and how it is programmed depends upon the kind of device. Each type has a control register which will allow the behaviour of the lines to be programmed and a status register to allow the status of the handshake to be monitored.

4.3.3 First-in, first-out buffers

Sometimes the recipient of data is unable to process the data at the speed at which it is being presented to the system. Whenever a 'bottleneck' of this kind occurs, the data must be placed into an orderly 'queue' awaiting processing.

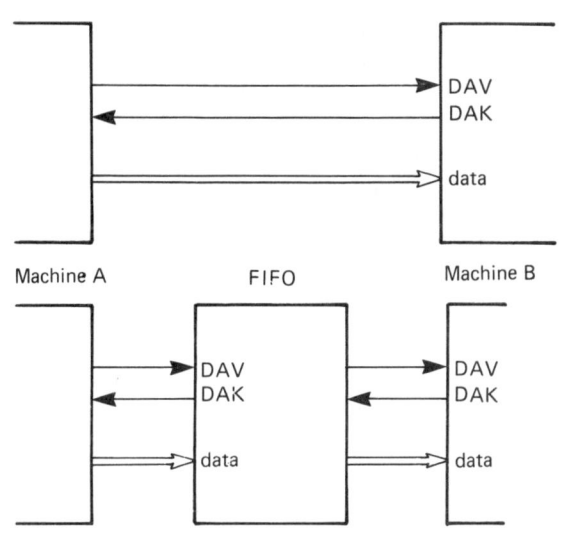

Figure 4.10 Using a FIFO to buffer two machines

The main characteristic of a queue is that data must be removed from it in the same order in which it was placed into the queue. This is the origin of the name **First-In, First-Out** which is usually abbreviated to **FIFO** and which provides a direct contrast with a stack in which the first piece of data to be placed in the stack is the last one to be removed. A FIFO can be implemented in software using indexed addressing, or using special hardware. In the former case the maximum speed of the FIFO is limited by the speed of the microcomputer to a few tens of thousands of words per second, while if special hardware is used, speeds up to tens of millions of words per second are feasible. A FIFO with a handshake at both the input and output ends may be used to connect together any pair of asynchronous processes, as shown in Figure 4.10. The effect of the FIFO is to allow a microcomputer (or other system) to accept data at a peak rate which is limited by the speed of the FIFO, provided that the average rate does not exceed that which the microcomputer can handle.

FIFOs are often included in the interfaces to communication links between microcomputers because they allow messages to be sent at a high speed which is limited by the capacity of the communications link and the properties of the FIFOs rather than the processing speed of the microcomputers.

4.4 Interrupts

The main problem with polled input/output is that the microcomputer's time is wasted. Often the time that the microcomputer spends in waiting for the appropriate flag bit to be set in a status register could be spent doing something more productive. One way to arrange this is simply to test the flag bit less frequently, but this is not possible if the input/output device requires a rapid response from the microprocessor. In the sort of real-time programs used in instrumentation and control, delays in responding to external flags must usually be minimised, and this means that either a short waiting loop or an interrupt system must be used.

The solution to this dilemma is to arrange a mechanism to allow **interrupts**. Interrupt mechanisms allow the flow of a program to be suspended when a designated flag is set so that a special **Interrupt Service Routine (ISR)** can be run to take care of the input/output device. When this routine has finished, the microcomputer resumes the instruction sequence which it was running when interrupted.

One way to visualise the sequence of events is to compare it with a telephone call arriving in an office. The person answering the call must stop what he or she is doing, find out the identity of the caller and what is required, and take action before resuming the work which was in progress when the bell rang. Corresponding steps occur during an interrupt in a microcomputer:

- When the interrupt occurs, details of the current register contents must be saved so that they may be reloaded later.
- The identity of the interrupting I/O device and/or the action to be taken must be determined.
- The appropriate interrupt service routine must be run.
- Finally, the original instruction sequence must be resumed.

The complexity of microcomputer-based systems requires considerable sophistication in handling interrupts, and many techniques have been devised for use with interrupts in microcomputers. These will be discussed in the next few sections of this chapter.

4.4.1 The interrupt mechanism

The first step in the sequence of events which makes up an interrupt is that the interrupting device must inform the microprocessor that it requires attention. In the analogy this corresponds to the telephone bell ringing. The request for an interrupt is made by changing the logic level on an **Interrupt Request (IRQ)** input to the microprocessor itself, which is monitored after the execution of each instruction. Sometimes there is more than one such input, as will be discussed later.

Next the microcomputer must make a note of the point in the program which it had reached when the interrupt request occurred. This is indicated by the contents of the program counter, which must be saved in some way. The normal method of doing this is to push it on to the stack in just the same way as when calling a subroutine. In fact, an interrupt service routine closely resembles a subroutine, except that the interrupt call is due to an external source. There is another important difference between the two. A subroutine frequently uses the registers of the microprocessor to pass parameters between the subroutine and the calling program. An interrupt service routine must not change the contents of any of the registers in the microprocessor unless it saves their contents at the beginning of the service routine and replaces the information before returning to the original instruction sequence. The occurrence of the interrupt routine must be made 'invisible' to the other program, otherwise they will interact in an unpredictable manner.

Some microcomputers save the contents of all the processor registers on the stack when an interrupt occurs, and return this information to the microprocessor registers after the ISR has finished. This is done using a **Return From Interrupt** (RTI) instruction which automatically reloads all the microprocessor registers including the program counter in just the same way as a 'return from subroutine' instruction returns control to the calling program by reloading the program counter from the stack. It is convenient to save all the registers on the stack because this allows the ISR to make use of all the processor registers, except the stack pointer. In fact the stack pointer can also be used provided that it is returned to its original state before returning to the other program.

Saving all the registers on the stack is attractive when the microprocessor has only a few registers, but in the case of the 6809, for example, the large number of register bytes involved means that saving all the registers takes about 20 microseconds, and a similar amount of time is needed to return from an interrupt. This time overhead may be excessive if the ISR does not make use of all the processor registers, and for this reason the 6809 allows another 'fast' interrupt which saves only the contents of the program counter and processor status register. When using the **Fast Interrupt Request (FIRQ)** input, the 6809 programmer must preserve the contents of any registers which are used by including appropriate instructions within the ISR.

A more radical approach to this problem of stacking the contents of the registers is to make use of another set of registers within the microprocessor. This is possible with a number of types of microprocessor, including the popular Z80, which has a duplicate set of working registers which can be selected by means of a software 'switch' controlled by special instructions. However, this second set of registers does not include a program counter, and when an interrupt occurs, the contents of this register are saved on the stack and they are recovered after the interrupt service routine has finished. If the ISR is written in such a way as to switch to the spare set of registers when it starts and switch back to the original set at the end of the ISR, there is no need to save many of the registers on the stack. Similarly, the 9900 and 8048 families of microprocessors make use of memory registers as working registers, and when an interrupt occurs another set of registers can be selected, avoiding the need to save the contents of the old ones. Problems arise when there is more than one possible source of interrupts, and these will be discussed in section 4.4.3.

4.4.2 Loading the interrupt vector

Once the contents of the microprocessor's registers have been saved, the program counter must be loaded with the starting address of the ISR; the information used to specify this starting address is called an **interrupt vector**. In principle calling an ISR is quite simple, since it is just like calling a subroutine, but in practice there are many ways of supplying the start address.

The simplest of these is to arrange for the ISR always to start at the same address in memory. This starting address is often at the lowest address in memory, and often the first instruction in the ISR is a jump to another address where the ISR proper begins. This allows the main ISR to start at any memory address, and many more modern microcomputer designs arrange to store the starting address, rather than the first instruction, of the ISR at a fixed place in memory. This is the system used by the 6809 and 6502 microprocessors, among others.

The 8085, on the other hand, uses a somewhat more indirect, but more flexible, system in which the interrupting device supplies the first opcode of the interrupt service routine. Normally this instruction causes the microcomputer to start executing an ISR at a prearranged address in memory. The Z80 also supports this mode of operation, as well as two other modes. In the first of these other modes the microprocessor automatically assumes that the address is stored at a prearranged place in memory rather like the 6809. The second is more complicated; here the Z80 receives a vector byte from the interrupting device and assumes that it represents an offset from the start of a table of addresses. This offset is combined with an address held in an **Interrupt Vector Register (IV register)** within the microprocessor to form a pointer to the address at which the ISR commences, as illustrated in Figure 4.11.

The pointed address is then loaded into the program counter and the microcomputer starts running the interrupt service routine. Allowing the interrupting device to indicate directly which ISR should be run saves valuable time when handling multiple interrupt sources, as will be discussed in section 4.4.3.

Sometimes, during parts of a program in which timing is critical, it is important that a program is not interrupted. To return to the analogy, this corresponds to leaving the telephone 'off the hook' so that it cannot ring. In the case of the interrupt, this is done by means of a 'mask' bit in the processor status register which, when set, prevents the signal on the IRQ line from causing an interrupt. Usually the mask is set automatically by the processor before executing the interrupt service routine. This 'deafens' the microprocessor so that it cannot 'hear' the interrupt, but by stretching the analogy further, we can imagine that a bell might sound for some very important reason, such as a fire. A correspondingly important event in the

4.4.2

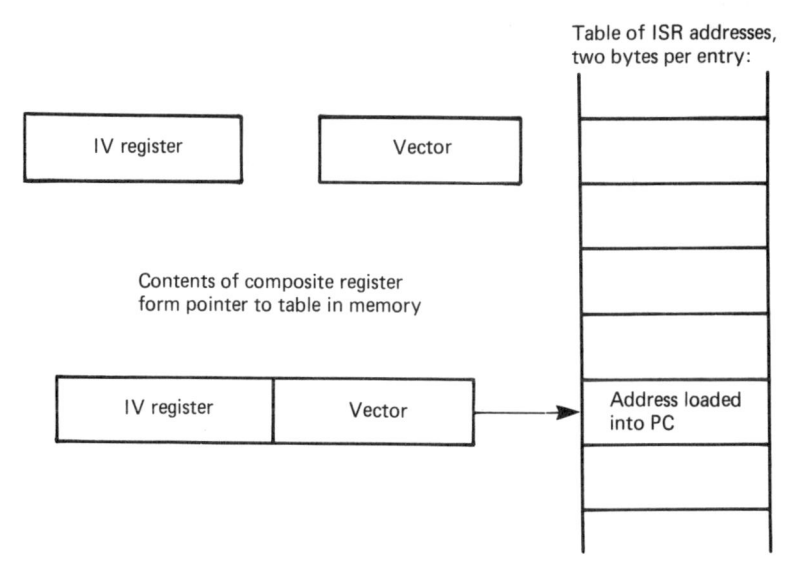

Figure 4.11 Z80 interrupt vector system

case of a microcomputer program might be a signal from the power supply unit that the electricity supply was failing. In this case the microcomputer might need to save all important information in some non-volatile form in which it could be retained until the power supply was established once more. For this reason, most microcomputers also have a **NonMaskable Interrupt (NMI)** which cannot be masked like the IRQ input.

Most input/output devices allow their interrupt outputs to be disabled by means of special **interrupt mask** bits in a control register. The VIA, for instance, has a register called the **Interrupt Enable Register (IER)** which contains a bit corresponding to each flag in the interrupt flag register. If an interrupt source is enabled by means of the IER, the corresponding flag bit will cause an interrupt if set to '1'.

4.4.3 Multiple interrupts

If more than one device is connected to the microprocessor, some provision must be made for differentiating between their interrupt signals. The easiest way is to use a separate connection for the interrupt request line from each device, but this often uses an excessive number of connections. A more compact approach results when all the IRQ lines are connected together with a 'wire-OR' arrangement so that the IRQ input of the microprocessor switches low if any of the interrupting devices causes an interrupt. **Daisychaining**, as illustrated in Figure 4.12, provides a more flexible arrangement in which each I/O device has an interrupt enable input and an interrupt enable output. The input and output are connected by internal logic circuitry which allows the device to generate an interrupt internally or to propagate an interrupt request from its IRQ input to its IRQ output. The IRQ output of the last device is connected to the IRQ input of the microprocessor. When the microprocessor responds to an interrupt request, it generates an **Interrupt Acknowledge (IACK)** signal which is passed back to the interrupting devices. The IACK signal is passed back along the chain until it reaches a device which is requesting an interrupt which then identifies itself to the microcomputer. In this way the IRQ and IACK signals form a handshake between the interrupting device and the microcomputer.

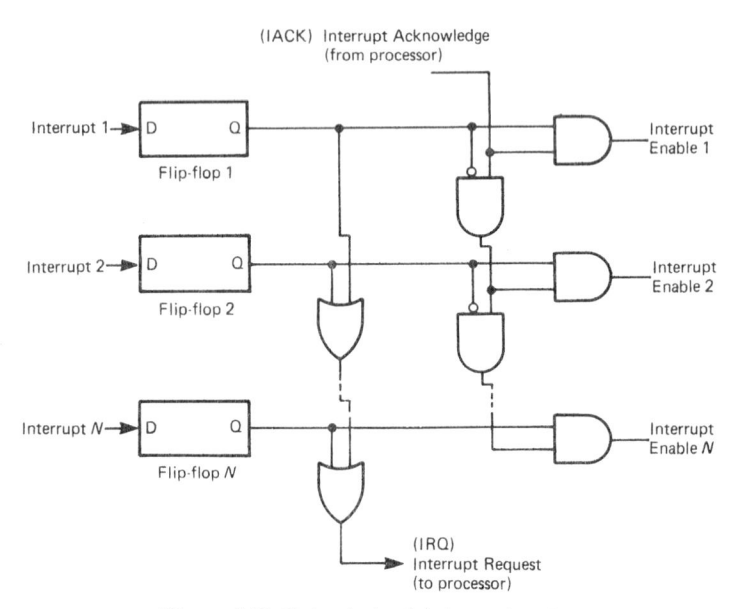

Figure 4.12 Daisychained interrupt system

Identifying the source of an interrupt
When there is more than one possible cause of an interrupt, the microprocessor must determine the source of each interrupt and the appropriate ISR for that source. The simplest method of doing this is **polling** in which the ISR interrogates each device in turn to ask if it caused the interrupt. The devices are 'polled' by reading their status registers and testing the flags to see if any are set. Once a flag has been found to be set, the corresponding piece of program can be run by the microprocessor. Unfortunately, the process of polling takes a long time in electronic terms, and the amount of time

taken could be excessive. For this reason, the more time-critical sources of interrupt are tested first, and less critical ones are left until later.

The fragment of program shown in Figure 4.13 shows how an interrupt handler might locate the source of an interrupt. First the IFR is tested to see if any flags are set; if the VIA is currently interrupting the processor, the most significant bit of the IFR will be set to 1. If the VIA is the source of the interrupt, the processor must next determine the source of the interrupt within that VIA. This fragment assumes that there may be several sources of interrupt from different devices, although later in the design example we shall see that this is unlikely to be the case in practice.

```
IRQHAN  LDA   IFR            Test IFR
        BPL   OTHER          Bit 7 set => flag set
        BITA  #%00000010     Bit 1 set => CA1 flag
        BNE   ISRCA1
        BITA  #%00010000     Bit 4 set => CA2 flag
        BNE   ISRCA2
        ...                  {Test other devices}

ISRCA1  LDA   ORA            Clear interrupt from CA1
        STA   FRED           by reading ORA

        ...                  {ISR routine for CA1}

        RTI                  Return to calling program when done

ISRCA2  LDA   BORIS          Clear interrupt from CA2
        STA   ORB            by writing ORA

        ...                  {ISR routine for CA2}

        RTI                  Return to calling program when done

        ...                  {Remaining ISRs go here}
```

Figure 4.13

A much faster technique is to arrange that each interrupting device has a separate code which it uses to identify itself when it causes an interrupt. The address of the ISR corresponding to that device can then be found from a table in the program and the ISR run immediately without further need for polling.

Interrupt priorities

Some of the sources of interrupt requests will almost certainly be more important than others. For instance, a request for service from an interface handling information for an electronic device must be handled more urgently than information being passed to a slow mechanical printer, while a power-fail interrupt is more important than either of them. In addition to the need to identify the source of an interrupt, the **interrupt priority** must be established. When an interrupt occurs, the interrupt priority system will arrange that the interrupt will be acknowledged only if the microcomputer is currently running a program of a lower priority. If the current program has the same or a higher priority, the interrupting device will have to wait until that program finishes. The software within the microcomputer can be visualised as a set of tasks, each with appropriate priority. Most of these will be interrupt service routines, but some subroutines where timing is critical may have a higher priority. Whenever a new task starts running, it sets an interrupt mask which prevents interrupts with a lower priority from having effect. This ability to have interrupts during interrupts, that is, **nested interrupts**, is one of the reasons why the stack is used in most microcomputer systems to handle subroutine linkage.

Sometimes it is useful to arrange variable priority for the same device, for example if it is sometimes, but not always, carrying out time-critical operations. If a number of identical devices are connected, it may be important that each stands an equal chance of being serviced, and one way to ensure that each device is treated fairly is to use **rotating interrupt priority** so that once a device in the group has been serviced it has the lowest priority, but that this priority gradually increases until it is serviced once again.

4.4.4 Interrupt controllers

The tasks involved with overseeing the various mechanisms involved in handling interrupts — the supervision of priorities, and selecting the ISR vector appropriate to each device — can become sufficiently complicated to warrant the inclusion of an extra integrated circuit to control interrupts. An interrupt controller has a number of separate IRQ inputs which can be connected to the various I/O devices in the system and one or more outputs which are connected to the interrupt inputs of the microprocessor. The IACK output of the microprocessor is connected to a corresponding input on the interrupt controller so that it can present the interrupt vector when the acknowledgement is received. In this way a number of devices can be connected, each with its own priority and vector, even when the microprocessor itself has a relatively simple interrupt handling mechanism.

4.4.5 Software interrupts

The idea of an interrupt, in which the normal operation of the program is suspended and special action is taken, is a useful one even when the interrupt is not initiated by an external event. The more general idea of an

exception covers all occasions when it might be desirable to suspend a program temporarily and run a special routine instead. Examples of such occasions are when a divide routine attempts to divide by zero or testing software, when it may be desirable to step the program one instruction at a time, to examine the contents of the processor's registers between each instruction. When an exception such as this occurs, the contents of the processor's registers may be saved in the same way as with an interrupt and the program counter is loaded with an exception vector just as if an external interrupt had occurred. The exception is then handled by an exception handling routine; for example, the exception handling routine used when tracing the operation of a program would cause information about the register contents to be printed or displayed.

Sometimes it is convenient to have exceptions called deliberately during the execution of a program. For example, if the character input and output routines were programmed in this way rather than as subroutines, there would be no need to specify the start addresses of the subroutines within the program. This gives a useful degree of independence between different computers, because even if the same microprocessor is used, the input and output arrangements can be very different. This method of calling a routine by means of a special instruction rather than with a subroutine call and a start address is known as using a **software interrupt** or a **trap**.

The exceptions can arise from the use of a machine code instruction byte that has not been specified or implemented for that type of microprocessor. For example, the 6301 causes a special software interrupt to be called whenever an illegal opcode is used. The 68000 series of microprocessors have opcodes specified for floating point data, but the 68000 itself does not have these instructions implemented. If any attempt is made to use these instructions, an 'unimplemented instruction' trap occurs. The service routine for this trap can consist of software for carrying out these floating point operations, so that software can be written which will operate without modification both on microcomputers which have floating point capability in their instruction set and those which do not.

4.4.6 Advantages and disadvantages of interrupts

The advantages of interrupts have already been pointed out: they allow a fast response from a real-time microcomputer system without wasting processor time. There are also some disadvantages which must be accepted when a system is designed to use interrupts. Such systems tend to be more complicated both in terms of hardware, and in terms of the software which is needed to handle interrupts. The interrupt routines which service the devices operate 'asynchronously' from the main program; that is, the software for the device is not called at a certain point in the program as is the case with a subroutine. This makes the testing of programs more difficult. Timing routines which make use of the microprocessor will be affected by the time which the interrupt service routine takes to execute unless the ISR is designed to interact with the main program and modify the values of the counts used in software, which is rather 'untidy' to arrange.

4.5 Counting

One task which microcomputers are frequently called upon to perform is to count external events such as objects passing on a conveyor belt or people passing through a turnstile. This can be done quite easily by monitoring a data input line for changes in logic level or by using a handshake input line to detect each time that the input signal changes from logic '0' to logic '1'. The method is quite satisfactory provided that the events being counted are relatively infrequent in electronic terms and that the microprocessor is not being used to anything like its full processing capacity, but if the input signal changes more than a few hundred times per second the microcomputer will need to spend most of its time monitoring the input line. This is a relatively trivial task for what is a very flexible piece of electronics, and the counting task can be done much more effectively by a separate counting circuit.

The simplest method of doing this would be to wire up a conventional electronic counter and then connect its output pins to the inputs of an I/O device so that it could be read by the microprocessor. This is not difficult to do, but it does use a large number of connections and it is much neater to incorporate the counter within an integrated circuit which can be read directly by the microprocessor without the added complication of using a separate input register. One of the major limitations to the complexity of a modern integrated circuit is the number of connections which are available on the package in which it is housed, and if the counter is within the same integrated circuit as the bus connections then there is no need to have separate pins for each output from the counter itself. All that the counter really needs is a single input connection to carry the electronic pulses which are to be counted.

A counter may also be used to measure time. If a signal of a fixed frequency is connected to the input of a counter the number held in the counter will increase at a fixed rate, and the program can measure the time interval between any two events simply by reading the counter on each occasion. The counter can be connected in the same way as any other kind of register, except that in this case the contents of the register will change independently of the microprocessor.

Many kinds of I/O device containing counters and timers are available, but the VIA will again be used as an example here. This has two counters, T1 and T2, which both count down towards zero, but which behave somewhat differently. T1 may only be used to count cycles of the microcomputer's internal clock signal, but the input of T2 can also be connected to bit 6 of port B in order to allow external pulses to be counted. Both T1 and T2 count down towards zero. When the counter reaches zero, it is reloaded from an associated register, after which it continues to count down once more. Thus if the register contains the number 123, the counter will count from 123 down to 0 and then be reloaded with 123. When the counter reaches zero, it also causes a flag to be set in the interrupt flag register. In this way time intervals may be preset by writing an appropriate value into the counter's latch register. For example, if the timer were required to cause an interrupt every ten milliseconds and the system clock frequency were 1.000 MHz, the counter would need to be programmed to cause an interrupt every 10 000 clock pulses. This number is too large to fit into a single 8-bit register, and in fact the 'latch' register used by T1 is 16 bits long, appearing to the microprocessor as a pair of 8-bit registers occupying two successive addresses.

```
LDA   #%01xxxxxx   (other bits would be specified
STA   ACR          in final program)
LDD   #9998        two-byte time interval
STB   T1LL         Write LSB to high byte of latch
STA   T1LH         Write MSB to low byte of latch
STA   T1CH         and to timer, to start it.
LDA   #%11000000   Enable timer 1 interrupts
STA   IER          by setting bit 6 of IER.
```

Figure 4.14

The behaviour of T1 is configured using a second control register, the **Auxiliary Control Register (ACR)**, bits 6 and 7 of which are concerned with T1. The program fragment of Figure 3.14 shows the steps necessary to configure timer 1 to generate interrupts at a fixed interval of 10 milliseconds. The system clock is assumed to have a frequency of 1 MHz so that the required time interval is 10 000 clock cycles. However, the VIA generates interrupts at intervals which are two clock cycles more than the number in the latch, so that the value to be used is 9998.

Notice that the double length accumulator of the 6809 is used in this example. With other 8-bit microprocessors which have only a single accumulator, it may be necessary to compute the more significant and less significant bytes of the 16-bit representation of 9998 separately by making use of the fact that

$$9998 = 39 \times 256 + 14$$

In this case the section

```
LDD   #9998   Two-byte time interval
STB   T1LL    Write LSB to high byte of latch
STA   T1LH    Write MSB to low byte of latch
```

must be replaced with the sequence

```
LDA    #14        Write less significant byte
STA    T1LL       to low byte of latch
LDA    #39        Write more significant byte
STA    T1LH       to high byte of latch
```

This sequence will program the VIA to cause interrupts every 10 milliseconds. An interrupt service routine of the type described in section 4.4.3 must be provided to handle these interrupts; the interrupt caused by T1 can be cancelled (somewhat arbitrarily) by reading the less significant byte of T1, TIL.

There are many other ways in which a timer can be used in a general purpose input/output device such as the VIA. For example, a timer could be used to measure the time interval between data being made available by an output and the acceptance of that data by another machine as indicated by the DAK signal. A program could test the timer and cause a message to be printed if a preset time limit was exceeded. An output is often provided which can be controlled by the counter to produce pulses independently of the microprocessor, although another input is often included to enable or disable the counting as required.

4.5.1 Measuring time intervals

Some transducers produce output signals which consist of sequences of pulses with a variable frequency or where the duration of a pulse is of significance. In such cases the microcomputer must convert frequency or time into numerical form.

This is done by counting. Time intervals can be measured by counting the number of 'ticks' of a clock waveform which occur during the pulse, while frequencies can be measured similarly by counting cycles of the input waveform during a fixed time interval. In each case two things must be monitored; elapsing time and the incoming signal. Time can be measured by counting cycles of a constant frequency, usually derived from the crystal clock of the microcomputer itself. The counters can be built in hardware or programmed in software, the approach taken depending upon the frequency of operation.

The measurement of time intervals is important because input voltages can be converted to time intervals and hence into numerical form. The hardware for measuring a time interval consists simply of a counter circuit which is connected to the input in such a way that it can count only when the input pulse is present. The counter can be set to zero at the start of the pulse and read again at the end of the time interval. However, it is easier with a microcomputer-based system, to read the counter at the start of the pulse and again at the end of the pulse, and then calculate the difference. The rate at which the counter increments should be known because it is derived from the system's own clock signal generator.

An important requirement in many control and instrumentation systems

4.5.1

is the ability to keep track of so-called **real time**, the time that an ordinary clock would show. This allows measurements to be made at predetermined times or operations to be carried out by a controller at certain times. The easiest way to incorporate a **real-time clock** into a system is to program one in software. All that is needed is a source of interrupts at regular intervals, which can be arranged by dividing the system clock using a counter circuit that raises interrupts every, say, tenth of a second. These interrupts are then counted; after every ten counts, another second has elapsed and a 'seconds' counter can be incremented; every sixty seconds a minute has elapsed and the seconds counter should be reset to zero and a 'minutes' counter incremented by one, and so on. The listing of Figure 4.15 shows a simple real-time clock based upon this principle; it could form part of an interrupt service routine called by the 10 ms interrupts generated by the VIA timer.

```
CLOCK   INC     TICKS           Count 10ms interrupts
        LDA     TICKS           and check to see if a
        CMPA    #100            second has elapsed
        BNE     EXIT

        CLR     TICKS           Reset interrupt counter
        INC     SECS            and count seconds
        LDA     SECS            Check to see if a
        CMPA    #60             minute has elapsed
        BNE     EXIT

        CLR     SECS            Reset seconds counter
        INC     MINS            and count minutes
        LDA     MINS            Check to see if an
        CMPA    #60             hour has elapsed
        BNE     EXIT

        CLR     MINS            Reset minutes counter
        INC     HOURS           and count hours
        LDA     HOURS           Check to see if a
        CMPA    #24             day has elapsed
        BNE     EXIT
```

... similar code can be used to count days, weeks, months, etc. Remember that months and years are not all the same length!!

```
EXIT    RTI                     Return from interrupt
```

Figure 4.15 Real-time clock routine

Practical real-time clock programs can be quite complicated, not only because of the peculiarities of our clock system, but because they may be programmed as 'alarm clocks' which must take special action at certain times.

Hardware real-time clocks are available in the form of single integrated circuits which can be connected directly to the bus of the microcomputer. These offer the advantage that they can carry on working even when the microcomputer is switched off, provided that a small battery is available to power them. It is even possible to arrange for the real-time clock to control the power supply to the rest of the circuit, switching on the microcomputer only when it is required, so that sophisticated instrumentation may be designed using microcomputers even when they are intended for use on sites where power is very limited.

4.5.2 Frequency measurement

Frequency measurement requires two counters. One is used to count a fixed number of clock pulses to provide a fixed measurement 'window' during which a second counter counts the number of pulses that appear on the input. This counter can be zeroed before counting and read afterwards, but again it is easier to read it before the start of the 'window' period and again at the end.

Special integrated counter-timer circuits are available for use with microprocessors. They contain counters which can be connected to count either cycles of the system clock or pulses from an external input. They can often be programmed to produce square wave output signals which have frequencies that are submultiples of the system clock frequency. Sometimes general-purpose input/output devices also contain counter-timers of this type, because they are so useful in interfacing.

4.5.3 Serial interfacing

When data is to be connected between machines which are not immediately adjacent to one another, the cost of providing parallel connections becomes prohibitive, because of the number of wires used. The solution is to convert parallel data to serial form, so that only one pair of wires need be used. This makes use of a **shift register** which is a register in which the data can be shifted 'sideways'. Serial interfacing can be used to save connections on integrated circuits and printed circuit boards as well, and is widely used for this reason.

The VIA provides a serial interface facility by means of the 8-bit shift register with associated logic. This can be programmed to act either as an input or as an output. When used as an output, the microprocessor loads the shift register with data by means of its parallel inputs just like any other kind of register. The shift register then shifts the data one bit at a time, the bit 'falling off' the end of the shift register appearing on one of the handshake lines of one of the parallel interfaces. The clock signal can be provided externally, or can be produced internally by the VIA, either directly from the system clock (but via a circuit which divides the frequency by two) or at a fraction of the system clock obtained by using timer T2. When all eight bits have been sent, a flag bit is set to logic 1 in the IFR; this bit will be cleared again when the microprocessor writes the next byte into the shift register. The input of serial data can be programmed in a similar fashion.

This serial interfacing technique allows information to be passed between VIAs by means of a pair of wires, or a VIA can be used to output data to another device containing a shift register by means of a pair of wires. This could be another VIA, but in the case to be studied in section 4.8 it will be used to send data to a circuit used to drive an LED display.

The transfer of data between microcomputers can be quite complicated, and Chapter 8 is devoted to the techniques associated with data communication between machines. Specialised interface devices are available for various types of data communication.

4.6 Direct memory access

Interrupts allow the processing power of the microprocessor to be used by an input/output device whenever the device needs attention. Frequently, however, the amount of actual processing needed by the device is minimal; it simply wishes to be able to access information held in the microcomputer's memory.

A more human analogy (Figure 4.16) may help some of the processes involved to be more understandable. Imagine that someone wishes to know the time at which a train is going to leave a station, and that the enquiry counter is not staffed continuously. The simplest thing to do would be to wait until a clerk appeared at the counter, but this might take a long time. The more positive approach would be to call for service using a bell; this would interrupt the clerk from whatever else he or she was doing and provide faster service. However, an even more direct approach would be to bypass the clerk and read the timetable oneself.

The three approaches correspond to handling the transfer of data between internal storage (the timetable in the analogy) and the input/output device (the enquirer). The first approach is the polled approach, in which the I/O device must wait until the processor tests to see if it requires service, and the second is the interrupt approach, in which the I/O device calls for attention by means of an interrupt. The third approach corresponds to **Direct Memory Access (DMA)** in which the device causes the microprocessor to momentarily relinquish control of the system buses so that it can access memory itself. This requires the use of a **DMA controller** which can generate addresses on the address bus in place of the microprocessor.

The DMA controller usually appears as a set of registers (Figure 4.17) which can be programmed as if it were an input/output device in its own

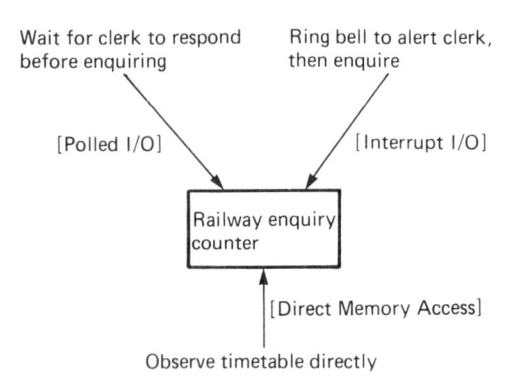

```
                Wait for clerk to respond    Ring bell to alert clerk,
                before enquiring             then enquire

                     [Polled I/O]             [Interrupt I/O]
                                 ┌──────────┐
                                 │ Railway enquiry │
                                 │ counter  │
                                 └──────────┘
                                      ↑
                                      │ [Direct Memory Access]
                                Observe timetable directly
```

Figure 4.16 Input/output analogy

right. Usually these registers consist of an **DMA address register** which holds the address in memory from which data is to be read and written, and a **DMA word count register** which holds a count of the number of words to be transferred. In addition to these registers the system will need a **DMA control/status register** to supervise and monitor its operation, and in some cases a **DMA data register** to temporarily hold the data being transferred.

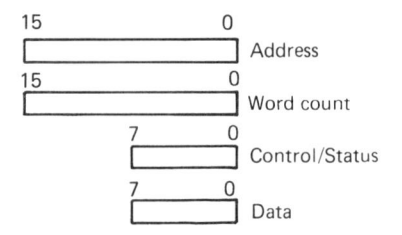

```
        15                0
        └────────────────┘  Address
        15                0
        └────────────────┘  Word count
             7            0
             └────────────┘  Control/Status
             7            0
             └────────────┘  Data
```

Figure 4.17 DMA registers

The DMA controller is connected to the I/O device, as shown in Figure 4.18, by a pair of handshake connections: DMA request and DMA grant, and to the microprocessor by another pair of handshake connections: bus request and bus grant. Whenever the I/O device wishes to read or write a word, it sends a DMA request to the DMA controller, which in turn requests access to the system buses from the microprocessor. When it is ready, the microprocessor disconnects its outputs from the system buses by switching them to their high impedance state and returns a 'bus available' signal to the DMA controller. This connects the address to be used for the DMA transfer to the address bus and a 'read' or 'write' signal on the control bus as appropriate. Then it sends a 'DMA Grant' signal to the I/O device which can then read data from the data bus or write data to the data bus as appropriate. When the DMA transfer has finished, the DMA controller removes the 'bus request' signal, disconnects its outputs by switching them to their high impedance state, and the microprocessor reconnects its outputs and continues.

The advantage of the DMA method over interrupts is that there is no need to save the contents of the microprocessor's registers because they are not used in the operation. A DMA is programmed by loading the DMA controller's address register with the starting address in memory to be used, the count register with the number of words to be transferred and the control register with the appropriate information for the type of transfer to be carried out. The microprocessor need then have no further part in the operation except to check by means of the status register that the previous transfer has finished before initiating another one.

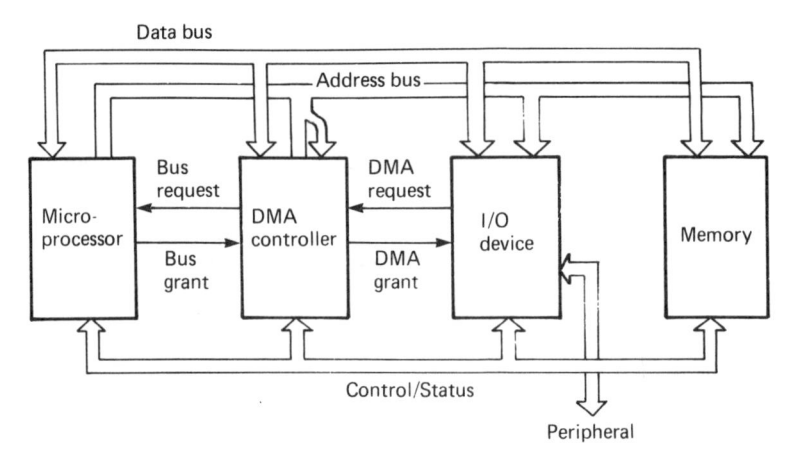

Figure 4.18 Connecting a DMA controller

Frequently a DMA consists of a sequence of several transfers, which may not all occur directly one after another. When the final transfer in a sequence has been carried out, a flag bit is set in the status register of the DMA controller to indicate the fact. The DMA controller can also be programmed to raise an interrupt at this point. Various types of DMA transfer sequencing exist. The most straightforward of these is the **burst mode DMA** in which the microprocessor is stopped and information is transferred continuously until the DMA operation is complete. This minimises the time taken for the transfer, but stopping the microprocessor for the time taken for a long input/output transfer can affect the operation of time-critical programs. In **distributed mode DMA** the transfers take place at time intervals so that DMA transfers are interleaved with normal microprocessor operation and the effect upon the program is reduced. A third method, **cycle-stealing mode DMA**, transfers data only during cycles when the microprocessor is not making use of the data bus itself. This allows the program to proceed at the same speed during a DMA transfer.

There is no reason why more than one device in a system should not be able to carry out direct memory access, and indeed most DMA controllers allow more than one I/O device to be used. Then the relative priority of each device must be defined just as with an interrupt controller. Some multichannel DMA controllers can be programmed to carry out block moves within memory, transferring a specified number of words from a block with the starting address specified using one channel of the DMA controller to a block with a starting address specified in another channel of the controller.

4.7 A comparison of input/output devices

Although the 6522 versatile interface adapter has been used as an example throughout this chapter, a very wide range of devices are available for digital interfacing of microcomputers. The main functions required by an interface circuit are:

- Input lines, the logic levels on which can be read by the microprocessor.
- Output lines, the logic levels on which can be set by the microprocessor.
- Data direction registers, if lines can be programmable as inputs or outputs.
- Handshake lines which can be used to supervise data transfers via the input/output lines when they are used as data ports.
- Serial data communication. The data can be converted from parallel to serial form by the program, or serialised by a hardware shift register in the interface device.
- Timing and counting functions; these may have sophisticated control functions associated with them.
- The interface devices require control registers to supervise their operation and contain status (flag) registers to allow the behaviour to be monitored by the microprocessor.

The Versatile Interface Adapter

Before briefly examining some other interface devices, it will be useful to review the VIA (Figure 4.19), the functions of which have been introduced at various points within this chapter. The 6522 is a general purpose interface circuit which provides two 8-bit parallel ports, each with a pair of handshake lines, and two counter/timer circuits. These counter/timers can be connected to certain of the input/output pins; additionally, one pair of handshake lines can be connected to a shift register within the 6522, which allows serial interfacing to external shift registers to be accomplished very easily.

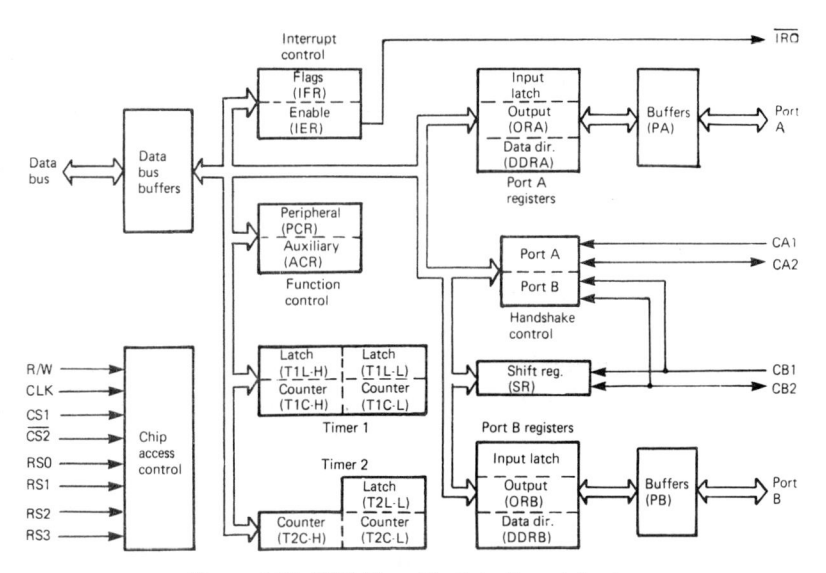

Figure 4.19 6522 Versatile Interface Adapter

Parallel input/output
All microprocessor families have parallel interface devices designed for use with that type of microprocessor. The 6820, 6821, and 6520 peripheral interface adapters contain two parallel ports, each with data direction registers and handshake lines, the functions of which can be programmed by means of a control register; these are purely parallel interface devices with no other functions such as timing or serial input/output. The 8255 is more sophisticated, with three input/output ports. One of these ports can be programmed to provide handshake ports for the other two ports; the 8255 also allows individual output bits to be set or reset with special modes of operation. The 68230 is yet another device which can be used as a pair of 8-bit parallel ports with handshake lines; this circuit is designed for use with the 68000 16-bit microprocessor.

Some microprocessors, such as the 8035 and 6803, have parallel input/output interfaces in the same package as the microprocessor itself. These are versions of single-chip microcomputers in which the internal buses are accessible, allowing them to be used as microprocessors with additional functions.

Timer/counters
As has been seen earlier in this chapter, the VIA contains two counters which can each be connected in various ways to other parts of the VIA. In this circuit, each counter decrements by one each time a clock pulse is counted, and when the count reaches zero, it is reloaded from the latch associated with it. In this way time intervals may be generated which are proportional to the number in the latch. The timers may be programmed to

generated repeated interrupts, or to create a single interrupt after a defined time interval. The 68230 is another example of an interface device which contains both input/output ports and a timer.

The 6840 and 8253 each contain three 16-bit counter-timers with sophisticated control functions. They may be used to generate output pulses in much the same way as timer T1 of the VIA, or to count input pulses like T2. The input pulses can be 'gated' by separate inputs so that counting can be controlled externally.

Each member of the 6801 family of microprocessors contains a timer with three special-purpose registers associated with it. The timer itself is a 16-bit counter which counts up at the system clock rate. Each time the number in the counter reaches the maximum possible value of 65535 (hexadecimal FFFF) it returns to zero (hex. 0000) on the next clock pulse and sets a timer overflow (TOVF) bit in the status register.

This **Timer Control/Status Register (TCSR)** has eight bits, and as its name implies, it is used to control the operation of the counter and its associated registers, and to report the status of the registers.

The **Input Capture Register (ICR)** has an input connected to a pin of the integrated circuit, which is monitored for changes in logic level in much the same way as a handshake input line. It waits for an 'active' transition, the direction of active transitions being programmable by means of the TCSR. When one is detected, the ICR is loaded with the current count in the timer and a flag bit is set in the TCSR. This event can also be used to cause an interrupt if it is enabled to do so by the appropriate bit in the TCSR.

The **Output Compare Register (OCR)** causes an 'output compare flag' bit to be set in the status register when the value in the timer is the same as that in the output compare register. This can also be programmed to raise an interrupt by means of a control bit in the TCSR, and to update the logic level on one of the output pins without direct software involvement.

4.8 Design example

In this section the input/output requirements of the controller circuit, used as one of our running examples, will be examined. To the user, the controller must appear to carry out several continuous operations simultaneously. In fact it will be required to carry out separate tasks at regular time intervals; these tasks can be grouped as follows:

- Every ten milliseconds or so, the keyboard must be scanned to determine if any more information has been typed in. Each time that a key is pressed, some special action may need to be taken, even if this consists only of updating the LED displays.
- At regular intervals which may depend upon the particular application of the controller, the controller must read the analog input to the system and compute a new value for the analog output using the three term control algorithm outlined in Chapter 2.

- The output from the controller may be in the form of a time delay from the last time that the mains voltage passed through zero. This means that the system must detect zero crossings of the mains voltage and cause an interrupt some specified time later.

Each of these tasks requires interaction with the external circuit. The keyboard scanning routine needs a parallel digital interface, while as will be seen later, the display can be interfaced using a serial technique. The analog input requires an analog-to-digital converter which can be interfaced to the microcomputer by means of a parallel port. However, the analog output will need only a single digital connection because the analog signal will be coded as a time delay from the last zero crossing of the mains waveform. A single digital input will be needed for the input from the zero crossing detector.

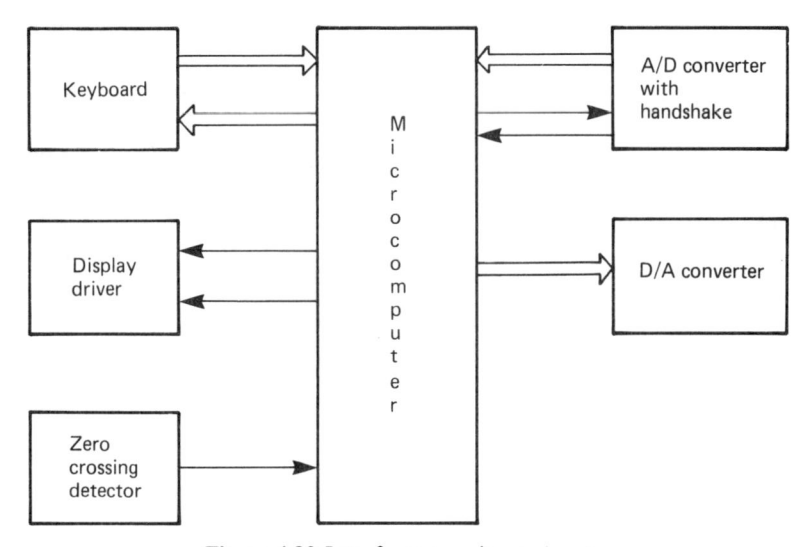

Figure 4.20 Interfaces to microcomputer

The input/output device or devices used by the controller must therefore provide the following functions, which are shown diagrammatically in Figure 4.20:

- parallel digital inputs and outputs for the keyboard interface.
- a parallel input port for the ADC. This will require a pair of 'handshake' control lines.
- a serial interface for the display interface. This needs two connections: one for the data and one for the clock. The data can be converted to serial form using either hardware or software within the program.
- an output connection for the analog output.
- an input connection for the zero crossing detector.

The input/output device or devices to be used in the design must provide sufficient connections to cater for all the interfaces listed in the paragraphs above. Note that all the connections required are essentially digital in nature, but that some control over time delays is necessary. A very large range of input/output devices exist for general applications such as this, where both digital input/output and timing functions are required. In this case the requisite functions can be carried out using a VIA (6522) and a 6803 microprocessor, which is a member of the 6801 family.

The 6522 has 20 input/output lines, while the 6803 has an 8-bit port without handshake and a further five input/output lines. One port of the VIA, together with its handshake lines, can be assigned to the ADC. The function of the handshake lines will be explained when ADCs are examined in Chapter 6. The digital-to-analog converter does not require a handshake, so that one of the two remaining 8-bit ports may be assigned to this. The zero-crossing detector produces a transition between logic levels which is to be used as the basis for some of the timing functions within the system. The input to the 6803's input capture register provides the most suitable connection for this, because the exact time of the transition can be 'logged' automatically by the ICR and read later by the processor. The LED displays are to be driven by special integrated circuits which are loaded serially, and the shift register of the 6522 provides a convenient method of outputting this serial data without the need to use software to send the individual output bits one at a time. The remaining input/output bits can be assigned to the keyboard; keyboard scanning will be discussed in Chapter 5.

Setting up time intervals

The timers in the VIA and the 6801 microcomputer family do not differ only in their manner of operation. The 65xx and 68xx families differ in the order in which the bytes of two-byte registers are stored. In the case of 65xx devices such as the 6522 VIA the less significant byte is stored at the lower address, while in the case of 68xx devices such as the 6803 the more significant byte is stored at the lower address. This fact must be borne in mind when using the VIA with the 6803; if a 16-bit number in the D accumulator of the 6803 is stored in one of the two-byte registers of the 6522, the more significant and less significant bytes will be transposed.

Timer 1 of the VIA can be programmed to cause interrupts at regular intervals quite easily. First the mode of operation of the timer must be configured using a second control register within the VIA called the **Auxiliary Control Register (ACR)** which controls operation of the timers and the shift register. Bit 7 of the ACR determines whether timer 1 will cause repeated interrupts (ACR6 = 1), or whether it will simply stop when it reaches zero (ACR6 = 0). If bit 7 of the ACR is 1, the output level on bit 7 of Port B will be controlled by timer 1. In this application timer 1 is configured by setting ACR bits 7 and 6 to logic 0 and logic 1 respectively.

The interval to be measured by timer 1 is determined by the number loaded into latch 1. If the time interval is fixed, it can be calculated when the program is assembled and the appropriate bit pattern loaded into the latch. However, if the time interval is to be variable, each time a new 16-bit value is calculated the two bytes must be stored separately.

The bits in the interrupt enable register of the VIA are set and cleared in a somewhat unusual fashion. When a byte is written to the IER, the most significant bit (1 or 0) of the byte is written into those bit positions where there is a logic 1 in the remainder of the byte; where there is a logic 0 in the byte being written into the IER the current value of the IER is unaffected. For example, the sequence

```
LDAA    #%10000011
STAA    IER
```

will set the two least significant bits of the IER while leaving the others unaffected, while

```
LDAA    #%00001100
STAA    IER
```

will clear bits 2 and 3 of the IER. Although this approach might seem rather strange at first sight, it allows sets of bits in the IER to be set or cleared, to enable or disable sources of interrupts within the VIA, very easily.

Each time that timer 1 generates a timeout interrupt, the corresponding interrupt handler must clear the interrupt. This can be done in various ways: by writing to the more significant byte of timer 1, by reading the less significant byte of timer 1, or by clearing the flag bit in the interrupt flag register directly. This last approach involves writing a byte containing a logic 1 in the bit position corresponding to the flag which is to be cleared. Once the flag has been cleared, the interrupt handler can continue with reading the analog input and the calculation of the next output sample. This will take an appreciable amount of time, and it may prove necessary to allow other sources of interrupt while this calculation is being carried out. Once the timer 1 interrupt has been cleared, the microprocessor's interrupt mask can be cleared to allow other interrupts to occur while calculating this sample.

4.9 Questions

1. The various methods of transferring data between a peripheral device and the microprocessor and/or memory of a microcomputer can be categorised broadly into three types:

    (a) Programmed input/output.
    (b) Interrupt-driven input/output.
    (c) Direct memory access.

    Explain carefully the differences between the three types, comment on their efficiency in terms of data transfer rate and processor utilisation, and briefly describe one typical application of each method.

2. What are the problems encountered when more than one peripheral device is allowed to interrupt a microprocessor? Discuss possible solutions to these problems and include details of any additional hardware or software which would be required.

3. Explain what is meant by the term 'handshaking' and describe one application of a handshaken data transfer.

4. Write a pair of Pascal procedures to simulate the input and output ends of a FIFO buffer. Rewrite your procedures using 6809 assembly language.

5. A set of model traffic lights is to be interfaced to a pair of 8-bit parallel ports with one output line controlling each lamp. Write a program in 6809 assembly language which will cause the model to turn the lights on and off according to their conventional sequence, with time delays generated using software.

6. Central heating controllers are usually designed with a clock which can switch the heating on and off at predetermined times during the day. In this way energy is not wasted in heating an empty house, but it is usually necessary to override this timer at weekends and during holidays when the house is occupied. Design an interface to allow a central heating controller with a built-in calendar to be programmed by the user. Remember that the interface should be as inexpensive and easy to use as possible. In what other ways could microcomputer technology be used to improve the performance of a heating system?

7. The frequencies of the notes on an even-tempered musical scale are arranged logarithmically, so that with twelve notes to an octave the frequency ratio of two adjacent notes is given by the twelfth root of two. This is an irrational number, which means that frequencies can only be approximated if they are produced by integer division from the same master clock. The table shows the frequencies (in hertz) of notes in the octave starting with middle 'C', tuned to concert pitch:

note	freq.	note	freq.
C'	261.63	F#	369.99
C#	277.18	G	392.00
D	293.67	G#	415.31
D#	311.13	A	440.00
E	329.63	A#	466.16
F	349.23	B	493.88

Design a simple monophonic organ which generates these frequencies as accurately as possible by dividing a system clock frequency of 1 MHz by an integer.

8. Explain how a block of data is transferred between a peripheral and memory using burst mode DMA. Describe how the transfer is initiated using waveform diagrams depicting the handshake signals.

9. Design a pulse generator which uses the timing functions of the VIA. What are the frequency limitations of such a system?

Chapter 5 DIGITAL INTERFACING

The signals used within microcomputer circuits are almost always too small to be connected directly to the 'outside world' and some kind of interface must be used to translate them to a more appropriate form. The design and selection of interface circuits is one of the most important tasks facing the engineer wishing to apply microcomputers, and two chapters of this book will be devoted to the subject of interfacing.

We have seen that in microcomputers information is represented as discrete patterns of bits; this **digital** form is most useful when the microcomputer is to be connected to equipment which can only be switched on or off, where each bit might represent the state of a switch or actuator. On the other hand, many of the quantities which the engineer may want to measure or control are continuously variable and must be converted to or from the discrete form used within the microcomputer. This important subject of **analog** interfacing will be postponed to the next chapter, and in this chapter we shall concentrate on the design of digital interfaces.

5.1 Transistors

Like all solid-state electronic circuits, microcomputers are based upon **transistors** which can be regarded for our purposes as very small electronic switches. There are two basic types of transistor: **bipolar transistors** in which the output current is controlled by the input current, and **metal-oxide semi-conductor transistors** or **MOS transistors** in which it is controlled by the input voltage. In fact, integrated circuits can be classified in terms of those which use bipolar technology and those which are based upon MOS transistors. A detailed understanding of either kind is unnecessary for our purposes because simplified **circuit models** can be used when designing interface circuits.

5.1.1 The bipolar transistor

This is a semiconductor device with three leads called the **emitter**, **base**, and **collector**. Figure 5.1 shows the symbols conventionally used for the two kinds of bipolar transistor, which are distinguished by the direction of the arrow in the emitter lead; this arrow shows the direction in which current flows when the transistor is operating. The circle represents the package containing the transistor and it is often omitted because it is not relevant to

5.1.1

Figure 5.1 Bipolar transistor symbols

the operation of the device itself. Although we shall restrict ourselves here to the *npn* **transistor**, the operation of the *pnp* **transistor** is similar except that the polarities of all the voltages and currents are reversed.

The flow of current between the emitter and collector is under the control of a small current flowing in the base. No current can flow in the base of an *npn* transistor if it is negative with respect to the emitter. If the base voltage is made increasingly positive negligible current will flow until the **base-emitter voltage** (V_{BE}) reaches a certain value characteristic of the type

(a) Input characteristic

(b) Output characteristics

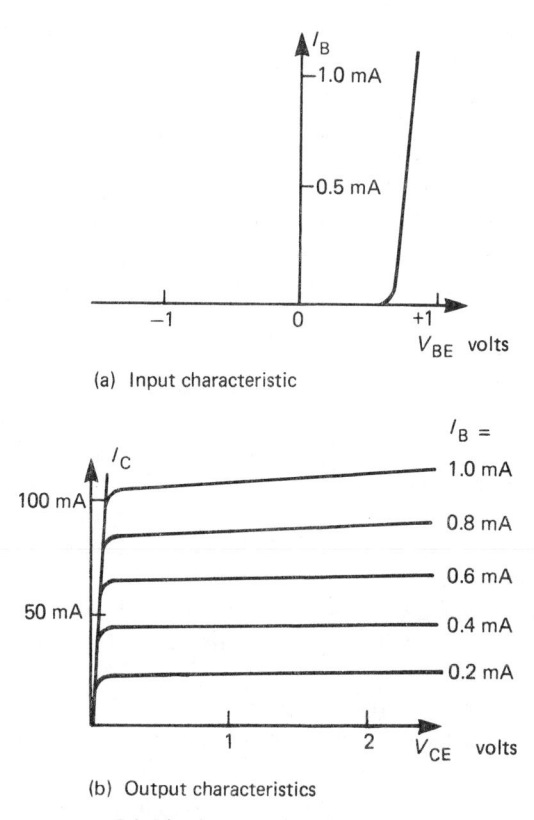

Figure 5.2 Bipolar transistor characteristics

of transistor, after which the current will increase very rapidly with a small increase in V_{BE} as Figure 5.2a shows. This condition is called **forward bias**, and it is usually adequate for the purposes of interface design to assume that the base-emitter voltage of forward biased transistors is about 0.7 V. Provided that the current is not limited by any external factor (such as a resistor, as is usually the case), the collector current is proportional to the base current. The constant of proportionality is called h_{FE} and is normally in the range 100 to 600 for low-current transistors, although larger transistors (which can handle currents up to tens of amps) often have much lower values of h_{FE}. The value of h_{FE} is specified by the manufacturer, and it frequently varies considerably between transistors of the same type, with a ratio of three to one between the highest and lowest figures being common. Figure 5.2b illustrates how the collector current depends upon the base current and collector-emitter voltage for a typical small transistor. There is a minimum **collector-emitter voltage** (V_{CE}) called the **saturation voltage** which will support a given collector current. For modern transistors this is a fraction of a volt, typically 0.1 V for a small transistor. The collector current is largely independent of increasing collector emitter voltage up to a maximum voltage rating which can be found in the manufacturer's data. If this limit is exceeded the current may increase to a value where the transistor is destroyed. Excessive negative voltages between the base and emitter can also damage the transistor; most transistors cannot tolerate a reverse base-emitter voltage of more than about 5 V.

There are three ways in which the transistor can be 'configured' in a circuit, distinguished by the choice of input and output connections. In **common emitter configuration** the base is used as the input lead and the collector as the output lead, with the emitter common to the input and output circuits as shown in Figure 5.3a. This allows a small base current to control the collector current which is h_{FE} times as large. Voltages much larger than the +5 V normally used for microcomputers can be controlled with this configuration.

The **common collector configuration** uses the base lead as the input and the emitter lead as the output. Since the base-emitter voltage is almost constant as long as base current is flowing, the emitter voltage varies in the same way as the base voltage, but because a larger current flows in the emitter ($1+h_{FE}$ times the base current) a useful current amplification is obtained. This connection is called the **emitter follower configuration** because of the way that the emitter voltage 'follows' the input base voltage, as shown in Figure 5.3b.

In **common base configuration** the emitter is used as the input lead. Almost all this emitter current appears at the collector, the small difference being due to base current. Although the current gain is actually less than 1, common base mode is also useful in the design of output interfaces because it enables voltage gains and shifts in voltage levels to be achieved (Figure 5.3c).

The design of microcomputer interfaces makes use of transistors in all

5.1.1

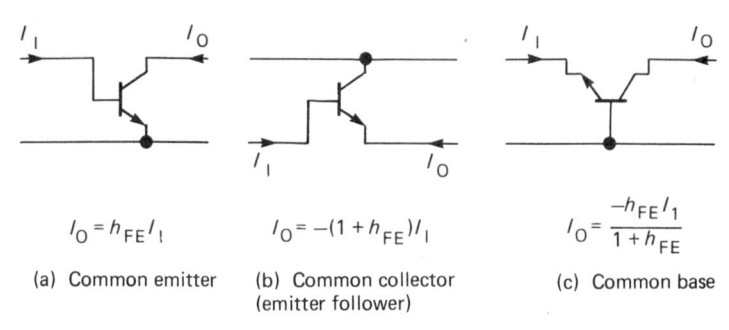

Figure 5.3 Three modes of operation

(a) Common emitter: $I_O = h_{FE} I_I$

(b) Common collector (emitter follower): $I_O = -(1 + h_{FE}) I_I$

(c) Common base: $I_O = \dfrac{-h_{FE} I_I}{1 + h_{FE}}$

three configurations. Usually the transistor is switched between two states: **cut-off**, where no current is flowing, and **saturation** where the current flowing between collector and emitter is limited by the resistance of an external load with only a very small voltage drop across the transistor.

Each configuration offers its own advantages. The common emitter configuration allows a large output voltage swing to be obtained with a relatively high input resistance (which leads to a small input current). If a low input resistance is needed for any reason, common base mode can be used, while emitter followers have high input resistances and low output resistances.

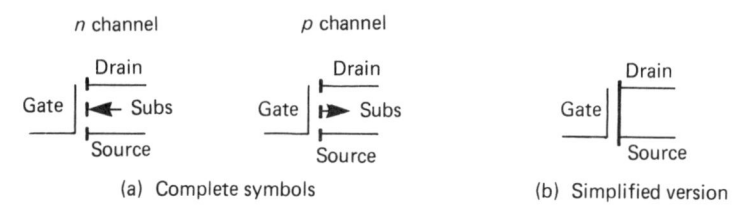

Figure 5.4 MOS transistor symbols

(a) Complete symbols (n channel, p channel)

(b) Simplified version

5.1.2 The MOS transistor

An MOS transistor also has three connections, but in this case they are called the **source**, **drain**, and **gate**. Current flowing in a channel from the source to the drain is controlled by the difference in voltage between the gate and source. *n*-**Channel transistors** are operated with the drain and gate positive with respect to the source; *p*-**channel transistors** behave in the same way but the polarities of the voltages and currents are reversed. The symbols for the two types of MOS transistor are shown in Figure 5.4, which also illustrates how the direction of the arrow in a fourth (substrate) lead is used to distinguish the types. This substrate does not take any direct part in circuit operation and frequently a simplified symbol is used for MOS transistors as shown in Figure 5.4b.

Figure 5.5 shows the characteristics of a typical small MOS transistor. If the gate-source voltage is increased from zero the channel between source and drain will remain closed to current flow until a threshold voltage V_T is reached, when a current will begin to flow. This current will rise nonlinearly with further increases in the gate-source voltage. Note that unlike the case of the bipolar transistor where the dependency of base current upon input voltage was shown, this figure illustrates how the output current depends on the input voltage. This is called the **transfer characteristic** and is used because the input current of an MOS transistor is extremely low and irrelevant to its operation.

Like their bipolar counterparts, MOS transistors can be connected in three modes: **common source**, **common drain** and **common gate**. Unlike the bipolar transistor, however, negligible input current is required when the gate is used as the input terminal and the principal advantage of the MOS transistor in interface design lies in its very **high input impedance,** which means that the amount of power required to control an output current is very small. MOS transistors are now available which allow currents of several amps to be controlled directly from a microcomputer output using only a single transistor. The same interface might require two or three transistors and a similar number of resistors if bipolar transistors were used.

When MOS transistors are switched on they behave as if they have a small resistance between source and drain, the value of this resistance depending upon the voltage between the gate and source and on the size of transistor. MOS power transistors typically have 'on' resistances of a fraction of an ohm, while the much smaller transistors used in integrated circuits may have 'on' resistances of several hundred ohms or more.

Many other semiconductor devices have been developed for use in microcomputer interfaces. They are mostly based on the two types of transistor already described and some will be introduced where appropriate in the remainder of this chapter.

5.1.3 Power dissipation

The power dissipated by a bipolar transistor is almost entirely due to **collector power dissipation**, the product of collector current and collector-emitter voltage. Similarly, the power wasted as heat by any electronic device can be calculated by multiplying the voltage dropped across it by the current flowing through it. It follows that if a device is used as a switch in which it either conducts a current with only a small voltage drop across it (the 'on' state) or conducts negligible current (the 'off' state), it can be made to waste only a small fraction of the power which it controls.

The power consumed by a transistor or integrated circuit ultimately appears as heat which must be removed to prevent it from getting too hot. Natural convection is sufficient to remove this waste heat from a small device, but with dissipations greater than about one watt further provision must be made to keep it cool. The power dissipation can be increased substantially by attaching the device to a metal **heat sink** which conducts the heat

(a) Transfer characteristic

(b) Output characteristics

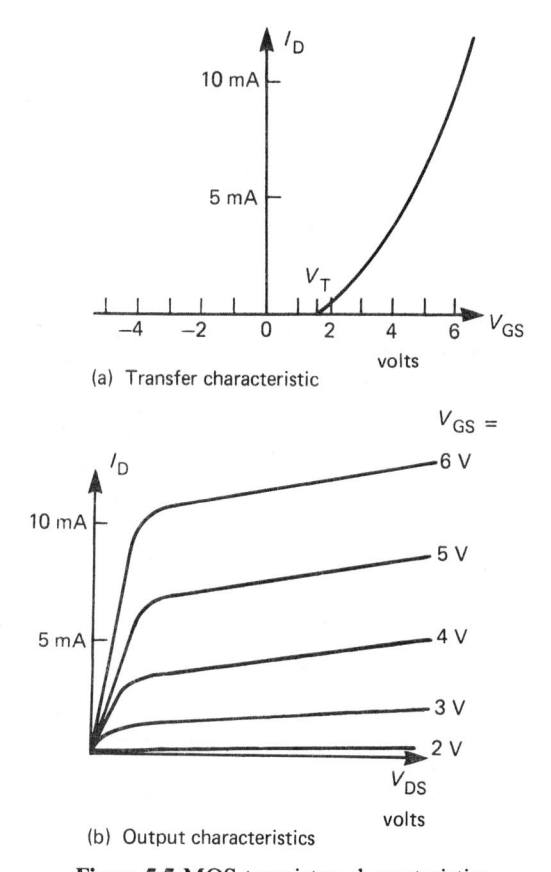

Figure 5.5 MOS transistor characteristics

away to surfaces which can be cooled by convection. At higher powers of tens of watts or more forced cooling by means of an electric fan is usually necessary.

5.2 Circuits and signals

Although it is convenient to think in terms of the flow of information from one part of the microcomputer system to another, this information is communicated using currents flowing in the circuits which make up the system. Current flows from a power supply to the system via a 'live' connection and back via 'zero volt' or 'earth' (or 'ground' in American usage) connection. The circuit's **live connection** is conventionally labelled V_{CC} (the 'C' standing for collector supply voltage) in bipolar circuits or V_{DD} ('D' for Drain) in circuits which use MOS devices, while the **earth connection** or **ground connection** is labelled **0V** or **GND**, or sometimes V_{SS} (the S standing for 'source') in MOS circuits.

5.2.1 Digital signals

At any instant a digital signal can take one of only a limited number of nominal values or **states**. A state cannot be represented by precisely defined voltages or currents in practice because of the effects of manufacturing tolerances and electrical noise, and instead it must be defined in terms of a permitted range of voltage or current values.

In principle a voltage or current could represent any number of possible states, but increasing the number of permitted states increases the complexity of the circuit required to distinguish them. It also reduces the voltage or current difference between states, increasing the system's susceptibility to electrical noise. For this reason the great majority of digital signals are two-valued or binary, despite the potential increase in information handling capacity of signals with more than two permitted states.

The choice as to which of these two states should represent binary '1' and which should represent binary '0' is arbitrary, although microcomputers normally output a high (more positive) state corresponding to logic '1'. This convention is known as **positive logic**; naturally enough, the opposite is **negative logic**.

5.2.2 Current sources and sinks

An output in the **high** state acts as a **source** of current which flows from the power supply via the output lead to the load (i.e., the immediate circuit which is being driven). Similarly, an output in the **low** state can **sink** current which flows into it from a load. The currents which a typical logic output can source or sink depend upon the type of logic in use: MOS logic outputs typically source or sink about 1 mA, while bipolar outputs can manage somewhat larger currents. Currents are conventionally regarded as having a positive sign if they flow into a terminal and negative if they flow out of it, which means that current sources are specified with negative current values and current sinks with positive values in manufacturers' data sheets.

5.2.3 Logic inputs

The inputs of a logic circuit must discriminate between the 'low' and 'high' states. MOS transistors are controlled by the voltage on their input electrode and can be designed to switch on at a specified voltage. An output driving an MOS transistor or integrated circuit frequently needs to source and sink only sufficient current to charge and discharge the small capacitance of the input because it has an extremely high input resistance. Bipolar transistors, on the other hand, are controlled by their input current which can exceed 1 mA.

5.2.4 Totem pole outputs

A logic gate can control an output voltage using transistors in its output circuit as switches, as shown in Figure 5.6.

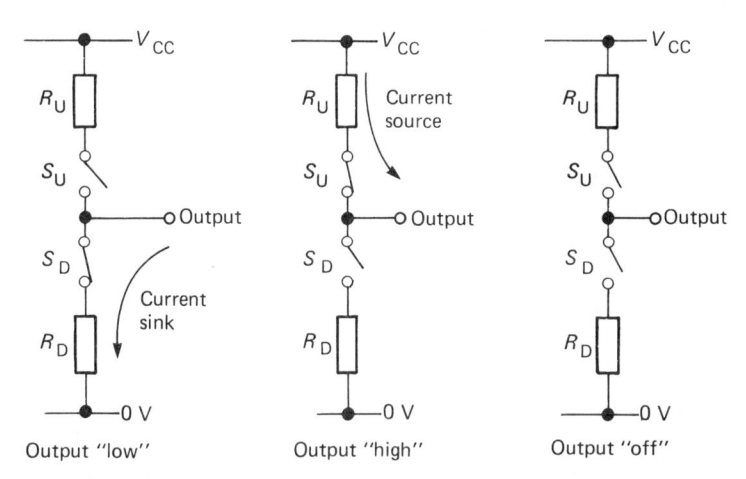

Figure 5.6 'Totem-pole' output with three-state capability

The output produces a 'high' state voltage by providing a connection from the supply voltage V_{CC} (or V_{DD}) by closing switch S_U to source current via resistor R_U. Similarly, a 'low' voltage can be produced at the output by closing switch S_D to provide a path to the zero volt line instead, allowing the output to sink current through R_D. The actual voltage which appears at the output depends upon the internal circuit of the logic device and the load to which it is connected. Most logic outputs are of this **totem-pole** or **active pull-up, active pull-down** type but the other types in common use are listed below. (The circuit is called a 'totem-pole' output because when drawn as a circuit with bipolar transistors it looks like a pair of faces, one above another like a Red Indian totem pole, given a vivid imagination!)

5.2.5 Three state outputs

Some outputs allow both the pull-up and pull-down transistors of a totem-pole output to be switched off at the same time, so that the output can neither source nor sink current. Logic outputs with this capability output are referred to as **three-state** or **tri-state** (this is the registered trade mark of National Semiconductor Corp.) outputs, the possible output states being high, low, and off (high impedance). Normally it is bad practice to connect logic outputs together because the currents that flow if one output sources current and another sinks current at the same time can cause overheating and possibly damage the output circuits. On the other hand, three-state outputs can be connected together provided that the designer ensures that all but one of the outputs are switched 'off' (and hence in effect disconnected!) at any time. This feature allows the wide-spread use of three-state logic in microcomputer systems where several outputs may be connected to a **common bus line**.

5.2.6 Open collector outputs

Some logic outputs omit the pull-up switch and resistor and are called **open drain** or **open collector** outputs. They can only sink current, and to source current they require an **external pull-up resistor**. When the pull-down switch is closed the output sinks current via S_D and R_D, which pulls the output voltage low if the external resistor is large compared with R_D. If S_D is open the output voltage rises to the supply voltage because the current is sourced via the external resistor. This type of output circuit with a pull-up resistor but no pull-up switch is called **passive pull-up**, and sometimes logic outputs are made with a passive pull-up resistor on the chip.

Several open-collector outputs can be connected together with a common pull-up resistor, so that if any of the outputs sinks current their common voltage will go low, but otherwise it will remain high. This has the effect of an 'AND' operation if logic 'high' is taken as equivalent to '1' and is sometimes called the **wired-AND connection** or **collector-AND connection**. Conversely, it has the effect of an 'OR' if the opposite convention of logic 'low' representing '1' is used, in which case it is referred to as the **wired-OR connection** or **collector-OR connection**.

5.2.7 Bidirectional input/output pins

If an output is neither sourcing nor sinking current it cannot affect the voltage on the output pin, which means that the same pin can be used as a logic input while the output is switched off. This is the principle of operation of the **bidirectional input/output ports** described in section 4.1.3. Most bidirectional ports use three-state totem-pole outputs, although the **quasi-bidirectional port** is a variant which is sometimes used. This uses passive pull-up but has a transistor in parallel with the pull-up resistor. The pull-up transistor is switched on momentarily when the output switches from the low to the high state, charging any capacitance connected to the output pin. The pull-up resistor then maintains this high level by sourcing a relatively small current.

5.3 Interconnecting logic circuits

Care must be taken when connecting logic circuits to ensure that their **logic levels** and **current ratings** are compatible. The output voltages produced by a logic circuit are normally specified in terms of worst case values when sourcing or sinking the maximum rated currents. Thus V_{OH} is the guaranteed minimum 'high' voltage when sourcing the maximum rated 'high' output current I_{OH} while V_{OL} is the guaranteed maximum 'low' output voltage when sinking the maximum rated 'low' output current I_{OL}.

There are corresponding specifications for logic inputs which specify the minimum input voltage which will be recognised as a logic 'high' state, V_{IH} and the maximum input voltage which will be regarded as a logic 'low' state, V_{IL}.

5.3.1 TTL logic levels

The first type of integrated circuit logic to be used widely was **Transistor Transistor Logic (TTL)** This widespread use of TTL led to the voltage levels used by this logic family becoming a standard used by most microcomputer components. In fact there are now many kinds of TTL available in addition to the various TTL-compatible MOS circuits on the market. Although they have widely varying input and output current ratings they all have similar input/output voltage levels and operate from +5 V power supplies. Table 5.1 shows the specifications for standard TTL inputs and outputs.

Table 5.1: Characteristics of Standard TTL

Input		Output	
V_{IH}	2.0 V	V_{OH}	2.4 V
V_{IL}	0.8 V	V_{OL}	0.4 V
I_{IH}	40 µA	I_{OH}	−400 µA
I_{IL}	−1.6 mA	I_{OL}	16 mA

For example, within a microcomputer the 'low' state is typically represented by a signal in the range 0.0 to 0.8 V, while the other 'high' state is represented by a voltage in the range 2.0 to 5.0 V. The states are distinguished by comparing this voltage with a test value which although nominally 1.4 V is also subject to manufacturing tolerances and temperature effects, which is partly the reason why voltages in the range 0.8 to 2.0 V are not used to represent a valid state.

5.3.2 Fanout

The maximum number of inputs which can be driven by a logic output is determined for bipolar logic by the ratio between the currents which an output can source and sink and those which the inputs can sink and source. Consider an output connected to N identical inputs; this is called a **fanout** of N.

When the output is in the 'high' state it can source a maximum guaranteed current of I_{OH} while each input sinks a maximum current I_{IH}. In order for the circuit to operate correctly the output must be able to source as much current as the total current sunk by the inputs, i.e.

$$I_{OH} \geq N \cdot I_{IH} \tag{5.1}$$

Similarly we require for the low state

$$I_{OL} \geq N \cdot I_{IL} \tag{5.2}$$

Clearly the maximum fanout allowable is the maximum value of N satisfying both equations 5.1 and 5.2. The data in Table 5.1 show that a maximum fanout of 10 is achievable with standard TTL, although there are special **bus-driving devices** available which are capable of sinking and sourcing higher currents, and low input-current devices which enable considerably larger fanouts to be obtained if needed.

The situation with MOS logic is rather more complicated. Although some MOS inputs have resistors connected between the input terminal and the V_{DD} power supply line, most appear as an extremely high resistance in parallel with a small capacitance of typically 5 to 10 pF. In such cases it is the capacitance which limits the fanout because of the need to charge or discharge the input when the voltage changes from one state to another. The small output currents available from most MOS circuits and the need to switch in a fraction of a microsecond typically limit the allowable capacitive load to about 100 pF, again dictating a maximum fanout in the order of 10.

5.3.3 Noise immunity of logic inputs

A major advantage of digital circuits over their analog counterparts is the ability to tolerate a small amount of electrical **noise** without affecting their operation. Noise can arise from many sources, most commonly taking the form of inductively and capacitatively coupled 'spikes' due to currents or voltages changing in adjacent wires.

Referring back to Table 5.1 again, we see that $V_{OH} > V_{IH}$ and $V_{OL} < V_{IL}$. Suppose that an output is producing a voltage V_{OL} which is connected to an input, and that a noise voltage V_{NL} is induced in series with this voltage, giving a voltage of $V_{OL} + V_{NL}$ at the input to another logic circuit. Provided that

$$V_{OL} + V_{NL} \leq V_{IL} \qquad (5.3)$$

the logic input will still interpret the input voltage correctly; similarly, if a voltage V_{NH} were induced when in the high state the circuit would still function correctly provided that

$$V_{OH} + V_{NH} \geq V_{IH} \qquad (5.4)$$

If we rearrange equations 5.3 and 5.4 we find that errors will not occur if

$$V_{NH} \leq V_{OH} - V_{IH} \qquad (5.5a)$$

$$V_{NL} \leq V_{IL} - V_{OL} \qquad (5.5b)$$

The quantities $V_{OH} - V_{IH}$ and $V_{IL} - V_{OL}$ represent the noise immunities (or margins) of the circuit in the 'high' and 'low' states respectively.

Almost all microcomputers are based upon MOS logic, but most use the same logic levels as bipolar TTL. Table 5.1 shows that the **noise margins** of TTL are 0.4 V in each state, adequate for logic connections within a microcomputer circuit. However, these logic levels are normally unsuitable for connecting directly to external equipment, and interface circuits must be used to change them to something more appropriate. Microcomputers are available which use **CMOS logic**, a development of MOS logic which offers a higher noise immunity, but nevertheless they still require interface circuits in most cases.

5.4 Interface circuits for digital inputs

Perhaps the main problem facing the designer of an **input interface** circuit is that of electrical noise. Small noise signals may cause the system to malfunction, while larger amounts of noise can permanently damage it. The designer must be aware of these dangers from the outset, and take precautions.

5.4.1 Protecting against excessive input voltages

All electronic equipment is susceptible to damage if the voltage on an input or output lead exceeds the maximum permitted value. Sometimes the damage is due to the breakdown of insulation at some point in the circuit, while in other cases failure is caused by overheating as a result of excessive current flowing. The very small dimensions of integrated circuits mean that they can be damaged by very brief overloads.

Most large-scale integrated circuits are vulnerable to **static electricity** because of the extremely high input impedance of MOS logic. The maximum voltage that an input can withstand is a few tens of volts, and since the input capacitance of an MOS device is so small, they can be damaged by charges as low as a few hundred picocoulombs. In practical situations such charges can arise during handling; the effect of excessive voltage is to cause the very thin insulating layers on the silicon chip to break down, irreversibly damaging the input transistors.

Practical MOS circuits incorporate protection circuits in parallel with their inputs to conduct away any excess charge. These contain a resistance in series with the input to limit the current which can flow in the input connection and a circuit in parallel with the input which conducts a relatively large current if a voltage larger than a specified value is applied. Normally the very high input resistance of an MOS input means that negligible current flows and virtually all the voltage on the input pin of the circuit appears on the input of the MOS transistor itself, but the protection circuit will prevent excessive voltages from appearing on the MOS input provided that the value of the series resistance is sufficiently large.

Circuits with this characteristic can be implemented quite easily. One method which is commonly used is the voltage clamp circuit shown in Figure 5.7. This consists of a pair of diodes connected to the power supply lines in such a way that they are always reverse biased in normal operation. If the input voltage attempts to exceed the power supply voltage or go negative, one or other of the diodes will conduct and allow a current to flow. This current will cause a voltage to be dropped across the resistor in series with the input, so that the voltage at the input will appear to be **clamped** to V_{CC} or zero volts. Another technique is to use a device connected between the input terminal and the zero volt line which will conduct if the voltage across it exceeds a certain value.

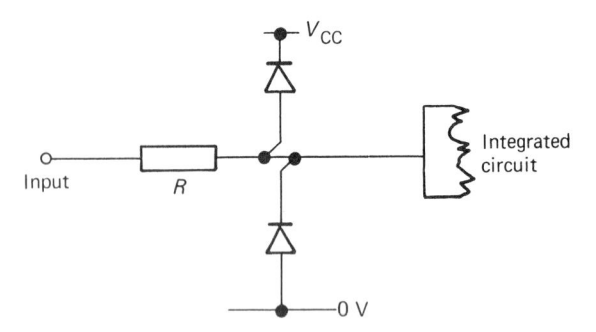

Figure 5.7 Diode clamp

These protection circuits make the integrated circuit less liable to damage from sources of 'stray' electrical charge such as the static electricity produced by artificial fibres in clothing and carpets. Damage can still occur if the static charge is especially large, however, because the protection circuit may not be able to discharge the input fast enough to prevent excessive voltages appearing. For this reason care should always be taken to avoid static electricity when handling MOS integrated circuits, by using electrically conductive packaging materials and ensuring that one's own body is earthed before touching the devices. They should be stored in special conductive foam rather than ordinary plastic foam, and specially treated plastic bags should be used for wrapping. Printed circuit microcomputers often have input/output lines connected directly to MOS devices, and they should be similarly handled with care.

Bipolar integrated circuits such as TTL usually have voltage 'clamps' built into their inputs too, although these are usually intended to prevent **ringing**, an oscillatory effect due to signal 'echoes' which appear when high-speed signals are transmitted on long lines. Integrated circuits with input clamp circuits are prone to damage due to excessive currents flowing in their inputs if the voltage ratings are exceeded for more than a fraction of a second. This is because of the small dimensions and limited current handling capacity of their protection circuits, and is usually more of a problem with TTL than with MOS logic.

5.4.2 Externally connected overvoltage protection circuits

The techniques already described can also be used to build external circuits to provide increased protection if necessary. If the input signal has a large amplitude it can be reduced by using a high value resistor and clamping diodes, or a **Zener diode** could be used instead of the two voltage clamping diodes to protect an input. Zener diodes are semiconductor diodes which conduct in the normal way when forward biased, but which also conduct in the other direction when a specified reverse voltage is applied. However, they also have large internal capacitances which combine with the series resistance to slow down the rate of response of the input to a change in logic level.

Nonlinear resistors are available which have a high resistance at low voltages but a much lower resistance at higher voltages. This effect can be used to protect logic inputs, but is also useful for protecting the circuit from damage due to excessive power supply voltages. The power handling ability of the resistors is higher than that of small semiconductor devices, and if the power supply voltage suddenly increases due to a fault or noise pulses carried by the mains supply, a nonlinear resistor connected across the power supply can prevent damage to the circuit.

5.4.3 Input isolation

Input signals must often be obtained from sources having a '0V' reference voltage which differs from that of the microcomputer; this can cause an undesired voltage to appear in addition to the signal. Even quite small voltages and currents present serious problems in analog interfacing, and some further techniques which do not require isolation will be described in the next chapter. For larger voltage differences the connection must be in a form which ensures that the input voltage to the system is **electrically isolated** from the integrated circuit itself.

One method of achieving this is to use a **transformer**, in which case only a.c. inputs can be coupled. However, transformer coupling still leaves an input vulnerable to noise 'spikes' which may be coupled inductively through the transformer or capacitatively between the primary and secondary windings.

A more reliable and versatile technique is to use a **Light Emitting Diode (LED)** and a **light-sensitive transistor (phototransistor)** so that the signal is coupled optically through an insulating barrier. This is the principle of the **optoisolator** in which the light source and detector are combined in the same package. The breakdown voltage of the insulator is usually a kilovolt or more, and the capacitance between the input and output is low enough to be negligible in almost all applications. LEDs are special diodes which emit visible or infrared light when they are forward biased. The voltage across a forward biased LED is more than the 0.7 V typical for silicon and depends to some extent on the colour of the light which the LED is designed to emit. Usually it is about 1.7 V and does not change much with current, although the amount of light emitted is approximately proportional to the current passing through the diode. The phototransistor is similar to a bipolar transistor except that the current flowing between emitter and collector is controlled by the amount of light falling on it instead of the base current. This means that the output current of an optoisolator is roughly proportional to its input current; the ratio between them is called the **Current**

Transfer Ratio (CTR) and is quoted at a particular input current. It is typically about 10% to 20% when a single phototransistor is used, although values of about 200-300% are possible with composite **Darlington phototransistors**.

LEDs gradually emit less light for a given current as they age, and as a result the CTR often decreases gradually over the first few hundred hours of operation. For this reason the circuit should be designed with an adequate margin of gain to ensure correct operation. Phototransistors switch on and off more slowly than ordinary bipolar transistors and many optoisolators are limited in their speed of operation to a few kilohertz. Although this appears to be a disadvantage, it should be borne in mind that the microcomputer is itself capable of coping with only a limited range of frequencies without external logic. This restricted bandwidth also tends to suppress short pulses which could be troublesome, and the bandwidth can further be limited by using external components. The LED should be protected by a diode connected in reverse parallel (D_1 in Figure 5.8) if there is a possibility of reverse voltages appearing on the input, because LEDs are capable of withstanding only quite low reverse voltages of about 3 V.

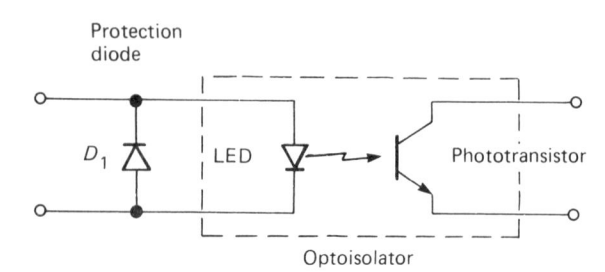

Figure 5.8 Optoisolator

5.4.4 Slowly varying input voltages

A logic signal must pass through the undefined range between high and low levels when changing state. If this signal is produced by a logic output connected in accordance with the fanout specification for the logic family being used, the transition time between states will be very short and no problems should result. If the signal varies more slowly difficulties arise not only because the input level is indeterminate, but also because input voltages between the defined states can cause excessive current flow in some types of logic gate as both output 'switches' conduct at the same time.

This problem can be overcome by processing 'slow' input signals to obtain fast switching versions. Figure 5.9a shows how the transfer function of a **voltage comparator** can be used to achieve this result. Voltage comparators are essentially special purpose amplifiers and closely resemble operational amplifiers, which will be introduced in section 6.2. If the signal voltage V_{in} is below the threshold voltage V_T the comparator will output a logic 'low' voltage level, while if $V_{in} > V_T$ the output voltage will be 'high'. The rate of switching of the comparator output still depends upon the rate of change of the input voltage, but is reduced by an amount approximately equal to the voltage gain of the comparator (typically about 10 000) for very slow signals. The response to faster input signals is limited by the maximum **slew rate** (rate of change of voltage with time) of the comparator output which is typically several volts per microsecond.

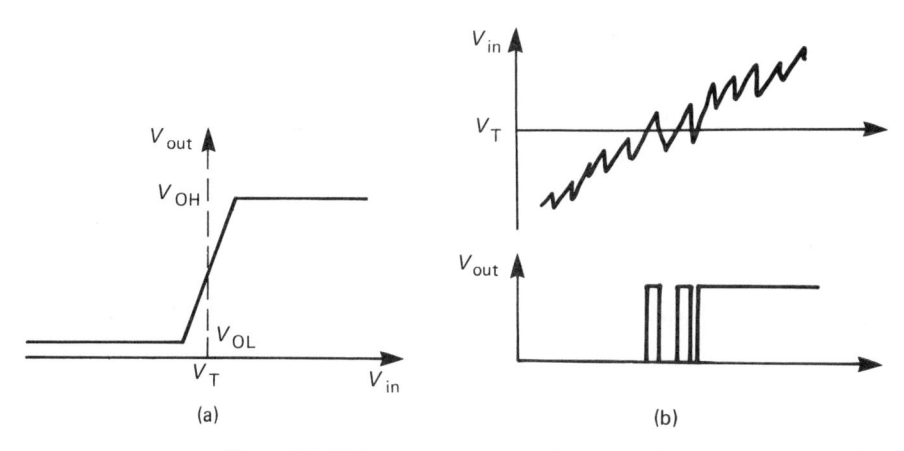

Figure 5.9 Voltage comparator characteristics

The main disadvantage of this technique lies in its vulnerability to noise which can cause the output state of the voltage comparator to change several times for a single transition of a noisy input signal, as shown in Figure 5.9b. If the microcomputer is programmed to respond to changes in input level rather than the level itself, the system will malfunction. For example, the input signal might be from a photodetector sensing objects interrupting a light beam as they pass along a conveyor belt. A microcomputer counting these objects by sensing logic transitions will miscount if noise causes multiple transitions for each object.

The classic solution to this problem is the **Schmitt trigger** circuit which if properly designed will produce a single abrupt change in logic level even when the input signal is slowly changing and noisy. It consists of a voltage comparator together with a circuit to feed back some of the output signal to the input in an arrangement called **DC positive feedback**. The effect of the feedback is to cause a square **hysteresis loop** to appear in the transfer characteristic, as may be seen in Figure 5.10 which shows that there are now two threshold voltages: V_{TH} and V_{TL}. When the input voltage is below V_{TL}, the output voltage is always 'low', whereas an input voltage greater the V_{TH} ensures a 'high' output voltage. If $V_{TL} < V_{in} < V_{TH}$, the output can be in either state.

The behaviour of the circuit can best be understood by considering an

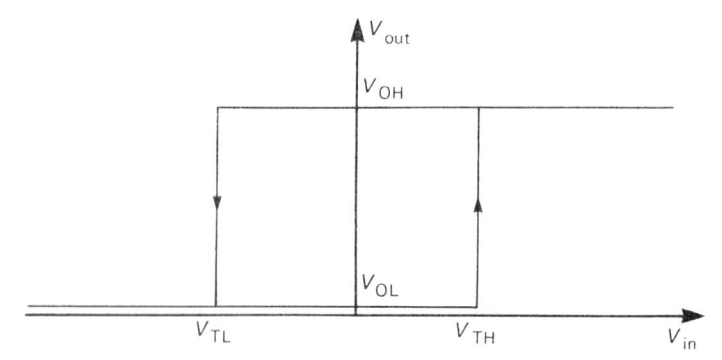

Figure 5.10 Transfer characteristic of Schmitt trigger

input voltage which increases from below V_{TL} to above V_{TH} and then falls back to its original value. Initially the output is 'low', and as V_{in} increases, the output remains 'low' until V_{TH} is reached when it is forced to switch to the high state. It will remain in this state until V_{in} again falls below V_{TL}, when it will be forced to switch low once more. When the input voltage is between threshold voltage levels the output voltage of a correctly designed Schmitt trigger will always remain stable, but if the input voltage passes a threshold level the output voltage will be forced to the appropriate output state. This characteristic makes a Schmitt trigger circuit insensitive to noise waveforms with a peak-to-peak voltage less than the difference between the threshold voltages. Note, however, that when the output voltage does change, it will switch between states very rapidly, irrespective of how slowly the input voltage changes.

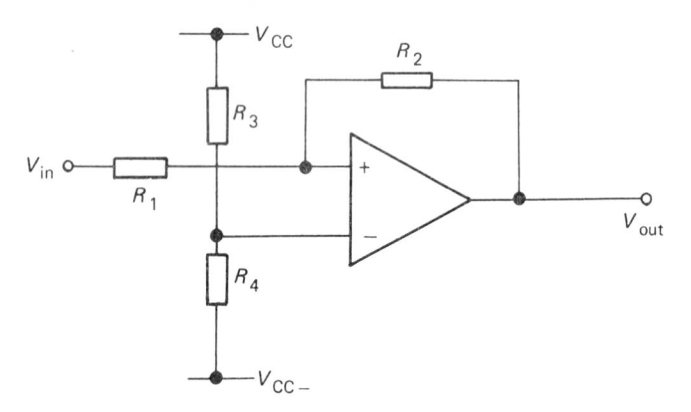

Figure 5.11 A simple Schmitt trigger circuit

The choice of threshold voltages depends upon the range of input voltages to be used. Logic gates are available with threshold voltages suitable for TTL logic levels and these are useful where the input signals are

5.4.4

nominally TTL compatible. Frequently, however, the input levels are from sources which produce other voltages in their low and high states, and then it becomes necessary to design a Schmitt trigger with appropriate threshold levels. Fortunately this can be done quite simply using a voltage comparator and resistors to provide suitable positive feedback, and Figure 5.11 shows a simple circuit. Here R_1 and R_2 provide the feedback, while two further resistors R_3 and R_4 are used to obtain a reference voltage from the supply voltage V_{CC}. This reference voltage is applied to the inverting $(-)$ input of the voltage comparator, and can readily be found by noting that R_3 and R_4 form a potential divider:

$$V_- = R_3 \cdot \frac{V_{CC}}{(R_3 + R_4)} \tag{5.6a}$$

The voltage on the noninverting $(+)$ input can similarly be found by considering R_1 and R_2 as a voltage divider:

$$V_+ = \frac{V_{in}R_2 + V_{out}R_1}{R_1 + R_2} \tag{5.6b}$$

When V_+ is greater than V_-, the comparator output voltage will be in its high state V_{OH}, while it will be in its low state V_{OL} if V_- is greater than V_+. The effect of the positive feedback from the output to the noninverting input is to cause the comparator output voltage to switch rapidly as V_+ increases or decreases past V_-. Thus switching happens when $V_+ = V_-$, and substituting in equations 5.6a and 5.6b gives expressions for the threshold voltages at which switching occurs:

$$V_{TL} = \frac{[V_-.(R_1 + R_2) - V_{OH}.R_1]}{R_2} \tag{5.7}$$

$$V_{TH} = \frac{[V_-.(R_1 + R_2) - V_{OL}.R_1]}{R_2} \tag{5.8}$$

In fact these equations assume that the voltage gain of the voltage comparator is very large compared with the ratio $(R_1+R_2)/R_1$ and that its input impedance is high compared with the values of the resistors used, both of which can usually be arranged quite easily. The low and high output state voltages V_{OL} and V_{OH} are quite closely defined for most voltage comparators, enabling the designer to obtain values for the threshold voltages V_{TH} and V_{TL} by a suitable choice of resistor values.

5.4.5 Using a low-pass filter to remove noise pulses

The noise experienced in digital interface circuits often consists of pulses which are short compared with the rate at which the input signal changes. In such cases the amount of noise can be reduced by using a **low-pass filter** with a bandwidth chosen to let the signal through with minimal distortion but attenuate the noise pulses. The waveform appearing at the filter output will have have slower rise and fall times than the original, but the fast transitions can be restored using a Schmitt trigger.

The effect of the filter is to 'spread out' the noise pulses so that they last longer but have lower amplitude. Figure 5.12a shows that if the peak value of the 'spread' noise pulse is not large enough to switch the output of the Schmitt trigger to the other state, the trigger circuit will remove the residual effects of the noise.

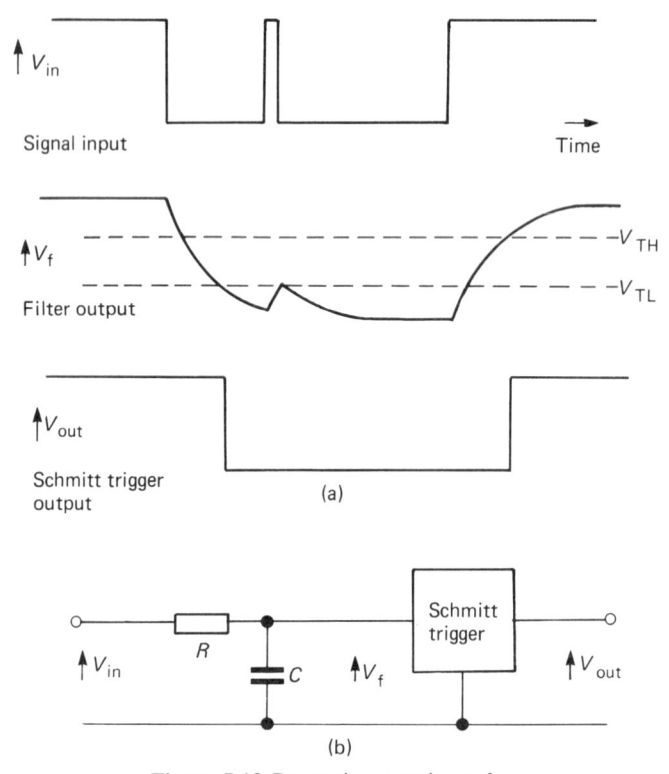

Figure 5.12 Removing a noise pulse

The low-pass filter can be made using only a resistor and a capacitor, as shown in Figure 5.12b. The effect of the capacitor is to slow down the rate at which the output voltage V_{out} changes in response to a change in the input voltage V_{in}. A step change in V_{in} will cause V_{out} to change in an exponential arc until a new steady value is reached. The rate at which this happens is determined by the product of the resistor and capacitor values used, which is called the **time constant** of the network. If the resistance is specified in ohms and the capacitance in microfarads, the time constant can be expressed directly in microseconds.

Unfortunately the time taken for the circuit to respond to a step change in input voltage is more difficult to calculate exactly. If the noise pulse is rectangular with an amplitude V_n and lasts for a time T_n, the maximum output voltage from the filter can be shown to be approximately

5.4.5

$$V_m \simeq \frac{V_n . T_n}{CR} \qquad (5.9)$$

where C and R are the capacitor and resistor values, the effects of any load connected to the filter output are ignored, and provided that T_n is much less than the product CR. Equation 5.9 will give results which are correct within a factor of two or three for most kinds of noise pulse, and although this tends to err on the pessimistic side it shows that short pulses can be substantially reduced in amplitude.

The voltage at the filter output is connected to a Schmitt trigger with threshold voltages of V_{TH} and V_{TL}. The time taken for this circuit to respond to a step change in input voltage can be found by substitution. In fact an exact solution of equation 5.9 is usually unnecessary and the speed of response can be assumed to be of the same order of magnitude as the time constant. The amount of delay which can be tolerated must be chosen by the designer taking into account the speed at which the interface is required to operate. Equation 5.9 can then be used to estimate the tolerance of the filter trigger combination to noise. If the response to the peak amplitude of the impulse is less than the difference between the threshold voltages, the effect of the input pulse can be removed. From equation 5.9, the input pulse will be removed if

$$V_n . T_n < (V_{TH} - V_{TL}) . CR \qquad (5.9a)$$

5.5 Switch inputs

Switches are often used as **input transducers** to microcomputers, for example in **keyboards** or as **limit switches**. The normal method of connecting a switch to a microcomputer input is to use a resistor to source current from the power supply and connect the switch between the input and the 0 volt line, so that it can sink all this current when the switch is closed. In this way the input can be made to 'see' a low level when the switch is closed and a high level when it is open.

The choice of switch is more limited than might appear at first sight. Switches intended for handling the low voltages and currents associated with electronic signals have contacts which 'wipe' over each other to counteract the effects of surface corrosion, but those designed for high power circuits are usually unsuitable for this type of application because they operate differently. Also, most switches do not open or close 'cleanly' because their contacts bounce as they meet. The result is that the switch does not cause a single transition from one level to another to appear at an input, but a rapid series of changes until it stabilises. This phenomenon, known as **contact bounce**, must be countered by suitable interface design as it can give rise to incorrect or erratic behaviour in the system as a whole.

5.5.1 Coping with contact bounce

Many methods have been evolved to counter this problem, either by assuming that any changes between the open and closed states that take place within a few milliseconds are due to contact bounce or by using more than one connection to the switch.

The short sequence of pulses can be distinguished from correct switch operation by the frequency of switching. One way to do this is to use a low-pass filter and a Schmitt trigger as already described, but the amount of interface circuitry required is prohibitive in most applications. Alternatively the switch contact can be **debounced** using software, in which case the only hardware required is the pull-up resistor on the input.

Because input switches are normally connected to sink current from an input when closed, they cause a logic 0 to be read when a switch is closed and a logic 1 when it is open. However, switches are conventionally regarded as being in the 1 state when their contacts are closed, so the input state must be complemented using software. If the input port has the label *inputs* the state of the switch detected during this scan could be assigned to a variable *this* (we shall ignore for now the problem of exactly how we define to the program the address of the input port, that is, the value of the label *inputs*).

this := **not** *inputs*

The debouncing subroutine can check that the switch state is stable by comparing its state this time (the value of the variable *this*) with its state during the last scan (the value of another variable, say, *last*). If the switch has been closed for two successive scans both *this* and *last* will have the logic value 1. A bit (called *switch*) used to model the state of the switch can be set when the switch has been set for two successive scans by the statement

switch := *switch* **or** (*this* **and** *last*)

and it can similarly be cleared when the switch has been open for two successive scans by the statement

switch := *switch* **and not** (**not** *this* **and not** *last*)

which simplifies using de Morgan's theorem to

switch := *switch* **and** (*this* **or** *last*)

The software to debounce a single input switch is thus quite simple, for example the following:

this := **not** *inputs*;
switch := *switch* **or** (*this* **and** *last*);
switch := *switch* **and** (*this* **or** *last*);
last := *this*

This can be written quite compactly in assembly language, as shown in Figure 5.13.

LDA	INPUTS	Read input port, complement pattern
COMA		so switches give a logic '1' when
STA	THIS	closed
ANDA	LAST	If the switch has been closed twice
ORA	SWITCH	in succession, set the corresponding
STA	SWITCH	bit of SWITCH
LDA	THIS	Check to see if the switch has been
ORA	LAST	open twice running
ANDA	SWITCH	If so, set the corresponding bit
STA	SWITCH	of SWITCH low again
LDA	THIS	Update record of previous
STA	LAST	input state...

Figure 5.13 Debouncing switches in software

Examination of this piece of program shows that the microprocessor operates simultaneously upon all the bits in a word, which means that an 8-bit machine can debounce eight inputs at the same time. If each line of the input port *inputs* is connected to a switch, the debounced status of each switch will be indicated by the state of the corresponding bit of *switch*.

Larger numbers of inputs could be handled by using indexed addressing to select different sets of input bits although the matrix technique described later in this chapter is normally preferred since it uses less interfacing hardware.

A shortcoming of this technique is that it assumes that the contact bounce time is short compared with the rate of switch operation, so that the programmer must know the characteristics of the switch being used. An alternative approach is to use a pair of inputs to provide more information about the internal behaviour of the switch. When a **Single-Pole Double-Throw (SPDT)** switch is operated, one pair of contacts usually opens a short time before a second pair closes. Provided that the design of switch ensures that this is the case, an SPDT switch can be interfaced via a simple logic circuit which removes the effects of contact bounce without recourse to time measurement. The advantage of this method is that it can correctly interpret the operation of a switch no matter how slowly it is operated, but it suffers from the twin disadvantages of requiring a more complicated

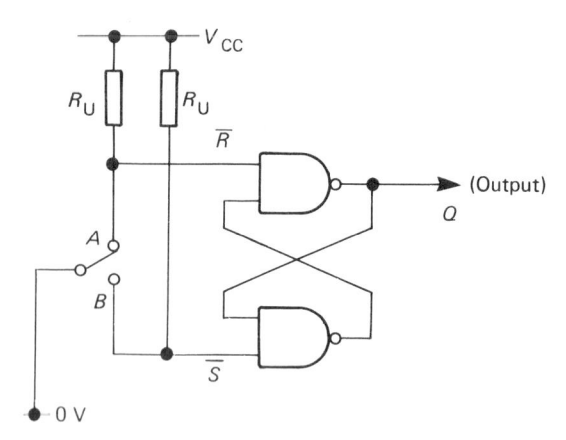

Figure 5.14 Debouncing an SPDT switch

switch and twice as many input lines.

Figure 5.14 shows such a debouncing circuit with a pair of NAND gates cross-coupled to form a **Set-Reset latch (SR latch.)** The output signal from this latch remains constant when both inputs are high, but it can be 'set' high by taking input $\bar{S}$ low and 'reset' low by taking input $\bar{R}$ low. When the switch is thrown from position A to B the first pair of contacts open and, as both inputs $\bar{R}$ and $\bar{S}$ are high, the latch output remains in the low state. The second pair of contacts then close momentarily at first as the contacts bounce. The $\bar{S}$ input of the latch switches low and the latch output is set high; it remains high as the contacts bounce. The output will switch low only when the switch is thrown back to the A position; when the first pair of contacts close, input $\bar{R}$ will go low and the latch will be reset.

5.5.2 Detecting switch closure

Usually it is the action of operating a switch which must be detected by an interface rather than its steady state. For example, it is the action of pressing a key on a keyboard which causes a character to be typed. In cases like this further hardware or software is needed to detect the operation of the switch.

External hardware can be used to debounce the switch signal with a connection via an **edge-sensitive input line** rather than a **level-sensitive input line**. These edge-sensitive inputs can be programmed to detect rising or falling edges, and optionally to cause an interrupt when a change in logic level is detected. A limitation of this method is the shortage of edge-sensitive inputs on most microcomputer input/output circuits, and although further hardware can be used to service several switches it is better to avoid the problem by using software instead.

The software used to debounce the switch inputs can be extended to detect changes in input state. The logical expression for the latch used in

5.5.2   DIGITAL INTERFACING   **209**

the SPDT switch debouncing method can be adapted to allow the condition '*S* AND NOT *Q*' to be tested. If this is true, then bit *Q* is about to be set, indicating that the switch has just closed. The time-sequence debouncing method is best modified by using a sampling interval which is longer than the bounce time of the switches. The program can then test for the sequence 'open, closed, closed' at successive sampling times. This technique requires three consecutive samples of the input signal; the current one *this*, the last one *last* and the previous one, say, *prev*. The switch closure can be detected by simple software such as that below:

>   *this*: = **not** *inputs*;
>   *switch* := *switch* **or** (*this* **and** *last* **and not** *prev*);
>   *prev* := *last*;
>   *last* := *this*

Note that no provision has been made for setting the value of *switch* back to 0 here; this is done by another program as described in the next section.

5.5.3 Using interrupts

It is not necessary to continuously monitor the state of switch inputs, because their mechanical operation is slow relative to the speed of a microcomputer. The delay can be generated using a software loop or a hardware timer, depending on whether the microprocessor can be employed for anything else more useful while waiting for a switch to close. If the switch inputs are used to control the behaviour of a continuously running program, for example to change set-point parameters in a controller, then a **hardware timer** should be used to generate interrupts at regular intervals. The debouncing software would then be incorporated in the interrupt service routine. When a switch closure has been detected by either method it must be reported to the main program, since switch closures are usually detected by interrupt routines. This is done by the **intercommunication register** *switch* which is used like an input/output register connecting the interrupt routine and main program. If the switch status is being reported, the bits are set by the interrupt routine when the switch closes and cleared by the interrupt routine when it opens. In the case of edge detection, however, the bit is set when the corresponding switch closes and must be cleared by the main program to acknowledge that it has responded to that switch closure.

5.5.4 Matrix connection of switches

Certain applications such as keyboards require large numbers of switches to be interfaced cheaply. The number of input and output connections, and hence the cost of the interface, can be reduced by connecting the switches between a matrix of wires connected in rows and columns as shown in Figure 5.15. Each row is switched low one at a time by software which allows it to sink current, and pull-up resistors source current to ensure that the outputs remain high as long as all the switches are open. When a switch closes it connects an input to an output and the program can identify which switch has closed by noting which pairs of input and output are low at the same time.

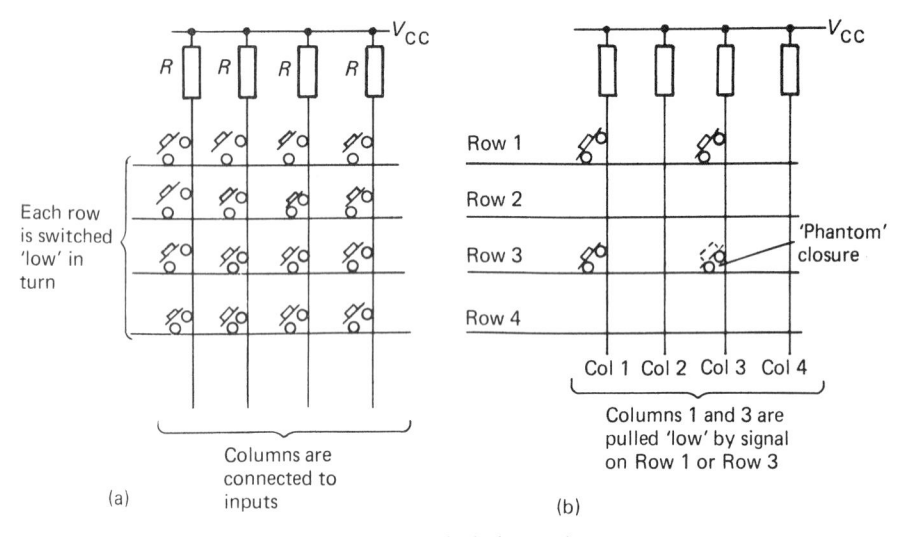

Figure 5.15 Switch matrix

This matrix technique becomes more attractive for larger numbers of switches because the number of input and output connections required rises roughly in proportion to the square root of the number of switches. In its simplest form it can reliably decode a maximum of only two switches closed at the same time, although this is adequate for small keypads where only one key is expected to be pressed at a time. Sometimes a second key may be pressed on a keyboard before the preceding one has been released **(rollover)** and this scheme is capable of handling 2-key rollover. If more than two keys are depressed at the same time there is a possiblility of interconnecting rows and columns to create 'phantom' key closures as shown in Figure 5.15b. Applications in which several switches can be closed at the same time need diodes to be connected in series with the switches to prevent interactions between the switches; examples are instrument panels and large keyboards where there is a possibility of **N-key rollover**.

The switch scanning routine must be called at regular intervals and is usually implemented as part of an interrupt routine which services a hardware timer. In the listing of Figure 5.16 each bit of each element of array *switch* is set to 1 if the corresponding switch is closed and reset to 0 if it is open. Of course, the main program must also have access to this array which appears to it as a set of status registers in a complicated input/output device.

The software for detecting switch closures can be modified in a similar way; bits are set in elements of the array *switch*, as shown in Figure 5.17,

```
row := 1;
for i := 1 to n do
  begin
    output := 255 - row; {switch one row output to '0'}
    row := 2*row; {next time, switch next row...}
    this := not inputs;
    switch[i] := switch[i] or (this and last[i]);
    switch[i] := switch[i] and (this or last[i]);
    last[i] := this
  end;
```

Figure 5.16

```
row := 1;
for i := 1 to n do
  begin
    output := 255 - row; {switch one row output to '0'}
    row := 2*row; {next time, switch next row...}
    this := not inputs;
    switch[i] := switch[i] or (this and last[i] and not prev[i]);
    prev[i] := last[i];
    last[i] := this
  end;
```

Figure 5.17

when the corresponding key is pressed. These bits must be cleared by the program reading the switch status.

5.6 Interface circuits for digital outputs

As we have seen, the output pin of a logic input/output device is connected internally to a circuit which is capable of either 'sourcing' an output current when in the logic 'high' state or 'sinking' a current when in the 'low' state. This information is usually presented in manufacturers' data sheets in terms of the parameters V_{OH}, V_{OL}, I_{OH}, and I_{OL}.

The values of these parameters are usually much too low to allow loads to be connected directly, and in practice an external circuit must be connected to **amplify** the current and voltage to **drive** a load. Although several types of semiconductor devices are now available for controlling DC and AC powers up to many kilowatts there are only two basic ways in which a switch can be connected to a load to control it: **series connection** and **shunt connection** as shown in Figure 5.18.

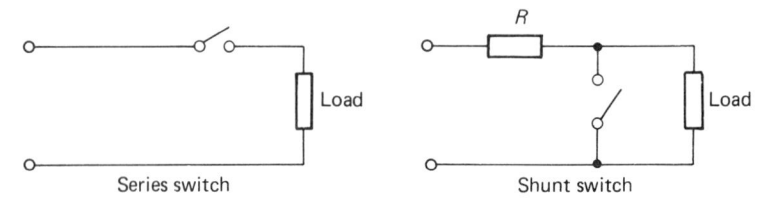

Figure 5.18 Series and shunt connection

With series connection, the switch allows current to flow through the load when closed, while with shunt connection closing the switch allows current to bypass the load. Both connections are useful in low-power circuits, but only the series connection can be used in high-power circuits because of the power wasted in the series resistor R. Interface circuits using more than one transistor can be designed in which each transistor controls the next either by series or shunt switching.

5.6.1 Operating the transistor from a logic output

MOS transistors, being controlled by their gate-source voltage, can be connected directly to logic outputs although they often need a larger voltage than the 2.4 V guaranteed to be available from a TTL output in the high state. This can be arranged by attaching an external pull-up resistor to the logic output to give a logic high voltage of +5 V, and indeed many microcomputer interface devices are designed with an internal pull-up resistor to provide this voltage when driving MOS devices. Thus an output circuit using an MOS device can be very simple as Figure 5.19a shows. However, MOS transistors tend to oscillate at high frequencies which shows up as unpredictable behaviour at the output and can damage the transistor. A simple solution which is almost always effective is to thread a ferrite bead on to the gate lead to reduce the high frequency gain of the amplifier.

Bipolar transistors are usually connected to logic outputs via resistors which allow their base-emitter voltage and hence base current to be controlled by a logic voltage. With appropriate design they can be switched between cut-off and saturation by the low and high state voltages of the logic output provided that enough current is available.

The properties of the transistor are matched to those of the driving circuit by a resistor network which allows us to specify the ranges of input voltage which our output interface circuit will recognise as 'low' and 'high', that is, we can choose values of V_{IH} and V_{IL} to suit our particular application. In Figure 5.19b, the resistors R_1 and R_2 form a potentiometer that reduces the input voltage and biases the base-emitter junction of the transistor.

When the input voltage is at V_{IL} or below, we want the transistor to be

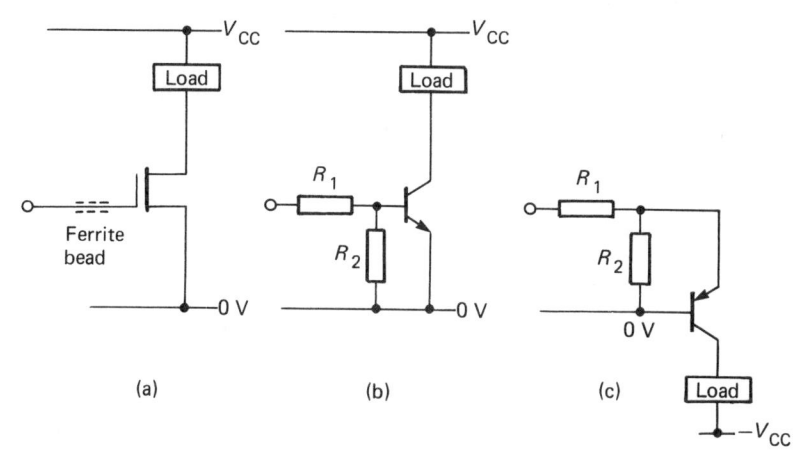

Figure 5.19 Simple interface circuits

cut off. In this case there will be no base current so the current in R_1 and R_2 will be the same and

$$\frac{(V_{IL} - V_{BE})}{R_1} = \frac{V_{BE}}{R_2} \tag{5.10}$$

If V_{IL} is below V_{BE} there is no need to reduce the input voltage by potentiometer action and R_2 can be omitted, so that the effective value of V_{IL} becomes V_{BE}.

On the other hand, when the input voltage is V_{IH} or above, we want the transistor to conduct the required collector current I_C. This requires a base current of

$$I_B = \frac{I_C}{h_{FE}} \tag{5.11}$$

to be provided by the driving circuit. Again the resistor network can be analysed, this time to give

$$\frac{(V_{IH} - V_{BE})}{R_1} - \frac{V_{BE}}{R_2} = I_B \tag{5.12}$$

These three equations can now be solved to give values for R_1 and R_2

$$R_1 = \frac{h_{FE} \cdot (V_{IH} - V_{IL})}{I_C} \tag{5.13}$$

$$R_2 = \frac{V_{BE} \cdot R_1}{(V_{IL} - V_{BE})} \quad \text{if } V_{IL} > V_{BE} \tag{5.14}$$

The design procedure for the common base circuit in Figure 5.19c is the same except that here the emitter is used as the input connection so instead of equation 5.11 we must use

$$I_E = \frac{h_{FE}.I_C}{(h_{FE} + 1)} \tag{5.11a}$$

We can assume to quite a good approximation that the magnitudes of the emitter and collector currents are the same because h_{FE} is much larger than unity for all modern transistors, so that the rule for R_1 becomes

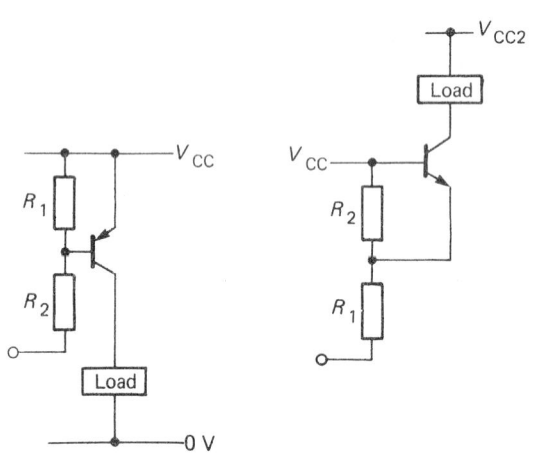

$$R_1 = \frac{(V_{IH} - V_{IL})}{I_C} \tag{5.15}$$

Figure 5.20 Further interface circuits

The same method can be used to design circuits of the type shown in Figure 5.20. Here the input will take current when it is 'low' but not when it is high; equations 5.13 and 5.14 can be used to calculate values for R_1 and R_2 except that because the common connection is now the power supply line at a voltage V_{CC} we must substitute $(V_{CC} - V_{IL})$ for V_{IH} and $(V_{CC} - V_{IH})$ for V_{IL}. Note that equation 5.13a is unaltered from 5.13

$$R_1 = \frac{h_{FE}.(V_{IH} - V_{IL})}{I_C} \tag{5.13a}$$

$$R_2 = \frac{V_{BE}.R_1}{(V_{CC} - V_{IH} - V_{BE})} \tag{5.14a}$$

We now have rules for designing the input networks for two useful output circuits, but there is another problem. When the input is in the high state it sinks an input current which can be easily shown to be

$$I_{IH} = \frac{(V_{IH} - V_{BE})}{R_1} \tag{5.16}$$

and the circuit driving the output must be able to source this amount. Similarly, when the input is low, the circuits of Figure 5.20 each source a current of

$$I_{IL} = \frac{(V_{CC} - V_{BE} - V_{IL})}{R_1} \qquad (5.17)$$

which the driving circuit must sink. This may be possible with common emitter output circuits, where the large value of h_{FE} means that an input current can control an output current up to perhaps a hundred times higher, but almost certainly it will not be the case with the common base mode. More complicated circuits are needed to achieve the output currents needed in most practical applications.

5.6.2 Series connection: Darlington transistors

One way to obtain a higher value of current gain is to connect two transistors together as shown in Figure 5.21a. This series connection is known as the **Darlington connection** and provides a composite transistor with a current gain which is approximately the product of the current gains (h_{FE}) of the two component transistors and a value of V_{BE} that is effectively twice that of a single transistor. Such composite 'Darlington transistors' thus have typical gains in the range of 1000-10 000 and values of V_{BE} of about 1.4 V.

The higher value of V_{BE} means that in most cases R_2 can be omitted and the Darlington transistor can be used to control currents of up to an amp or more from a logic output. In fact, the outputs of many input/output chips are specified not only in terms of their TTL compatibility, but also in terms of the minimum and maximum current that they will source when in the high state and connected to the base of a Darlington transistor, so that R_1 can sometimes be left out as well.

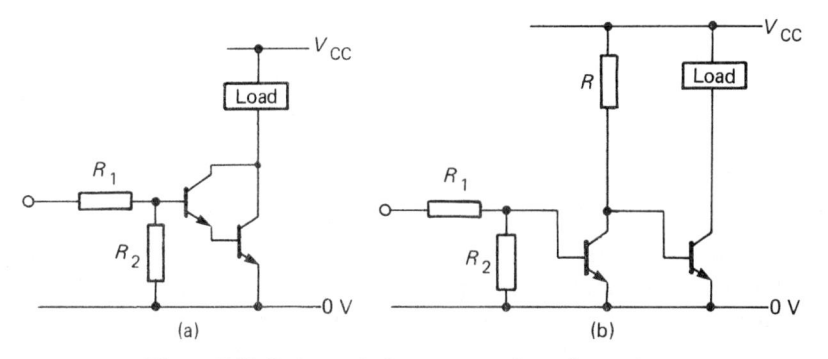

Figure 5.21 Series and shunt connection of transistors

5.6.3 Shunt connection of transistors

Figure 5.21b shows two transistors connected in such a way that if the first one conducts it deprives the second one of base current. When the left-hand transistor is cut off the base current available to the other transistor is

$$I_B = \frac{(V_{CC} - V_{BE})}{R} \tag{5.18}$$

There is no reason why the second transistor should have to be connected in common emitter mode. If a *pnp* transistor is connected in common base mode instead of the common emitter transistor shown, a useful current can be controlled in a load connected to a negative supply.

5.6.4 Characteristics of the load

We have assumed so far that the output circuit is required simply to source or sink a specified current via a load. In fact the characteristics of the load must also be taken into consideration, since the load is rarely a simple resistance.

Some devices, for example LEDs, require the current through them to be defined by the circuit which is driving them, so that the output interface must be more than a simple logic-controlled switch. In the case of LEDs this is easily achieved by the use of **current-limiting resistors** in series with the diodes; the value of resistance needed is found by subtracting the forward voltage drop across the diode from the supply voltage available and dividing by the desired current. Special LED drive circuits are available with internal current limiting.

An incandescent lamp has a much lower resistance when its filament is cold, with the result that a relatively large 'inrush' current can flow at the instant of switching on. The designer must make sure that this current will not damage the driving circuit by using a transistor which can cope with these transient increases in collector current. There is also a possibility of overheating because the transistor power dissipation is increased considerably during these transients.

A similar problem arises with capacitive loads. When the voltage across the load is changed any capacitance shown by the load must be charged or discharged. This is not a common problem with microcomputer interfaces, but it can be overcome by designing an output which is capable of sourcing and sinking relatively large transient currents. Long cables carrying signals sometimes show troublesome capacitance, but special drive circuits are available for this application.

Inductive loads, for example relays, solenoids and electric motors, are encountered much more commonly in interface design than capacitive ones. In such cases the current cannot increase instantaneously due to the effect of the inductance, but in the steady state the current in the inductive coil must be limited by the interface circuit or the resistance of the coil. Fortunately these devices are usually designed so that they have enough resistance to limit the current when they are operated from their specified supply voltage. Unfortunately, inductive loads present another problem to the circuit designer. When the current is switched off, a voltage appears across the load which is proportional to the rate of change of current. This voltage will almost always destroy the driving transistor unless protective circuitry is added to allow the current to die away more slowly in the load. A very effective way to arrange this is to connect a diode in parallel with the load, wired in such a way that it is reverse-biased while current is flowing in the drive transistor. When the current from the drive transistor ceases the voltage across the load will reverse polarity and the diode will conduct. Initially the current in the diode will be the same as that which was provided by the transistor, and the diode must be chosen to cope with this current and the reverse voltage rating which it will experience. This **flywheel current** in the diode will decay in a fraction of a second through resistive losses in the load coil, and the actuator will switch off quickly enough for almost all applications.

5.6.5 Controlling AC power

A very common requirement is for an output circuit to control the power being supplied to a load from AC mains. In this case electronic switching is still possible but special power switching devices must be used. Furthermore, the circuit being controlled must be isolated from the microcomputer or there is a strong likelihood that the microcomputer will be damaged.

Thyristor (SCR)

Thyristors are semiconductor devices which operate in one of two states: conducting and nonconducting. In one direction they are incapable of conduction, while conduction in the forward direction is under the control of a **gate** lead. Thus the thyristor behaves as a rectifier which can be controlled by its gate lead, and another name for the thyristor is the **Silicon Controlled Rectifier (SCR)**. Unlike the base or gate of a transistor, the gate lead of an SCR cannot control the amount of current passed as in the case of a transistor, and after the device has been switched on the current must be limited by resistances (or reactances) in the rest of the circuit. It is triggered by allowing a current (usually of a few tens or hundreds of milliamps) to flow in the gate lead, the voltage drop between gate and **cathode** being typically one or two volts. Once triggered in this way, a thyristor will continue to conduct current flowing from the **anode** to the cathode leads (the forward direction) for as long as that current exceeds a given threshold value. Again, this current is usually some tens or hundreds of milliamps, according to the type of SCR.

The only way to switch an SCR off again is to reduce its anode current below the threshold level. This **commutation** of the anode current is difficult in DC circuits, and the SCR is normally used only in AC circuits where the current falls to zero during each cycle. The SCR can be triggered from a simple microcomputer interface provided that the power circuit is completely isolated from the microcomputer. This requires the use of either a transformer or an optoisolator. The use of a transformer avoids the need to have a DC power supply for the output side of the optoisolator, and for this reason SCRs are normally triggered using small pulse transformers to provide isolation.

Special problems arise with SCRs due to the high voltages and currents at which they operate: hundreds of volts and several amps are quite typical. If the voltage across the SCR increases too rapidly, due to a noise spike for example, the device may be switched on by accident. To reduce the possibility of this happening, **snubbing circuits** are normally used to help absorb rapid transients. In their simplest form these consist of a resistor and capacitor in series across the SCR. If the SCR switches on too rapidly, the current may rise before the whole device has time to start conducting. Consequently a large current will be concentrated in a small part of the SCR, causing permanent damage. Here the solution is to rely upon the circuit inductance to limit the rate of rise of current, or to use an SCR type with a high di/dt rating.

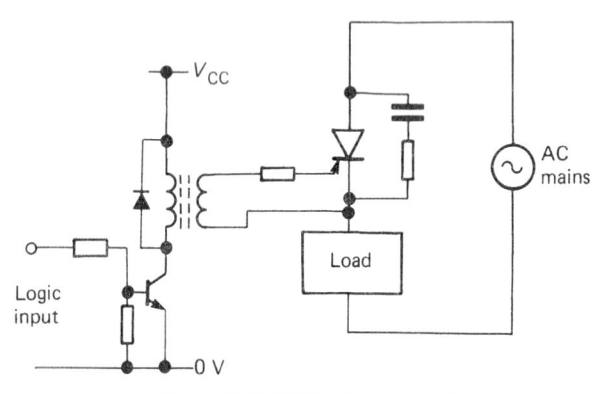

Figure 5.22 SCR trigger circuit

In Figure 5.22 a circuit is shown for a SCR switch controlled from a logic output. The driving circuit must provide pulses which can be coupled through the isolating pulse transformer. These pulses must trigger the SCR as soon as possible during each cycle that it is required to conduct. One approach is to use pulses which are synchronised with the mains, but delayed slightly to allow the mains voltage to rise to the point where sufficient current is available through the circuit to bring the SCR current above its threshold level. Another is to pulse the circuit at a high frequency so that when required to do so it switches on as soon as possible.

If an optoisolator is used to provide the isolation, the SCR can be switched on by a steady current flowing in the output of the optoisolator, so that the problem of relating the timing of trigger pulses to the zero crossings of the mains does not arise.

Triacs
A major shortcoming of the SCR is that it is capable of controlled operation in one direction only. If full control of the AC power being supplied to a load is required, two SCRs are needed, each with a separately isolated triggering circuit, to handle both positive and negative half cycles of the

waveform. The **triac** is essentially an SCR which is capable of operation in both directions, so that AC power can be controlled with only a single isolating circuit. The triac can be triggered by a pulse applied to the gate lead in much the same way as the SCR, except that it can be switched into conduction during both positive and negative half cycles, and the triggering pulse can be of either polarity. Again, snubbing circuits are normally used to prevent undesired triggering of the triac by rapid transient changes in voltage.

Gate turnoff thyristor (GTO)
Unlike SCRs, **Gate Turnoff Thyristors (GTOs)** can be turned off quite easily, by applying a pulse to the gate lead with the opposite polarity to that used to turn it on. These devices are capable of switching on and off in about a microsecond, comparable with a power transistor, but they can handle much more power than a transistor of comparable price. The BTW58-1300R, for example, is a GTO which can control currents up to 6.5 A at continuous voltages up to 750 V, which means that it could be used to control DC resistive loads of up to more than 4 kW. Moreover, it is capable of tolerating repetitive peak values up to 25 A and 1.3 kV, so that it can cope with transients in AC circuits and can be protected by a simple fuse.

The GTO can control current in one direction only, which is not a problem in DC circuits, but in AC applications the GTO must be protected by a rectifier diode in series with it to prevent conduction in the reverse direction. The GTO is switched on by a current in the gate lead, typically in the order of a few hundred milliamps; the corresponding voltage between gate and cathode would be 1.5 V to 2 V. This current can be applied continuously without ill effects; in fact it will reduce the forward voltage drop in the circuit being controlled and hence reduce the power dissipation of the device. Switching off is achieved by applying a pulse of reverse polarity lasting a few microseconds and a few volts in amplitude.

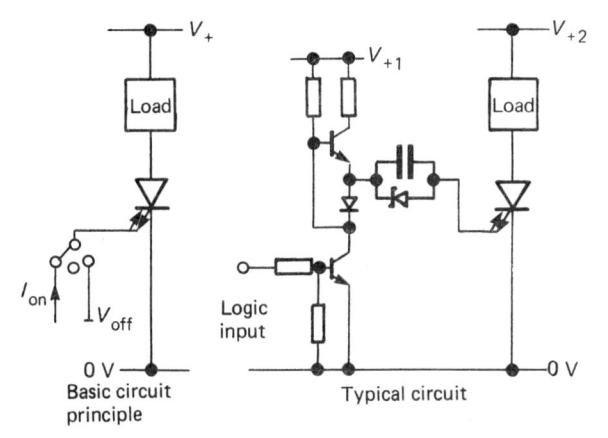

Figure 5.23 Controlling a gate turnoff thyristor

Figure 5.23 shows a control circuit for a GTO. The DC supply voltage for the drive circuit must be isolated from the microcomputer and may be derived from the mains supply by a transformer and simple power supply circuit. The control circuit is operated via an optoisolator, the output of which controls the base current to a Darlington transistor T_1. When this conducts, no base current is available to the other transistor T_2 and the output is close to zero voltage. When T_1 is turned off, the output is no longer connected to the zero voltage line, and because base current can now flow to T_2 the output of the drive circuit will rise close to the supply voltage. Now gate current is available for the GTO and it conducts, even if the gate current is removed.

Solid state relay (SSR)
The design of interfaces for controlling AC power requires specialised expertise and there is a danger of damage to the rest of the system if such interfaces fail. Fortunately, packaged circuits known as **solid state relays (SSRs)** are available, which, as their name suggests, behave quite like electromagnetic relays: a small current in the input circuit, usually a few milliamps at three to five volts, allows a much larger mains current to be switched on and off, while the input is isolated electrically from the output. Internally they consist of an optoisolator, triggering circuit, triac, and snubbing circuit. For many applications SSRs provide a more convenient and safer method of controlling high power AC loads. Unfortunately, they usually require rather more current to operate their inputs than can be provided directly by the output of an MOS interface circuit, but a TTL gate can be used to provide increased current drive.

5.6.6 Integrated interface circuits

After the foregoing, it may come as something of a relief to discover that many of these interface circuits are now available in integrated circuit form. These include the input level defining resistors, drive transistors, and sometimes the flywheel diodes as well. Usually several outputs are available in one circuit, and in some cases latches or shift registers are provided to ease interfacing.

The disadvantage of using a general purpose output driver is that the designer is less likely to be able to obtain exactly the characteristics that he or she requires than if the circuit had been constructed from discrete components. However, this is usually more than outweighed by the savings in space, printed circuit complexity and assembly cost that result from using integrated circuits. Also, integrated circuits allow certain functions to be achieved which would be impracticable with circuits made from discrete components.

The range of integrated circuits for interfacing outputs is evolving rapidly and any survey of available devices rapidly becomes obsolete. For 'one-off' applications it is frequently quicker to design an interface circuit rather than try and locate an integrated circuit with the precise specification required. With larger quantities the benefits of using the integrated approach dictate that suitable integrated circuit drivers should always be used when available.

5.7 Multiplexed outputs

Output connections can also be used more effectively by **multiplexing**. Any output device which can be driven with an intermittent signal can be operated in multiplexed mode, for example incandescent lamps and resistive heaters. In practice, however, multiplexed outputs are used almost exclusively with displays.

Information displays require a relatively large amount of input data but can exploit the fact that the human eye is insensitive to changes taking place at rates of more than about 50 times per second. Thus a television picture can be drawn by a single dot travelling very rapidly across the screen. As with switches, the display elements should be wired into a matrix with each element being connected between a different row and column. Each row is then selected in turn, for example by connecting it to a current source, and selected elements in that row are turned on by sinking current through them. (Note that the terms 'row' and 'column' are arbitrary and are used here simply to distinguish between the two sets of inputs to the matrix.) Although the principle is the same, there is one important difference between the multiplexing of switches and of displays: in the latter case appreciable power may need to be switched.

The calculation of the voltage and current required depends upon the type of display being used. Perhaps the easiest type of display to multiplex is the LED which emits light at an intensity proportional to the current flowing through it. Thus in a multiplexed display where each row is selected for $1/N$ of the time, the current which must flow when a given LED is selected must be N times that which would be needed for the same light intensity if it were continuously excited. The designer must make sure that the driving transistors can provide the current needed and put resistors of suitable value into the circuit to limit the current.

Incandescent lamps, on the other hand, emit light at an intensity which depends in a highly nonlinear way upon the power being dissipated. For a given brightness the power input must be the same in the multiplexed case as in the steady case, which means that if a lamp is selected for only $1/N$ of the time the power input must be N times as much — and the voltage and current larger by the square root of N.

Another type of display which is in fairly widespread use is the **fluorescent display** which is actually a latter-day valve, or tube. The brightness of these devices increases in proportion to the current flowing which depends in turn upon the supply voltage raised to the power 3/2. The supply voltage must be increased for multiplexed operation, although since these displays are usually intended for use in this mode their data sheets normally specify the voltage to be used.

Gas discharge displays require high voltages for their operation and hence special drive circuits. Their brightness also varies in direct proportion to the current, which can usually be set by means of a resistor; the manufacturer's data sheets usually give detailed explanations of how to set the current.

Liquid Crystal Displays (LCDs) are different from all the other types described here in that they operate by changing the way in which they reflect or transmit light instead of emitting light themselves. This offers two major advantages: they need have only a very small power consumption because they are not radiating light energy, and they can be read even in conditions of high ambient light. Their disadvantage is that driving LCDs requires the use of special drive techniques to ensure that no DC voltage appears across any element of the display at any time, or electrochemical effects will damage it. Special integrated drive circuits are now available for LCDs which implement quite sophisticated multiplexing schemes, and there is little point in attempting to use any other method to drive them. In fact, most LCD displays are now supplied with integral drive circuits.

5.7.1 Hardware or software multiplexing?

The multiplexing can be carried out either by software within the microcomputer or by external hardware. Where possible, software is usually preferable because it is cheaper, but care must be taken to ensure that damage to the display cannot result from malfunctions in the program. If the multiplexer stops scanning the display elements, the excessive DC current that flows in those elements that remain permanently selected may cause failure. For this reason the software must be reliable and in applications where the microprocessor could be halted or otherwise prevented from scanning the display, provision must be made to blank the display if scanning stops, or other techniques must be used. However the multiplexing is carried out, further savings in input/output connections are often possible by using the same set of outputs to scan a switch matrix and a matrix of display elements.

5.7.2 Software multiplexing

The first decision which the designer of a software multiplexed display must take in programming is to fix the **multiplex rate**. If it is too slow, troublesome flicker will become apparent. This is less of a problem where incandescent lamps are used because of their relatively slow speed of response, but ordinarily a rate less than 50 complete scans per second should not be used. If the rate is excessively high, more processor time than necessary will be dedicated to this rather mundane task. The timing for the multiplexer can be provided by a timer programmed to cause interrupts at appropriate intervals.

If LEDs or other relatively fast-acting displays are used there is a danger of **ghosting**. This occurs when elements which should be dark are seen to glow faintly and is due to the limited speed of response of the drive circuitry. The solution to this problem is to switch off the previous row of elements before selecting which outputs are going to be illuminated in the next row, then switch on the next row. If a suitable time interval is left between each step ghosting should not occur; a delay of a few microseconds is usually sufficient, and this can be provided by the execution time of the program.

The short program of Figure 5.24 shows how a multiplexed display with six 'rows' of eight 'columns' each can be programmed. In practice each 'row' might select a different digit and each 'column' might define the state of a segment in a seven-segment representation of that digit. In fact this needs eight outputs, the eighth one being used for the decimal point:

---

column := *alloff*; {*turn off all segments*}
$i := (i+1)$ **mod** 6; {*select next digit*}
row := *digit*[*i*]; {*turn on digit driver*}
column := *segments*[*i*]; {*turn on selected segments*}

Figure 5.24

Each time that the fragment of program of Figure 5.24 is run it selects the next row by incrementing the row counter modulo 6 (the number of rows) and outputs the pattern *digit*[*i*] to select row *i*. Before the row is changed all the column outputs are disabled by outputting a fixed pattern called *alloff* (which will consist entirely of 1s or entirely of 0s) to the column select outputs *cols*. Once the new row has been selected the outputs are energised according to the pattern in the *i*th element of the array *segments*.

The output buffer *column* contains an image in binary form of the pattern of elements to be illuminated. It is written by the main program, to which it appears as a set of output registers. If the display is going to display 7-segment numerical data we shall also need a subroutine to translate numerical information presented by the main program into patterns. This is easily done by using a lookup table, an example of which is included in the design example of section 5.8.2.

5.7.3 Hardware multiplexing

The alternative is to implement the multiplexing logic in hardware, a task which is made considerably easier by the availability of special purpose integrated circuits. These contain a set of data registers which contain an image of the data to be displayed, the function which *columns* performed in the program. Each register is connected to the column enabling outputs in turn and the corresponding row output enabled; the output waveforms are timed by a clock input or by an on-chip oscillator and are designed to avoid the danger of 'ghosting'.

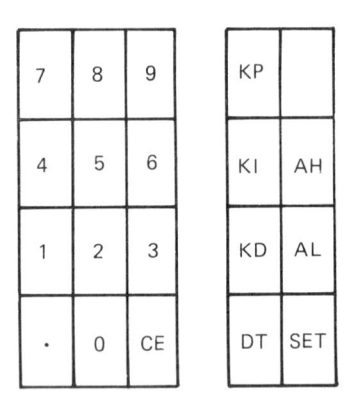

224 MICROCOMPUTERS IN ENGINEERING AND SCIENCE 5.8

7	8	9	KP	
4	5	6	KI	AH
1	2	3	KD	AL
.	0	CE	DT	SET

Figure 5.25 Keypad layout

5.8 Design examples

The design examples in this chapter are also concerned with aspects of interfacing the controller, because this provides several examples commonly encountered in engineering applications.

5.8.1 Scanning the controller keypad

Information is entered into the controller by means of a keypad on the front console of the machine. The layout of the keypad is shown in Figure 5.25; the keys are in two groups — one group allows numbers to be typed in, while the other group contains keys which are used to set parameters such as the setpoint, alarm levels, gains and sampling interval.

The keys are connected as a matrix, with four row and five column connections. This means that nine connections are required; if only one port can be spared for this interface, external logic will be required. An external three-to-eight line decoder will be used in this example because it demonstrates how larger keyboards can be interfaced. The eight output lines of the decoder allow keyboards with up to 40 keys to be scanned with the use of a single 8-bit port.

Figure 5.26 shows a subroutine, SCANKB, which scans the keypad; it is effectively a rewrite of the Pascal program in Figure 5.17 in assembly language, except that it codes the row and column numbers to form a single code which is returned in location KEYBUF. In operation, a separate output of a 3-to-8 line decoder is connected to each row and an input to each column. Each output is taken low in turn and the inputs monitored to determine if any key is currently being pressed. The keystrikes are debounced and decoded with the aid of tables showing the states of the column inputs the last time that the subroutine was called, LAST, and the state on the time immediately previous to that when the keypad was scanned, PREV.

If a keystrike is detected, the code for the key is worked out as a code

SCANKB	LDAA	#BOTTOM	Start with bottom row
	STAA	ROW	
	LDX	#LAST	Table shows history of keystrikes
SCANK0	LDAA	ROW	
	COMA		
	STAA	PORTA	Output row number
	LDAA	PORTA	Read in five columns
	COMA		Invert so that logic 1 => key pressed
	PSHA		and keep a copy of it
	ANDA	0,X	Check with values last time
	ANDA	8,X	and with previous values.
	BEQ	SCANK2	if a strike is found, work out code:
	LDAB	ROW	Work out position number
	ASLB		of key; start by multiplying
	ASLB		row number by five (number of
	ADDB	ROW	columns)
	DECB		Allow for INCB that follows
SCANK1	INCB		Now increment by an amount
	LSRA		that depends upon column number.
	BCC	SCANK1	
	PSHX		Save pointer and look up in table...
	LDX	#LAYOUT	Offset allows for codes starting at 8!
	ABX		
	LDAB	0,X	Appropriate ASCII code is in acc-B.
	PULX		Recover pointer.
	STAB	KEYBUF	save this code
	SEC		Set flag to indicate that
	ROR	KEYFLG	a character has been read.
SCANK2	LDAA	0,X	Update PREV record
	COMA		with complement of LAST (We're
	STAA	8,X	looking for sequence 0,1,1!) & store...
	PULA		Recover keystrike pattern and update
	STAA	0,X	LAST entry.
	INX		point to next history byte
	DEC	ROW	on to next row
	BPL	SCANK0	until all rows have been scanned
	RTS		

Figure 5.26 Keyboard scanning subroutine

number that depends upon the position of the key. This code is then translated into the internal representation by means of the lookup table LAYOUT (Figure 5.27) which contains a representation of the layout of the keypad.

```
LAYOUT  FCB   '7','8','9',KP,CAN
        FCB   '4','5','6',KI,AH
        FCB   '1','2','3',KD,AL
        FCB   '.','0',CE,DT,SET
```

Figure 5.27

This is organised with one character for each position on the keypad to correspond to the layout of the characters on the keys. The appropriate entry in the lookup table is placed in location KEYBUF, and location KEYFLG is set to a nonzero value to indicate that a character is in KEYBUF.

These two registers, KEYBUF and KEYFLG, appear to the main program as data and status registers, just as if they were in a physical input device. When the main program is waiting for another input character it will test the contents of KEYFLG to determine if a character is in KEYBUF; if KEYFLG is zero, KEYBUF is empty. When the main program wishes to read a character, it calls subroutine READKB (see Figure 5.28) which waits for the contents of location KEYFLG to become nonzero, indicating that SCANKB has detected a key being pressed on the keypad. When this happens, subroutine READKB reads KEYBUF and then resets KEYFLG to indicate that the character has been taken.

```
READKB  TST   KEYFLG    Test to see if KEYBUF is full
        BEQ   READKB    and wait until it is
        LDA   KEYBUF    Now take the character
        CLR   KEYFLG    and clear the flag.
        RTS             Return with character in acc.
```

Figure 5.28

The circuit of the keyboard scanning subsystem is shown in Figure 5.29; note that only four of a possible eight column lines are used.

5.8.2 Display scan

Information from the controller is displayed on an eight-digit light emitting diode display. The displays are driven by serial driver chips type 5450, each of which can drive up to 34 light emitting diodes or four seven segment displays. Similar display drivers are available for other types of display such as LCDs, and all that would be necessary to change the type of display would be to use the appropriate type of display driver circuit.

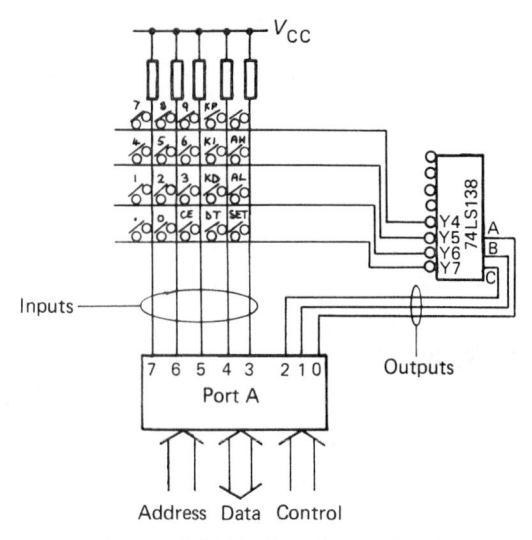

Figure 5.29 Keyboard scan circuit

The eight digits in the display are therefore presented as two groups of four digits. The display drivers need the data to be formatted in serial form with a start bit and a stop bit as shown in Figure 5.23. This means that six bytes must be sent to the display drivers: a 'start' byte which prepares the display driver to accept the data which follows, followed by four bytes each containing a seven-segment pattern, followed by a 'stop' byte which completes the transmission of serial data.

When a digit is to be displayed, it must first be translated into the appropriate seven segment pattern. This is done using a subroutine, TRANS7 (Figure 5.32), which finds the appropriate entry in a lookup table TABLE7 as may be seen in the listing of Figure 5.31. TRANS7 is called as required by another subroutine, DISPLY, also shown in Figure 5.32, which

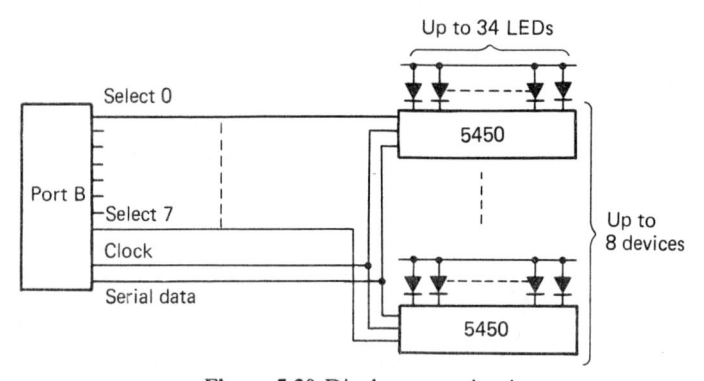

Figure 5.30 Display scan circuit

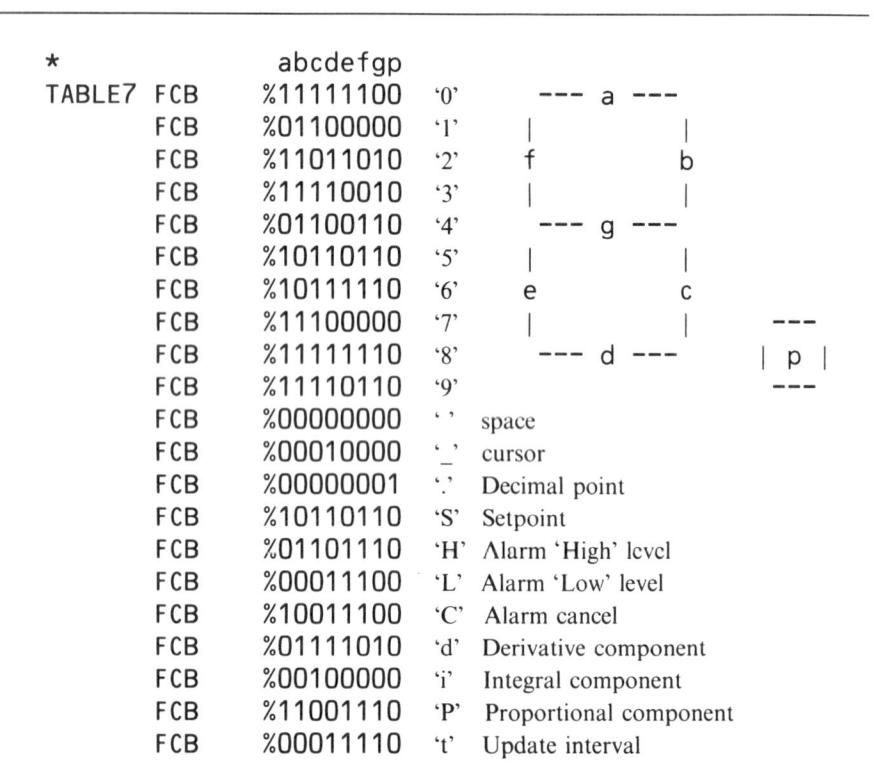

```
*                         abcdefgp
TABLE7   FCB    %11111100    '0'
         FCB    %01100000    '1'
         FCB    %11011010    '2'
         FCB    %11110010    '3'
         FCB    %01100110    '4'
         FCB    %10110110    '5'
         FCB    %10111110    '6'
         FCB    %11100000    '7'
         FCB    %11111110    '8'
         FCB    %11110110    '9'
         FCB    %00000000    ' '   space
         FCB    %00010000    '_'   cursor
         FCB    %00000001    '.'   Decimal point
         FCB    %10110110    'S'   Setpoint
         FCB    %01101110    'H'   Alarm 'High' level
         FCB    %00011100    'L'   Alarm 'Low' level
         FCB    %10011100    'C'   Alarm cancel
         FCB    %01111010    'd'   Derivative component
         FCB    %00100000    'i'   Integral component
         FCB    %11001110    'P'   Proportional component
         FCB    %00011110    't'   Update interval
```

```
    --- a ---
   |         |
 f           b
   |         |
    --- g ---
   |         |
 e           c
   |         |    ---
    --- d ---    | p |
                  ---
```

Figure 5.31 Character lookup table for display

---

outputs a block of four digits starting with the digit pointed to by the index register X when the subroutine is called. These digits are sent to all the serial display drivers, but only one of them will be enabled by the decoder chip to accept this information. The information as to which one of these displays should be selected is passed to DISPLY in accumulator B. First the 'start' pattern '00000001' is output using the shift register in the VIA by means of subroutine SEROUT. Then four successive bytes are read from memory starting from the address in X; each byte is translated to 7-segment form by TRANS7 and sent to the display driver by SEROUT. Finally the 'stop' code '00000000' is sent using SEROUT.

5.8.3 Isolated inputs and outputs

The controller needs to determine the time at which the mains voltage passes through zero. The problem of electronic design is made slightly more difficult by the need to cater for two different mains standards: 230 V 50 Hz, and 110 V 60 Hz. In order to isolate the mains input from the circuitry of the microcomputer, an optoisolator will be used. In fact this is not strictly necessary, but it provides an increased safety factor for anyone wishing to

5.8.3

```
        DISPLY  LDAA    #1              Output start code
                BSR     SEROUT
                LDAA    #4              Number of digits in section
        DISPL1  PSHA                    on stack
                LDAA    0,X             Get customer number
                BSR     TRANS7          Translate digit and
                BSR     SEROUT          output it.
                INX                     Bump pointer to next entry
                PULA                    Get count from stack
                DECA                    and decrement
                BNE     DISPL1          until done
                BSR     SEROUT          output zeroes to finish loading
                RTS

        TRANS7  PSHX                    Save contents of X at entry
                LDX     #TABLE7         point to table
                ABX                     offset to 7-seg code
                LDAA    0,X             and get it into acc-A.
                PULX                    Recover contents of X register.
                RTS

        SEROUT  STAB    PORTB           select digit
                STAA    SHIFTR          and output it...
                RTS
```

Figure 5.32 Display drive subroutine

---

construct the circuit.

The current flowing in the LED input of the optoisolator must be limited by an external resistor. Again, in order to 'play safe', two resistors will be used, one in each mains lead. Two factors must be considered; the current rating of the LED and the power dissipation rating of the current limiting resistors. If the power dissipation of each resistor is limited to 0.25 watt when 250 V AC (the maximum likely voltage) is applied to the two resistors in series, the rms voltage across each resistor will be 125 volts, and the value of the resistor must be no more than the value R given by:

$$\text{Power} \simeq 0.25\text{W} \simeq \frac{125^2}{R} \tag{5.19}$$

This value is 62.5 kΩ, which suggests that the next standard value of 68 kΩ should be used, giving a total resistance in the input circuit of 136 kΩ. The maximum instantaneous voltage corresponding to an rms voltage of 250 V is approximately 350 V, so that the maximum instantaneous current in the LED will be 350/136 or approximately 2.5 mA.

Optoisolators with single phototransistors (that is, those which do not use Darlington phototransistors) typically have current transfer ratios of about 20%, so that the peak current which can flow in the output in our example will be about 0.5 mA. This output must be connected to an amplifier which can detect the very small current which will flow just after zero crossing occurs. If the circuit is to switch at about 1 degree after the zero crossing, the input voltage at that instant will be about 1.75% of the maximum value, or about 6 V. The forward voltage drop across the LED is about 1.7 V, leaving roughly 4.3 V across the two input resistors, enough to cause a current of about 6 µA in the output if the 20% CTR of the optoisolator is maintained at this low current.

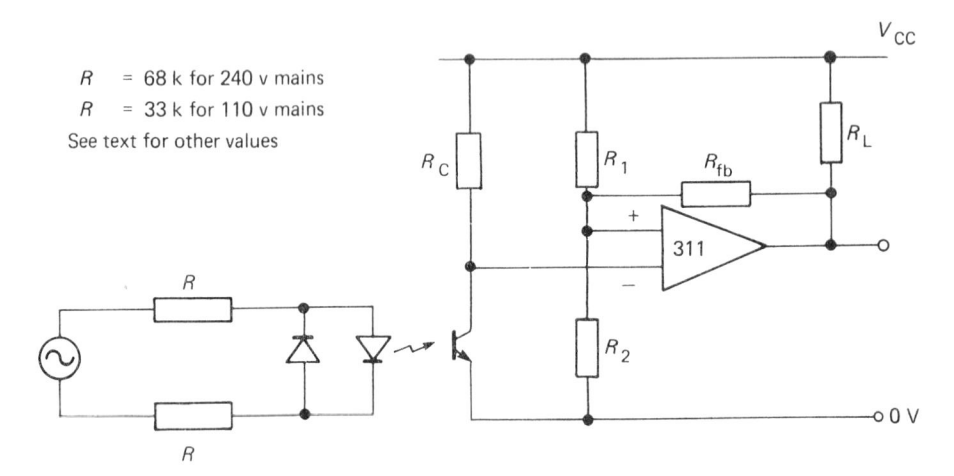

R = 68 k for 240 v mains
R = 33 k for 110 v mains
See text for other values

Figure 5.33 Isolated input circuit

This small signal must be amplified, and the amplifier must also act as a trigger circuit to give a 'clean' transition between logic levels when the mains voltage changes polarity. In the circuit of Figure 5.33 the feedback resistor causes the comparator to act as a trigger circuit. The voltage on the noninverting terminal of the comparator can be found using Kirchoff's current law — the sum of the currents flowing to this terminal will be zero, since the bias current taken by the 311 is very small, typically 0.1 µA. Thus

$$\frac{(V_{CC}-V_+)}{R_1} + \frac{(0-V_+)}{R_2} + \frac{(V_O-V_+)}{R_F} \simeq 0 \qquad (5.20)$$

When the output of the comparator is in the 'low' state, the voltage on its output terminal V_O will be approximately zero, so that the voltage on the noninverting input V_+ will be approximately

$$V_+ \simeq V_{CC} \left[\frac{\frac{1}{R_1}}{\frac{1}{R_1}+\frac{1}{R_2}+\frac{1}{R_F}} \right] \qquad (5.21)$$

5.8.3 DIGITAL INTERFACING

When the output of the comparator is in the logic 'high' state its output terminal will switch off and current will flow from the power supply line via R_L and R_F, and the voltage on the noninverting input will be

$$V_+ \simeq V_{CC} \left[\frac{\frac{1}{R_1} + \frac{1}{R_F + R_L}}{\frac{1}{R_1} + \frac{1}{R_2} + \frac{1}{R_F + R_L}} \right] \tag{5.22}$$

It is relatively easy to make R_F much larger than R_L so that the latter can be neglected in equation 5.22. If the three resistors R_1, R_2, and R_F are made identical, equations 5.21 and 5.22 then simplify dramatically to give

$$V_+ \simeq \frac{V_{CC}}{3} \tag{5.21a}$$

and

$$V_+ \simeq \frac{2V_{CC}}{3} \tag{5.21b}$$

If the comparator is to switch at 1.7 V when 6 μA is flowing in R_C, the voltage across R_C must be 3.3 V, suggesting a value of 470 kΩ for this resistor.

5.8.4 Triac output

The AC power supplied to a high power load can be switched on and off by means of a triac. Again, this must be isolated from the microcomputer, both for its safety and that of the user. The optoisolator is a convenient isolating device, as shown in Figure 5.34. The triac is switched on by applying a positive voltage on its gate with respect to the common cathode terminal, and this positive voltage is derived from the AC mains supply by means of a simple rectifier circuit. The resistor in series with the rectifier is used to limit the current, and hence voltage, appearing at the output of the rectifier circuit. Enough gate current (up to 80 mA is required by common types of triac) must be available to trigger the triac, and the capacitor used must have enough capacitance to hold sufficient charge to provide the triggering current during negative half cycles of the mains voltage. Only 1.6 mA is available from a standard NMOS logic output, so that a current gain of 50 must be provided. A Darlington optoisolator is used here to help provide this current gain; the 6N139 type used has a minimum current transfer ratio of 200% (i.e. 2), so that a further gain of 25 must be provided by another bipolar transistor. The transistor amplifier used in the design of Figure 5.34 is based upon one shown in Figure 5.19.

Some other components appear in the circuit shown. A resistor is placed in series with the LED in the optoisolator to limit the current flowing through it, while those resistors associated with the gate of the triac limit the triggering current and provide a parallel conduction path to minimise the risk of the triac triggering when not required. The resistor and capacitor

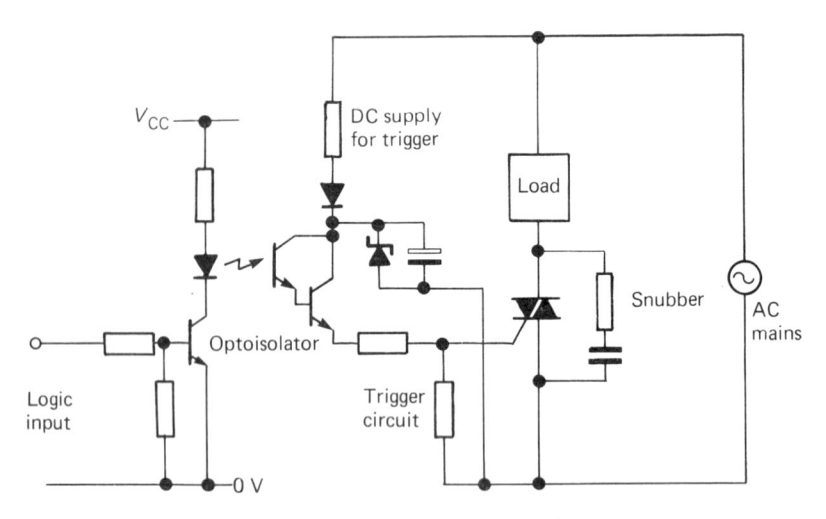

Figure 5.34 Triac drive circuit

across the triac form a 'snubbing network', intended to prevent damage due to voltage noise 'spikes' on the mains supply.

5.9 Questions

1. A jukebox which contains 100 records is controlled by a 6809-based microcomputer. The records, which are numbered 00 to 99, are selected by means of a small keypad and the number of the record selected is displayed on two seven-segment LED display modules.

    (a) Describe one method by which the binary coded decimal (BCD) equivalents of the decimal digits 0 to 9 may be generated from a 4×4 matrix non-encoded keyboard. Explain how the problems of rollover and switch bounce could be overcome.

    (b) The number of the record selected is stored in BCD format, in memory location RECD. The two 7-segment displays are connected in parallel directly to a VIA port, the symbolic address of which is defined as OUTD. The seven segments occupy bits 0 to 6 inclusive, and bit 7 is used to determine which display module is addressed; a logic '0' enabling the least significant decimal digit.

    Decide which segment is to be connected to which output bit of the output port, and write a 6809 assembly code subroutine which causes the number stored in RECD to be displayed in decimal form. You may assume that the VIA port has been initialised, that a binary 1 illuminates the display, and that additional latch enable signals are not required.

2. Design a Schmitt trigger circuit to operate from a single power supply of +10 V which will switch output state when the input voltage rises above two-thirds of the supply voltage or falls below one-third of it.
3. Many types of CMOS logic can operate from voltage supplies from +3 V to +15 V. Input voltages from 0 V to one-third of the supply voltage are interpreted as logic 'low', while input voltages above two-thirds of the supply voltage are regarded as logic 'high'. Design a circuit to interface a TTL output to a CMOS circuit operating from a +12 V supply.
4. The actuating coil of a magnetically operated valve may be represented as an inductive load with a resistance of 100 Ω which operates from a supply voltage of +24 V. Design an interface to allow the valve to be driven using a PIA output.
5. An electronic display is to be constructed from an array of seven rows of 80 LEDs, which will allow several characters to be displayed simultaneously. Messages of arbitrary length can be displayed by 'scrolling' the displayed patterns left one LED position at a time. Each LED will glow with adequate brightness if a forward current of 20 mA is allowed to flow through it and can tolerate peak currents up to 200 mA.
   (a) Design a drive circuit for the LED display. Remember that as few connections as possible should be used in the interests of cheapness, compactness, and reliability.
   (b) Write an assembly language subroutine which will produce suitable signals for your drive circuit.
   (c) Write a subroutine which will accept ASCII characters, convert them to a pattern suitable for display, and call the subroutine of section (b).
   (d) Modify your subroutines to allow the display to 'scroll' left one LED position when required.
6. A simple keyboard instrument has a three-octave keyboard (12 keys per octave) which must be scanned by a microcomputer within the keyboard to determine which keys have been pressed. Design an interface to carry out this function; remember that more than one key may be pressed simultaneously.

Chapter 6 ANALOG INTERFACING

In the 'real world' outside the microcomputer there are many kinds of continuously variable quantities, but within the 'electronic world' of the microcomputer itself there are only electrical waveforms. Nevertheless, these are variable voltages or currents which can change at variable frequencies or time intervals.

It is interesting to note that despite the fact that most quantities met in everyday life are continuous, we have to express these quantities digitally in order to handle them properly. We do this quite unconsciously, as when quoting a weight as a number of pounds or kilograms, and the same conversion from a continuously variable quantity such as weight to numerical form must be done when using a microcomputer. Within the microcomputer, quantities such as this are always represented numerically.

This chapter is concerned with the methods by which voltages and currents may be generated by a microcomputer interface and the techniques for converting analog electrical signals into digital form. The external transducers and actuators commonly used to provide and make use of the analog signals will be discussed in the next chapter.

6.1 The structure of an analog interface

The expression **analog interface** is loosely used to describe all parts of a system positioned between the transducer or actuator and the digital input/output section of the microcomputer. It cannot, however, be considered in isolation since the input/output section and the controlling software are integral parts of the data conversion process. Consider for example the simple feedback control system illustrated in Figure 6.1.

The input signal conditioning section transforms the range of voltage or current signals from the transducer to a range acceptable to the **Analog-to-Digital Converter (ADC)**. The ADC in turn converts the scaled analog signal into a binary code in a form suitable for presentation to the microcomputer, either directly to the system buses or via a digital interface circuit. In either case it appears to the microcomputer as the contents of one or more registers.

The **input subroutine** is an integral part of the analog interface which in some ways complements the input signal conditioning section. It **reads** the converted data and scales and sometimes linearises the numbers obtained

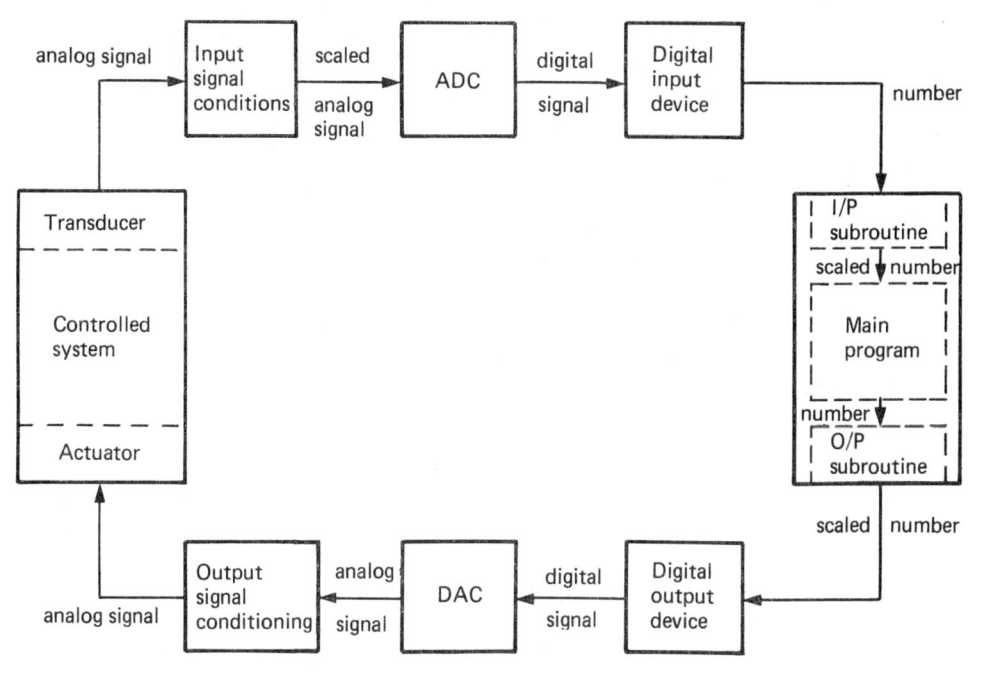

Figure 6.1 Generalised analog interface

before presenting the data to the main program. As an example, the ADC input signal may consist of an analog voltage which is related to the transducer's temperature in some non-linear way. Provided that the combined characteristics of the transducer and signal conditioning section are known, the input subroutine may translate the digitised input directly into units of temperature in, say, degrees Celsius. In fact, this process outlines one of the significant advantages of using a microcomputer to control equipment. Cheaper transducers often have characteristics that vary slightly from sample to sample of the same type. If the transfer function of each transducer is measured, a correction can be incorporated into a table used by the input subroutine of the microcomputer to which it is attached.

A secondary function of the input subroutine relates to the control of the analog interface. Most analog-to-digital converters require one or more **ADC control signals** for each data conversion. The generation of these signals and the monitoring of the **ADC status** can be handled within the input subroutine, which enables the main program to deal directly in terms of the quantity being measured, without the programmer being concerned with the detailed operation of the interface. The number and frequency of the control signals required will depend both on the sampling period and the type of ADC employed. Consequently the choice of interface components will not only be influenced by the overall system specification but also by a **hardware/software tradeoff** factor. This is the extent to which control logic can be implemented in software weighed against the additional hardware required for the conversion to be mainly independent of the input subroutine. Later in this chapter it will be shown that much of the logic needed in an ADC can be written in the form of software.

An analog output interface performs the opposite function to the one just described since information passes from the program to the outside world. A **Digital-to-Analog Converter (DAC)** is the counterpart of the ADC and, in addition to amplification, the output signal conditioning section may be required to remove any high frequency components from the generated signal.

The function of a DAC is to decode the digital number presented to it and to generate an analog output which is as closely proportional to this number as possible. If the number consisted of three bits, the analog current or voltage could have eight different values, as shown in Figure 6.2.

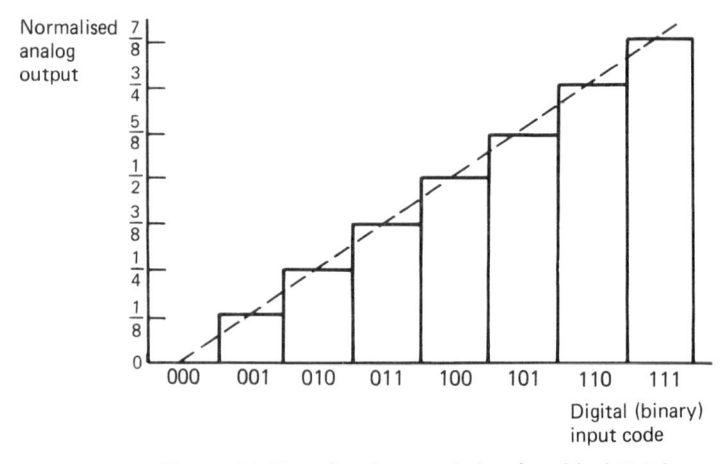

Figure 6.2 Transfer characteristic of an ideal DAC

This very generalised example of an interface has assumed that the analog signals have been voltages or currents. The microcomputer program has been concerned with a digital representation of their **amplitude**, either as input or output data. In many applications it is not the amplitude of the analog quantity which is of interest, but the signal frequency or the time between signal transitions. In some cases the measurement or generation of analog signals can be carried out most effectively by measuring or generating frequencies or time intervals. Sometimes an interface requires a waveform to be produced in which both frequency and voltage are controllable.

6.1.1 Terminology

Often there are many ways of converting a quantity from its physical form to a number within a program, and the designer must make a decision between the 'quality' of the measurement to be made and the amount of

6.1.1

money which can be allowed for this operation. There are several factors to be considered when quoting this 'quality'. In discussing these points the transducer and analog-to-digital converter will be considered together, since they both affect the conversion from a physical quantity to a number.

First there is the **precision** or **resolution** of the measurement to be made. This indicates the minimum difference between two values which can be represented, and it is not the same thing as the **accuracy** of the measurement. The accuracy is a measure of how closely the reading obtained from the transducer and its associated interface corresponds with the actual value of the quantity being measured. The difference between the indicated and actual values is the **error** in the transduction process.

If the transducer were accurately calibrated, the error might be known as a function of the reading being taken, making it possible to correct for this error within the microcomputer. Even if this were done, however, there would still be some **uncertainty** in the measurement due to changes in the parameters of the transducer, to uncertainties in the reference against which the transducer was calibrated, and arising from the calibration process itself.

A digital watch provides one example of a digital measuring device with which we are all familiar. If the watch displays the time to the nearest second, then it has a precision of one second, even if it is actually running one minute slow, so that its accuracy is only one minute. The watch will also gain or lose time, so that even if the watch had been checked and was known to be a minute slow, the error of one minute would become more and more uncertain.

The length of an engineering component could be measured using a metre rule to a resolution of $\pm$ 0.5 mm or with a micrometer to a resolution of $\pm$ 0.001 mm. If the micrometer measures values between the limits of 0 and 25 mm, it will allow the length to be expressed as one of 25 000 values. Similarly, the metre rule would allow it to be expressed as one of 2000 discrete levels within the range 0 and 1 metre. The **ranges** or **full-scale values** of the two measuring devices are 1 m and 25 mm respectively. The number of values which can be resolved in a measurement is calculated simply by dividing the range of the measurement by its resolution; the greater this number, the more bits will be needed to represent it within the microcomputer.

Another important factor when taking measurements is the rate at which it is changing. Each measurement of a time-varying quantity represents a **sample** of its waveform. Nyquist's criterion (Nyquist, 1928), well-known in electrical theory, states that to represent a signal accurately, samples must be taken at a rate at least twice as high as the highest frequency component in that signal. Provided that measurements are made quickly enough, accurately enough, and to an adequate resolution, there should be no overall loss in fidelity when employing digital techniques.

All these factors must be considered whenever a signal is to be converted from one form to another, and the separate components of an interface must also be considered in relation to one another. For example, the precision of the transducer is related to the number of bits which the ADC must handle, while the accuracy can be improved if the error is known by using a correction subroutine.

6.1.2 Parameters

The choice of ADC and DAC must be made in the light of the application being considered. Both types of converter are subject to the various shortcomings listed in the previous section, to which must be added the effect of **quantisation**, illustrated in Figure 6.3, which arises because the digital representation of the quantity can have one of only a limited number of values, even though the physical quantity being represented may be free to vary continuously.

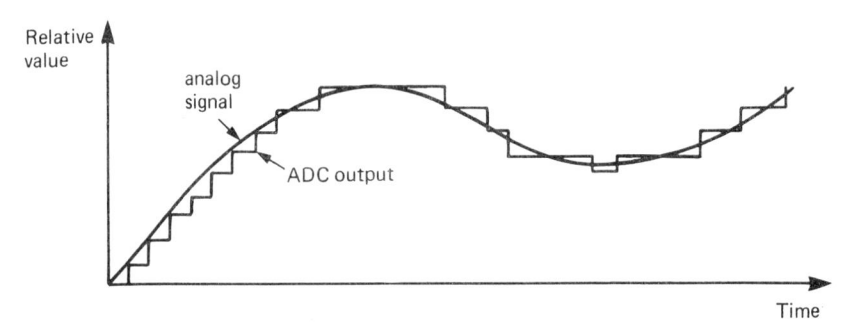

Figure 6.3 Effect of quantisation on analog signal

Conversion speed

In fact there are two ways in which we can regard an analog signal. The first is to consider its instantaneous value, measured in as short a period as possible, while the second is its value averaged over a period of time to remove irrelevant or undesired variations. It is possible to convert either instantaneous or average values, but the techniques involved are for the most part very different, as later sections of this chapter will show.

Resolution

The resolution of a DAC is related to the number of different digital values which can be distinguished by the converter, and is simply two to the power of the number of bits being used. Thus one of the first decisions to be made when selecting a DAC or an ADC is the number of bits to which the conversion should be carried out.

Each of the bits in the digital representation used by a converter corresponds to a voltage 'weighted' by a corresponding power of two. If, in an 8-bit converter, the analog voltage equivalent of the most significant bit (MSB) were 5.0 V, then the voltage equivalents of the other seven bits would be respectively one-half of this value, one-quarter, one-eighth, and so on down to the least significant bit (LSB), as shown in Table 6.1.

Table 6.1 DAC voltage levels

Bit	Voltage
(MSB) 7	5.0
6	2.5
5	1.25
4	0.625
3	0.312 5
2	0.156 25
1	0.078 125
(LSB) 0	0.039 062 5
Total	9.960 937 5

The total is included because it is the largest voltage which can be represented in this way, corresponding to the binary number 11111111. This is the range, or **full scale output**, of a unipolar converter.

Devices with conversion accuracies of between 6 and 18 bits are available. Table 6.2 shows the corresponding numbers of different possible digital values and the resulting resolutions as percentages of the range.

Table 6.2 Converter resolution

Bits	Levels	Resolution
6	64	1.5625%
8	256	0.3906%
10	1 024	0.0977%
12	4 096	0.0244%
14	16 384	0.0061%
16	65 536	0.0015%
18	262 144	0.0004%

Needless to say, converters with higher resolution cost more! Note that the resolution, being the smallest change that can occur in a digital system, corresponds to changing the least significant bit (LSB) of the digital number.

Input codes

It has been assumed so far, without any very good reason, that the converters are being used to handle voltages which are always of the same polarity, and that the corresponding numbers are unsigned binary. In fact there is no reason why voltages of either polarity cannot be used, with sign-magnitude or (more commonly) two's complement number representation.

Accuracy

The accuracy of a converter is usually taken to mean the largest deviation of any voltage level from its nominal value, and it is usually expressed as a percentage of the full scale output. The accuracy can be affected in a number of different ways, as Figure 6.4 shows.

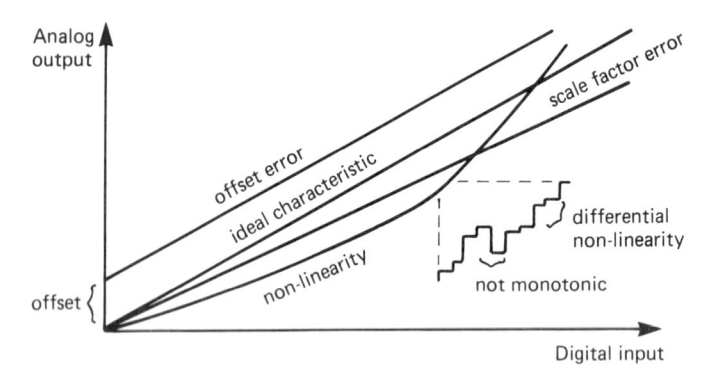

Figure 6.4 Errors associated with DACs

Zero offset
The transfer characteristic of the converter may not pass through the origin, that is, the whole graph may be displaced so that applying the number zero to the input of the converter does not produce zero volts at the output. The voltage produced by a DAC when zero is applied to its input, or the voltage that must be connected to the input of an ADC for it to produce an output of zero is called its **zero offset** voltage.

Scale factor errors
The converter will be designed to have a particular relationship between the digital numbers at one side of the converter and voltages at the other. This means that the gradient of the graph relating one quantity to the other should have a defined gradient or **scale factor**. In practice small deviations from the desired scale factor can occur, which will affect the accuracy of measurements.

Relative accuracy
This is determined by the **linearity** of the DAC, and is a measure of the maximum deviation from a straight line passing through the endpoints of the converter's transfer function. It is measured after adjusting for zero offset and scale factor errors, and again, is normally expressed as a percentage of the full scale range. Although most converters are designed to be linear, there are some applications, such as the recording and playback of digitally encoded audio signals, which require nonlinear devices. Converters are also available for use in the telecommunications industry which have nonlinear conversion laws that conform to the international standards for these applications.

6.1.2

Monotonicity

Another essential property of a DAC is that it should be **monotonic**. As the digital input is incremented from its minimum to its maximum value, each successive analog output level should be greater than the previous one. This is particularly important if the DAC is part of a feedback control system. Later in this chapter it will be shown that many of the analog to digital conversion techniques in common use rely upon DACs being connected in feedback systems. If a DAC is nonmonotonic, then an increase in the input number will cause a fall instead of an increase in the output signal at certain points in its characteristic; the negative feedback will become positive.

Differential nonlinearity

This is the difference between the measured change in the output signal and that which would ideally result from changing the input number from one value to the next. The differential nonlinearity of any converter should be no greater than an amount equal to the analog equivalent of the LSB to ensure that its characteristic is monotonic.

Another choice facing the designer is that between **unipolar converters**, which handle voltages of a single polarity and unsigned numbers, and **bipolar converters**, which can handle voltages of either polarity and which use signed numbers. Converters are also available which handle nonbinary codes such as binary coded decimal. BCD can be useful in simple applications because it avoids the necessity for binary to decimal conversion before displaying the result.

6.1.3 Signal conditioning

The electrical signal produced by a transducer can take various forms, and although some transducers produce a voltage suitable for direct connection to a digital-to-analog converter, most do not. Often, as in the case of a microphone, the signal needs to be amplified before it can be converted to digital form. In other cases the signal from the transducer is a current rather than a voltage, and a special type of amplifier must be used to convert it into a voltage for analog-to-digital conversion. Many transducers produce a variable resistance output, while others produce a variable capacitance or inductance.

For this reason it usually proves necessary to 'condition' the signal first. 'Signal conditioning' is the process of converting the electrical output of a transducer to a form, usually a voltage, which can be handled by the analog-to-digital converter.

The action of the signal conditioning section can be described mathematically by an expression which shows how the signal at the output of the section depends on the one at its input. The mathematical operation needed to do this can be performed by a circuit that acts as a simple analog computer, computing the output signal in terms of the input. Although analog computers are themselves obsolete, their basic 'building brick', the operational amplifier, is still widely used in analog circuit design. Analog interfaces use operational amplifiers in many ways, and we shall take a closer look at these useful components before going on to describe the principles of analog interfacing in greater detail.

6.2 The operational amplifier

This is an amplifier in which the ratio between a change in voltage at the output terminal and the corresponding change in input voltage which caused it, the **voltage gain**, is extremely high. In fact an operational amplifier has two input terminals and it is the difference between the voltages on these two terminals (the **differential input voltage**) which controls the output voltage. Ideally an operational amplifier would have infinite gain so that if the voltage on the output terminal were finite, the differential in input voltage would be zero. The voltage gains of practical operational amplifiers are finite, but nevertheless they are very large (typically in the order of 100 000) so that the assumption of zero differential input voltage is reasonably accurate.

Further assumptions are made about the ideal operational amplifier which can only be approached in practice. The first of these assumptions is that the differential input voltage is zero when the output voltage is zero. Unfortunately this is not the case in real operational amplifiers and there is always a small **voltage offset** present at the input if the output voltage is zero. The size of this voltage offset depends upon the type of operational amplifier; in most cheap types it is in the order of a few millivolts, but offsets as low as a microvolt are possible in special designs.

Another assumption made about ideal operational amplifiers is that the currents flowing in their input terminals are always zero. Practical operational amplifiers consist of integrated circuits which use transistors, and if bipolar transistors are used a small current (usually only a fraction of a microamp in modern designs) will inevitably be found to flow in each of the input leads. This is called **bias current**. Even operational amplifiers based upon field effect transistors draw some input current, although it will normally be extremely small.

6.2.1 Voltage amplifiers

Voltage amplifiers are often used for conditioning the small signals produced by transducers. The design of voltage amplifiers using operational amplifiers is quite straightforward provided that we assume that they behave like their ideal counterparts. The technique is based upon the idea of negative feedback; a small fraction of the output signal is compared with the input signal and the operational amplifier acts in such a way as to keep the two the same.

There are two basically different ways of doing this with an operational amplifier. The first of these is called **series feedback**, in which a voltage derived from the output signal is compared with the input voltage. In the second type, called **shunt feedback**, a current derived from the output signal is compared with the input current.

6.2.2 Series negative feedback

Here the analysis of the circuit's operation is based upon comparing voltages. The voltage connected to the inverting input of the operational amplifier is

$$V_- = \frac{R_1 + R_2}{R_1} \cdot V_{out} \qquad (6.1)$$

while that connected to the other input is the input voltage. Since we are assuming that the gain is infinite, the difference in input voltages will be negligible provided that the output voltage is finite. This means that

$$V_+ = V_- \qquad (6.2)$$

$$V_{in} = \frac{R_2}{R_1 + R_2} \cdot V_{out} \qquad (6.3)$$

which after rearranging gives us a formula for the **voltage gain** of the series feedback amplifier shown in Figure 6.5.

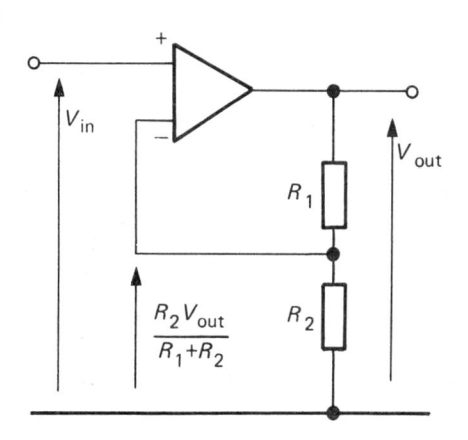

Figure 6.5 Series feedback

$$\frac{V_{out}}{V_{in}} = \frac{R_2 + R_1}{R_2} \qquad (6.4)$$

Equation 6.4 shows us that we can design an amplifier with a closely defined voltage gain provided that the operational amplifier itself has an extremely high voltage gain before the resistors are added. Series feedback, as used in this amplifier, allows us to design amplifiers with extremely high input impedances.

6.2.3 The voltage follower

A special case of the series feedback amplifier arises when the output is connected directly to the inverting input of the amplifier without the resistors R_1 and R_2 being used to form a potentiometer. In effect R_1 has become

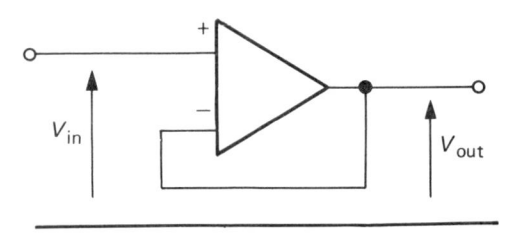

Figure 6.6 Voltage follower

zero and R_2 has become infinite, so that equation 6.4 becomes

$$V_{out} = V_{in} \tag{6.5}$$

Now the output voltage is the same as the input voltage, and it might seem that the circuit does not carry out any useful function. There is, however, an important difference between the input and output connections. The input resistance of the amplifier is extremely high, while the output resistance is quite low. This allows connections to be made to circuits which have very high resistance without unduly affecting their operation. This circuit is known as a **voltage follower** and is illustrated in Figure 6.6.

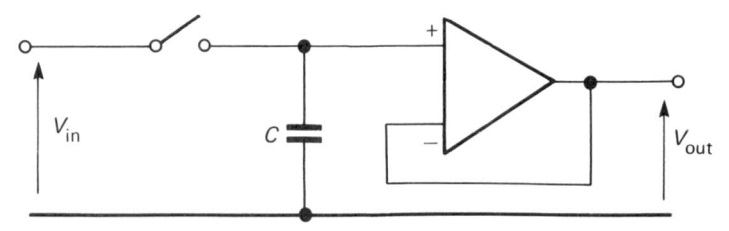

Figure 6.7 Sample and hold circuit

6.2.4 The sample and hold circuit

The very high input resistance of the operational amplifier can be exploited to make a circuit which can take a sample of an analog waveform in a very short time and present this sample as a steady voltage level on its output terminal. Such an arrangement is known as a **sample and hold circuit** or sample and hold amplifier, and is shown in Figure 6.7. All that is required is a capacitor with a low internal current leakage (this can be found from the manufacturer's data), an electronic switch such as an MOS transistor, and an MOS operational amplifier. When the MOS transistor conducts, it connects the capacitor to the waveform so that it can charge up to the signal voltage at that instant. The time for which the switch is closed is called the **acquisition time** of the circuit. This is limited by the speed of the switch and the time taken for the voltage on the capacitor to reach the same voltage as the input waveform, which is affected in turn by the resistance of the transistor switch when it is conducting.

When the MOS switch is turned off again by means of its gate voltage, the capacitor is disconnected from the input waveform but remains connected to the input of the voltage follower. The reason for choosing an MOS amplifier is that the input current of an MOS transistor is extremely low. The current flowing in the input of the amplifier will tend to discharge the capacitor, a phenomenon known as **droop** because of the way in which the waveform at the output 'sags' with time.

6.2.5 Shunt feedback

The second kind of feedback, shunt feedback, operates by comparing two currents rather than the two voltages which were assumed to be equal in equation 6.5.

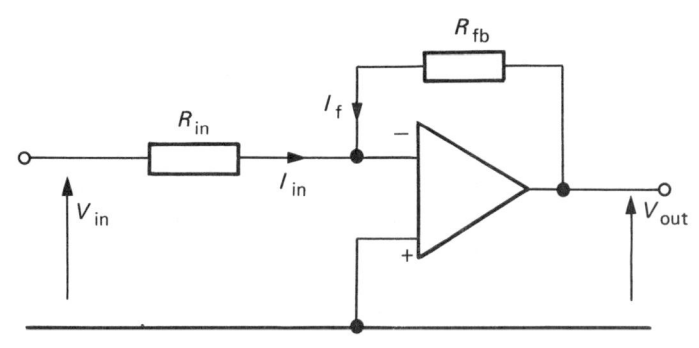

Figure 6.8 Shunt feedback amplifier

The starting point for the analysis of this type of amplifier again centres on the two inputs to the operational amplifier itself.

If we make the usual assumption that the voltage gain is extremely high, then the difference in voltage between the input terminals will again be extremely low and we can assume that the voltages on these inputs are approximately the same. The noninverting input is connected to the earth line which ensures that it is at zero volts, and so the inverting input will also be at zero potential. This point in the circuit is called the **virtual earth** because it is maintained at earth voltage even though it is not directly connected to the earth lead.

There are three connections to the virtual earth in the diagram of Figure 6.8. Kirchoff's current law states simply that the sum of the currents flowing into any point in a circuit, such as the virtual earth in this case, must always be zero. The connection to the operational amplifier carries only the bias current of the amplifier and this is almost always small enough to neglect. The currents in the other two resistors are easily calculated using Ohm's law, bearing in mind that the voltage at the virtual earth is zero

$$I_{in} + I_{fb} = 0 \tag{6.6a}$$

$$\frac{V_{in}}{R_{in}} + \frac{V_{out}}{R_{fb}} = 0 \tag{6.6b}$$

$$\frac{V_{out}}{V_{in}} = -\frac{R_{fb}}{R_{in}} \qquad (6.6c)$$

Note that this type of negative feedback produces an **inverting amplifier**; a positive-going change in input voltage will produce a negative-going change at the output.

6.2.6 Current-to-voltage conversion

Some kinds of transducer produce an output signal in the form of a current rather than a voltage. A current can be converted to a voltage quite easily by means of a resistor, possibly wired in series with the transducer. However a more complicated circuit is usually needed because the transducer may have a very nonlinear current-to-voltage relationship, or it may need to be operated with a fixed voltage across it.

The analysis of the shunt feedback amplifier provides us with a simple method of converting a current to a voltage without varying the voltage across the transducer. Equation 6.7a shows that the current in the feedback resistor is the same as the current in the input. If the input resistor were removed, the current flowing from the transducer into the virtual earth would still cause a voltage to appear at the output terminal of the operational amplifier. Because this output voltage is related to the input current by a constant resistance, this circuit, shown in Figure 6.9, is often called a **transresistance amplifier**.

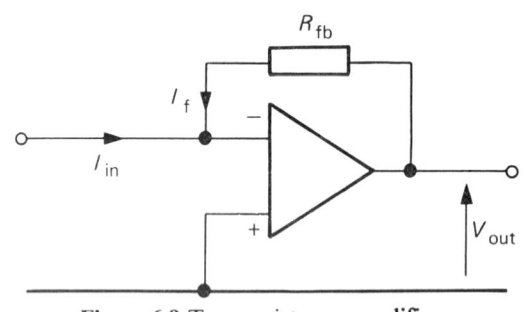

Figure 6.9 Transresistance amplifier

$$I_{in} + \frac{V_{out}}{R_{fb}} = 0 \qquad (6.7a)$$

$$V_{out} = -I_{in} \cdot R_{fb} \qquad (6.7b)$$

6.2.7 Differential amplifiers

An amplifier that converts a differential input voltage to an output voltage which is referenced to earth can be built quite easily using an operational amplifier and four resistors as shown in Figure 6.10. Two of the resistors provide shunt negative feedback around the operational amplifier, while the other two are wired as a potentiometer.

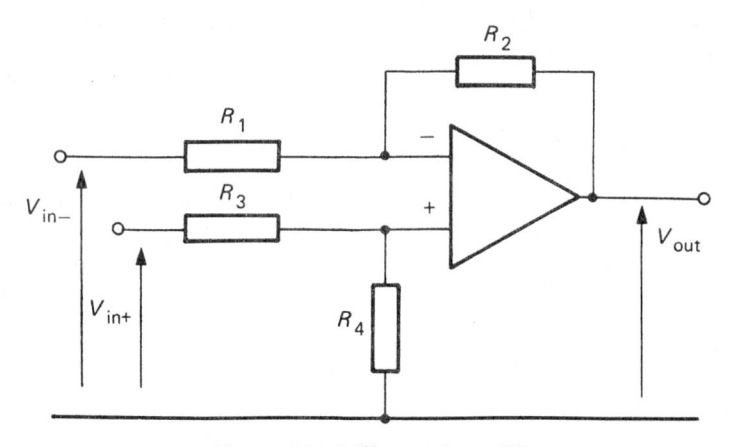

Figure 6.10 Differential amplifier

By considering the two resistors R_3 and R_4 as a potentiometer, we see that the voltage on the noninverting input is

$$V_+ = \frac{R_4}{R_3 + R_4} V_{in+} \tag{6.8}$$

while that on the inverting input must be almost exactly the same if the gain of the operational amplifier is very high. This voltage can also be expressed in terms of the other input voltage and the output voltage

$$V_- = V_{in-} + \frac{R_1}{R_1 + R_2}(V_{out} - V_{in-}) \tag{6.9a}$$

Rearranging gives

$$V_- = \frac{R_2}{R_1 + R_2} V_{in-} + \frac{R_1}{R_1 + R_2} \cdot V_{out} \tag{6.9b}$$

and equating the voltages on the two inputs of the operational amplifier results in the expression

$$\frac{R_4}{R_3 + R_4} \cdot V_{in+} = \frac{R_2}{R_1 + R_2} V_{in-} + \frac{R_1}{R_1 + R_2} \cdot V_{out} \tag{6.10}$$

If the resistor values are chosen so that $R_1 = R_3$ and $R_2 = R_4$, this equation can be rearranged and simplified to give an expression for the output voltage in terms of the differential input voltage.

$$V_{out} = \frac{R_2}{R_1}(V_{in+} - V_{in-}) \tag{6.11}$$

Note that this output voltage depends only upon the difference between the input voltages and not upon the voltages themselves. Any voltage component, such as electrical noise picked up by connecting wires, which is common to both of the inputs will be ignored. This ability to reject undesired signals is often exploited when interfacing analog inputs or when communicating between a pair of microcomputers using a pair of wires.

6.2.8 The integrator

Another useful trick which can be performed using the operational amplifier is to use the relationship between the charge in a capacitor and the voltage across it to produce a circuit (Figure 6.11) in which the output voltage is related to the input voltage by a mathematical expression involving integration. Among other things, this forms a mathematical basis for averaging, and it is used in 'averaging analog-to-digital converters'.

The voltage across the terminals of a capacitor at any time is proportional to the charge which it contains, and this charge represents the integral with respect to time of the current which has been flowing in it. This current can in turn be related to an input voltage by means of a resistance.

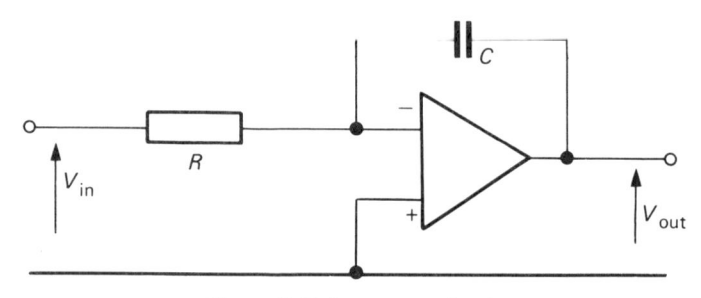

Figure 6.11 Integrator circuit

The current flowing in the resistor at any time is

$$I_{in} = \frac{V_{in}}{R} \tag{6.12}$$

and if the current flowing in the input terminals of the operational amplifier itself is ignored, all this current must flow into the capacitor, by applying Kirchoff's current law again. The voltage on the capacitor is

$$V_{out} = -\int \frac{V_{in} dt}{CR} \tag{6.13}$$

Substituting for the input current in this equation gives

$$V_{out} = -\int \frac{I_{in} dt}{C} \tag{6.14}$$

6.2.9 Summing amplifiers

If more than one input is connected, each through a separate resistor, the total current flowing into the virtual earth will be the sum of the input currents and the feedback current. Each current can be calculated quite easily by dividing the corresponding voltage by the value of the resistor used, because the voltage at the virtual earth is zero. For this reason the virtual earth is sometimes called a summing junction. This circuit can be analysed very simply by using Kirchoff's current law again at the summing junction to give

$$\frac{V_1}{R_1} + \frac{V_2}{R_2} + \cdots + \frac{V_n}{R_n} + \frac{V_{fb}}{R_{fb}} = 0 \qquad (6.15)$$

which can be rearranged to give an expression for the output voltage

$$V_{out} = -R_{fb} \cdot \left[\frac{V_1}{R_1} + \frac{V_2}{R_2} + \cdots + \frac{V_n}{R_n} \right] \qquad (6.16)$$

When all the resistors have the same value the output voltage has the same magnitude as (but opposite polarity to) the sum of the input voltages, which is the origin of the name 'summing amplifier'. This circuit, shown in Figure 6.12, provides the basis for a simple digital-to-analog conversion technique.

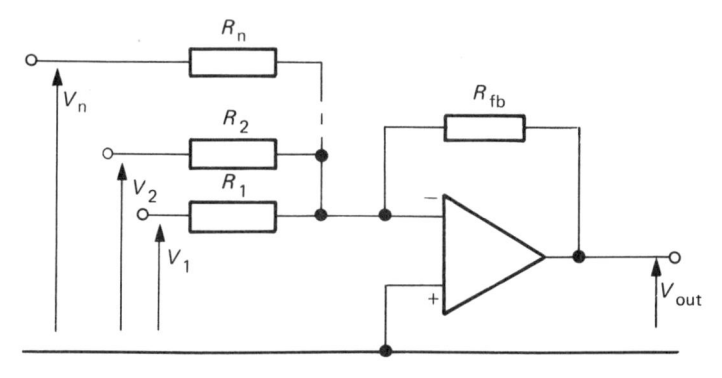

Figure 6.12 Summing amplifier

6.3 Digital-to-analog conversion

In this section various techniques for generating output signals will be introduced. Digital-to-analog converters can be divided into two main categories: 'instantaneous' types which produce samples of an output waveform and 'averaging' types in which some external filtering is used to smooth the output signal.

6.3.1 'Instantaneous' digital-to-analog converters

Nearly all commercially available DACs consist of four elements; a voltage or current reference source, an array of semiconductor switches, a resistor network and an operational amplifier. The simplest in concept is the **binary weighted resistor DAC** shown in Figure 6.13, which uses a voltage reference, and resistors with values inversely proportional to the binary weighted values of the corresponding input bits. This is just a special case of the summing amplifier circuit shown in Figure 6.12, but with switches added which can be used to connect or disconnect the reference voltage to each of the inputs. This is done according to the state of the corresponding bit of the digital input; the voltage source is connected if the controlling bit is 1 and the input is connected to ground if it is 0. The analog output voltage is proportional to the total current taken by the resistors.

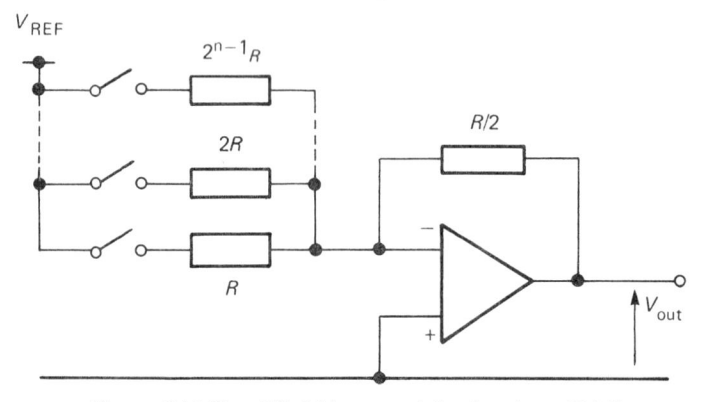

Figure 6.13 Simplified binary-weighted resistor DAC

If each of the input switches are connected to the constant voltage source, corresponding to a binary input of all 1's, the output voltage is given as follows

$$V_{out} = -\frac{R}{2} \cdot \left[\frac{V_{REF}}{R} + \frac{V_{REF}}{2R} + \frac{V_{REF}}{4R} + \cdots \right] \quad (6.17a)$$

$$V_{out} = -V_{REF} \cdot \left[\frac{1}{2} + \frac{1}{4} + \frac{1}{8} + \cdots \right] \quad (6.17b)$$

If the binary input consists of a pattern of bits ABC..., the output voltage will be

$$V_{out} = -V_{REF} \cdot \left[\frac{A}{2} + \frac{B}{4} + \frac{C}{8} + \cdots \right] \quad (6.17c)$$

In other words, the current in each of the input resistors is summed to produce an analog output voltage, V_{out}, equal to the reference voltage, V_{REF}, multiplied by an amount proportional to the binary number input to the converter.

6.3.1

The main disadvantage of the weighted resistor DAC is that the accuracy of the conversion process is not only dependent on the stability of the voltage reference but also upon the accuracy of the resistor values. If binary weighting is used in an 8-bit converter, the resistor corresponding to the least significant bit (LSB) must have a value 128 times greater than that for the most significant bit (MSB). If the LSB is to have a meaningful effect on the analog output signal, the MSB resistor must be very accurate and stable. Let us consider the case when the input word changes from 01111111 to 10000000, which should result in the output voltage changing by an amount equivalent to the LSB. The analog equivalent of the MSB must differ from the summed currents of all the other bits by the current equivalent of the LSB. The ratio of the MSB current to the LSB current, and hence the ratio between the highest and lowest resistor values, rises rapidly with the number of bits. The effective values of the resistors include the resistances of the switches in series with them, and it becomes increasingly difficult to produce resistor 'weights' in the correct ratio as the number of bits increases. The problem is difficult enough even for an 8-bit converter where the ratio is 128:1, but for larger numbers of bits it becomes virtually impossible and more practical alternatives must be sought.

R-2R ladder networks

The problem of the widely varying resistor values of the weighted resistor DAC can be overcome by using an **R-2R ladder network**, as shown in Figure 6.14. The current associated with each individual bit has the appropriate binary weighting applied to it at the output end of the ladder. This can be easily proved using the Superposition and Thévenin's theorems, which can be found in most textbooks on electrical circuit analysis. The currents generated are not proportional to the resistance values but depend only on the $R:2R$ ratio. With just two resistance values, the problems of resistor matching are considerably eased, and since the entire ladder is normally fabricated on a single silicon chip all the resistance values track with temperature variations.

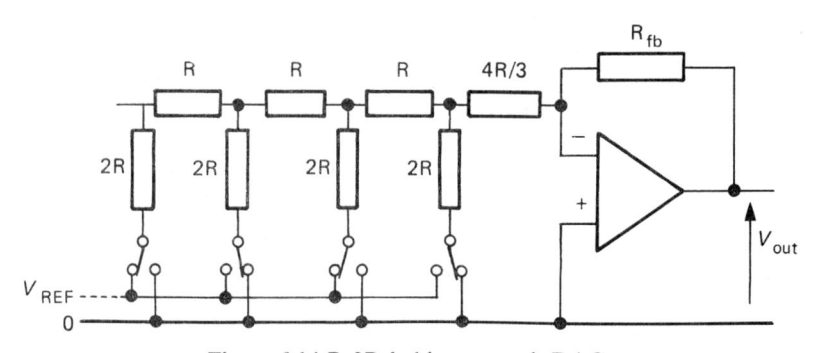

Figure 6.14 R-2R ladder network DAC

Resistive DACs with up to twelve bits of input can be manufactured in the form of a single silicon chip and are therefore relatively inexpensive. With larger numbers of bits, the prices of DACs tend to increase with resolution, and they are still usually fabricated as **hybrid integrated circuits**. These are built using more than one chip mounted within a single package, and tend to be more expensive, because of the extra assembly work required in their manufacture. However, using more than one chip allows chips of different types to be used, which is important when some parts of the circuit are used to handle digital signals and other parts to handle analog signals. The converters so far described tend to be slow in operation, due to the **settling time** of the output operational amplifier and the switching speed of the transistor switches. When voltages are switched, the signals take time to respond because of the need to charge and discharge the capacitances in the circuit, while if currents are switched the corresponding changes in the circuit's voltages are much smaller and the circuit can be operated at a higher speed. Hence an increase in speed can be obtained by using a resistive DAC with a current output, in which the output amplifier is omitted and switches which operate in the current switching mode are employed. In this case, a continuously generated current for each input bit is connected either to a common line which forms the output of the DAC, or to earth, depending on whether the input bit is 1 or 0.

All DACs use either a **current reference source** or **voltage reference source**. This is because they convert from a pure number to a voltage or current, and this conversion must be made to a reference value of some sort. It is now usual for voltage references to be included on the same chip as the other DAC elements. The voltage produced by the reference is defined by some property of the silicon. More than one physical property is available to be exploited, and a considerable amount of effort has been expended by manufacturers to develop voltage references which are as independent of temperature and manufacturing variations as possible. Earlier converters employed external voltage sources which used Zener diodes, and external voltage references are still used where higher accuracy is required.

In some cases the voltage reference is deliberately omitted from the package in order that the converter can operate as a **Multiplying DAC**. This is particularly useful when the reference voltage is to be varied, possibly even changing polarity. If a DC or AC signal is connected to the reference input, the output voltage or current from the converter will be the product of the input signal and the number on the digital input. This number is treated as a binary fraction in the same way as when a fixed reference is used. If the digital input is always regarded as a positive number, then **2 quadrant multiplication** is achieved. If the digital input represents a bipolar quantity, then **4 quadrant multiplication** can be obtained. An obvious application of a multiplying DAC is in a digitally controlled attenuator; an AC signal connected to the reference input will be attenuated by an amount depending on the polarity and magnitude of the number input to the DAC. Many single-chip converters also contain a register which is used to latch the digital input word before its subsequent conversion. These converters are usually arranged to be addressable and interfaced directly to the microcomputer buses as memory-mapped I/O devices.

6.3.2 Generating waveforms

Microcomputers can be used to generate voltage or current waveforms in which the instantaneous values change as a function of time. The changes in level can be generated directly by the program as it runs, or the task of waveform generation can be 'delegated' to a special output device. Variable output waveforms can be used both to control the speed of operation of external equipment such as motors, and as an alternative technique for digital to analog conversion.

Frequency division

The simplest way to generate a variable waveform is to divide the constant frequency 'clock' signal by a variable amount. If the constant frequency is f and the divisor is N then the frequencies which can be produced in this way are of the form f/N, where N is an integer. This technique can be implemented very simply by counting N pulses and then generating a pulse on the output pin, either by software or hardware. In each case the process is carried out most easily by counting down to zero and reloading the counter with the number N each time zero is reached. This method produces time intervals which are proportional to N and hence frequencies which are inversely proportional to this number. Its major disadvantage is that it cannot generate frequencies at regular intervals, but it is useful where a time interval in the output waveform is being controlled or where the exact interval between frequencies is not important. One example is in electronic music instruments where the interval between successive notes on a keyboard corresponds to a frequency ratio given by the twelfth root of two. This interval cannot be expressed exactly as the ratio between two integers, so that an approximation must suffice however the notes are generated.

If the input to a DAC is changed whenever a time interval is completed, a more general waveform synthesiser can be constructed in which the voltages are controlled by the numbers on the DAC input and the frequency by the rate at which the input to the DAC is changed.

Modulo counting

A second method of producing frequencies is rather less straightforward but it allows frequencies to be generated which are equally spaced. This involves repeatedly adding a number, say N, to an n-bit register. After 2^n additions the total value, if such a number could be represented in n bits, would be $N \times 2^n$. This means that the most significant bit of the register must have completed N cycles during these 2^n clock cycles, and a carry output would have produced a logic '1' each time a cycle was completed, i.e., on N out of 2^n additions. The proportion of times that a logic '1' is generated on the carry output will be $N/2^n$, and if additions are performed at a constant frequency f, the mean frequency of the '1's on the output will be $Nf/2^n$.

Note that after 2^n additions, the number in the register will have returned to its initial value.

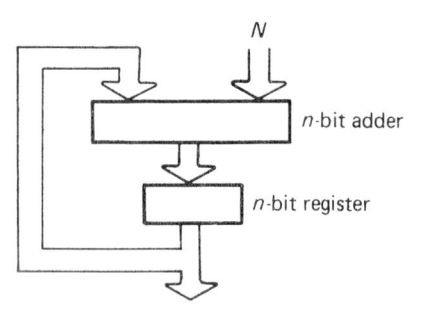

Figure 6.15 Modulo-2^n adder

Perhaps this can be made a bit clearer by considering a rather strange clock which normally moves its seconds hand by one division every second, so that after a minute it returns to the top of the dial again. If it moved the seconds hand two divisions each second, it would reach the top again after only thirty seconds, and after a minute it would have completed two complete circuits of the dial. If it increased by N divisions on the dial every second, it would complete N sweeps per minute. The same would be true of a digital clock; the number of times that the seconds display passed zero each minute would be proportional to the number being added each time.

Next consider an 8-bit register to which a number N is added each millisecond. After 256 (2^8) milliseconds the count would return to zero again after having counted through N cycles. If the level of the most significant digit of the register were observed, it would be seen to have completed N cycles in that time. The individual cycles might vary slightly in duration because the system is restricted to updating its output every millisecond, but the average frequency would be $N/256$ kHz.

Note that in this case the output frequency is proportional to the input number, and not its reciprocal as in the previous case. This makes calculations involving changing frequency much easier. Suppose, for example, that a stepper motor was being driven. Stepper motors have inertia, which means that there is a limit to the maximum angular acceleration to which they can be exposed, or they will 'stall' and not rotate at all. The speed of rotation at any time is proportional to N, and the acceleration is determined by the rate of change of N with time, which can be calculated quite easily. However, the method has an inherent limitation that it cannot in general produce waveforms in which all the cycles have exactly the same waveform, but in many applications this is not a severe problem. It offers the advantage that the frequencies can be generated to any desired resolution. If, for example, a sixteen bit register were used instead of an eight bit register in the example above, the output frequency would be adjustable in steps of $1/65\,536$ kHz instead of $1/256$ kHz.

6.3.2

Techniques for faster operation

Sometimes frequencies higher than the highest frequency in the microcomputer itself must be generated by the system. This can be achieved by using waveform sources whose frequency can be 'tuned' by means of an externally applied voltage; these are called **voltage-controlled oscillators**. The frequency of a voltage-controlled oscillator (VCO) can be set directly by means of a voltage generated by a DAC. However this is rarely done in practice because high frequency VCO's do not usually have stable enough output frequencies, and some form of feedback is needed to increase their frequency stability. This could be done by measuring the frequency using another interface to the microcomputer and calculating a correction for the voltage on the tuning input. However, a more common form of feedback to control the frequency is provided by the **phase-lock loop (PLL)** in which the output signal from the voltage-controlled oscillator is connected to an integrated circuit which divides its frequency by an integer N; in effect this is simply a counter. The resulting signal is then compared with a reference signal of an accurately known frequency using a circuit which compares their phases. The output from this **phase comparator** is a voltage signal which is filtered to remove the effects of 'noise' and then used to control the VCO's frequency. The effect of this circuit is to vary the frequency of the VCO until the output from the frequency divider has exactly the same frequency as the reference signal, which is usually based upon a quartz crystal oscillator, and is hence very stable. When this occurs, the phase-lock loop is said to be 'locked', and the VCO output frequency is N times the reference frequency; this arrangement is illustrated in Figure 6.16.

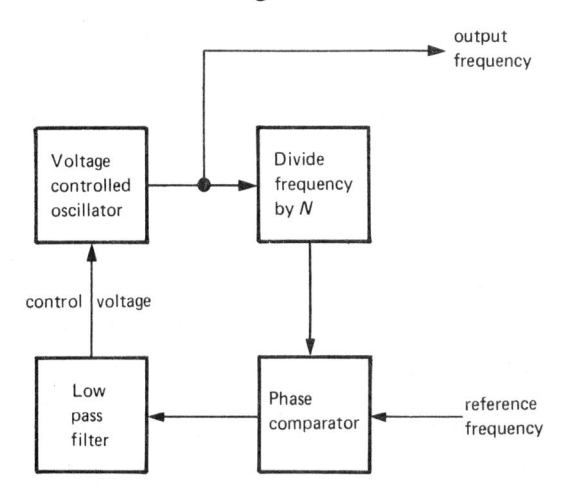

Figure 6.16 PLL frequency synthesiser

The frequency can be controlled by changing the value of N used in the frequency divider, so that a range of equally spaced frequencies can be produced. The separation between these frequencies is equal to the reference frequency used in the phase comparator. This is the normal mode of operation of a phase-lock loop **frequency synthesiser** but increased flexibility is possible by modifying the value of N used between each cycle of the output of the frequency divider, in which case the effective division ratio is the average value of N and the output frequency is N times the reference frequency. For example, if alternate division ratios of N and $(N+1)$ are used, the average division ratio will be $(N+0.5)$, and the output frequency will be $(N+0.5)$ times the reference frequency.

The value of N can be adjusted in software using a modulo-2^n adder. Each time a carry output is produced, the ratio in the frequency divider should be set to $(N+1)$, while the rest of the time it should be N. The average division ratio, d, will then be given by equation 6.18.

$$d = N + M \cdot 2^{-n} \tag{6.18}$$

As M is increased from 00...0 to 11...1, the effective division ratio increases from N to almost $N+1$, and by choosing a large enough value of n, the frequency steps can be made as small as desired. The logic for controlling the value of the division ratio can be built quite simply using a few logic circuits, but if processor time is available the software required, which is best implemented as an interrupt service routine, such as that shown in Figure 6.17, is almost trivial.

NEWM	LDA	REGSTR	Get modulo-256 counter
	ADDA	M	add in increment
	STA	REGSTR	and store result
	LDA	N	Get division ratio for counter
	ADC	#0	and add the carry bit if any
	STA	RATIO	Result is new division ratio
	RTI		End of interrupt

Figure 6.17

When the counter carrying out the frequency division has counted M cycles of the VCO output signal, it will reload and cause an interrupt to the microcomputer. The microcomputer must then run this interrupt service routine before the counter causes the next interrupt.

Variable duty cycle outputs
In addition to being able to change the frequency of an output waveform, it is possible to change its duty cycle. Suppose that each time the addition of the last section generated a 'carry output', a pulse of one clock period duration were generated. The number of such pulses per unit time, and hence the proportion of time that the output spent at logic '1', would be

proportional to the number N being added. This proportion is called the **duty cycle** of the waveform and for an n-bit register it is $N/2^n$.

6.3.3 Averaging digital-to-analog conversion

If a waveform consisting of just two possible levels is averaged over time, for example by filtering it to remove the fluctuations between levels, the resulting average will be proportional to the duty cycle of the original waveform. Because the duty cycle can be arranged to be proportional to the number N in our program, this method provides us with a technique for digital-to-analog conversion.

This technique differs fundamentally from the 'instantaneous' DACs described in section 6.3. It cannot be used when rapidly changing output voltages are required because of the need to filter out the fluctuations in the waveform to leave an average controlled by the duty cycle. Nevertheless, it is far from useless, because it provides a technique for controlling actuators which are slow in electronic terms but which use large amounts of electrical power, such as motors and heaters.

'Instantaneous' DACs and voltage amplifiers use transistors to control the voltage from their outputs. In doing so, these transistors, whether they are incorporated into integrated circuits or exist as separate components, must dissipate an amount of electrical power which is comparable with that applied to the load. Although this can be acceptable when small output powers are involved, it is obviously not the case when the relatively high power needed by, for example, an electrical heater is involved. Averaging DACs avoid this problem by using the variable duty cycle waveform to control an electronic switch in series with the load. The proportion of time for which this switch is closed determines the amount of power which flows to the load and yet the power dissipated in the switch can still be quite low.

It is not usually necessary to use electronic filters to obtain an averaging effect. Electromechanical actuators such as motors have mechanical inertia, while the temperature of a heater cannot vary rapidly because of its thermal 'inertia': it takes considerable time to heat or cool. Another advantage of this switching type of DAC is that it can be used to control the power being supplied to the load directly. A DAC which produces a voltage proportional to the number on its input will cause a nonlinearity to appear in a system controlling an electrical heater because the heating effect is proportional to the square of the voltage.

When used with DC outputs, the averaging technique requires high power transistors to switch the current flowing to the load. Bipolar or MOS transistors can be used, the latter offering the advantage of faster operation. The method is also suitable for controlling the AC power supplied to a load by means of SCRs or triacs. Here the situation becomes more complicated because the nature of the AC waveform must be taken into account. The simplest approach is to switch the power on or off for complete cycles or half cycles, which has the advantage that the amount of electrical interference produced by switching a heavy current is minimised because the output can be switched when no current is flowing. This technique is called **burst firing** of the SCR or triac.

One disadvantage of burst firing is that if the averaging interval required for the signal is too long, the variations in the output may become intolerable. For example, if AC lighting is controlled by either passing or stopping current for complete mains cycles, an annoying flicker can result. An alternative is to produce a waveform with a constant frequency but variable duty cycle. Several methods are available for doing this, but essentially they all start a timer at the same time that the output pulse starts, and end the pulse when the time delay finishes. The SCRs and triacs used to control AC power remain switched on once they are triggered until the current passing through them falls once more to zero. Thus if a device of this type is triggered part-way through a cycle it will conduct for an average amount which depends upon the point in the cycle at which it was triggered. Unfortunately the expression which relates the triggering time and the average power is very complicated because the input voltage varies sinusoidally with time. Equation 6.19 shows the expression for the case where a resistive load is being driven; for a reactive (inductive or capacitative) load the expression is more complicated because of the phase difference between the voltage and current in such circuits.

$$P = \frac{V^2}{\pi R} \int_{\phi}^{\pi} \sin^2 \omega t \; d(\omega t) \tag{6.19}$$

However, it is possible to correct for this pronounced nonlinearity by means of software. The easiest way to do this is to use a look-up table containing the time delay to be used as a function of the output power required.

6.4 Analog-to-digital conversion

The most direct method of converting a signal from digital-to-analog form is to use a large number of voltage comparators, one for each possible quantised input level. This is the basis of the **parallel converter**, or **flash converter**, which earns its nickname from the fact that it is capable of making a conversion in a very short time, some types taking well under 100 ns. Each comparator compares the input voltage with a voltage set to the corresponding quantisation level; those with a test voltage less than the input voltage will output a logic '1' while the remainder will output a logic '0'. The outputs of all the comparators are taken to a logic circuit which determines where the boundary falls between the two groups and hence derives a digital number corresponding to the input voltage; in fact this is the 'serial number' of the highest numbered comparator which is outputting a logic '1'.

The problem with the parallel converter is that it is complicated, which leads to its being expensive and using a lot of power. It is required in some applications such as the conversion of television and radar signals, but for most other applications alternative techniques are fast enough.

6.4.1 Feedback methods

Most practical designs for 'instantaneous' ADCs make use of fast DACs and feedback circuits. In each case the feedback circuit uses a voltage comparator to compare the converter's output voltage with the input voltage and adjusts the digital input to the converter accordingly.

Ramp feedback

The simplest method of determining the input voltage is to try each possible value in turn, starting with the lowest. The microcomputer outputs each possible value in turn, and tests the comparator output level to find out if the DAC's output voltage is higher than the input voltage. It carries on until this voltage exceeds the input voltage, after which the process stops. The input voltage now corresponds to a level somewhere just below the number currently being applied to the input of the converter. The output voltage from the DAC consists of a ramp 'staircase' which continues until it reaches the input voltage. If hardware logic is used to generate the sequence of numbers presented to the DAC, it can be made using a counter with some logic to stop it when the comparator output voltage changes. The counter should preferably have some means for resetting it to zero before the start of each conversion sequence. A ramp feedback converter is illustrated in Figure 6.18.

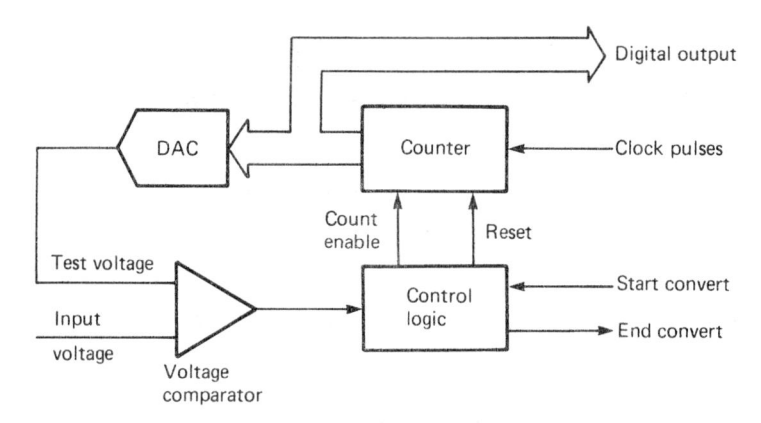

Figure 6.18 Ramp conversion method

The logic can also be programmed quite simply as an input subroutine. The programming examples used throughout this chapter will make use of 6809 assembly language, because they are all quite simple, and because assembly language is a convenient method for controlling the interface circuits used. For the purpose of these examples, assume that the DAC is an eight bit type controlled by an output register called OUTPUT and the comparator output is connected to the most significant bit of the register INPUT, so that a TST instruction can be used to test the state of this input line. If the comparator output is connected to another input line, it tested quite simply by means of the sequence

LDA	INPUT	Read the input register
BITA	#MASK	mask off a single bit
BNE	...	and test the Z flag.

The code necessary to program a ramp converter is thus quite simple, as shown in subroutine RAMP of Figure 6.19.

RAMP	CLR	OUTPUT	Set the output to zero
LOOP	TST	INPUT	Test the comparator output
	BPL	DONE	and exit the loop when done
	INC	OUTPUT	meanwhile try the next value
	JMP	LOOP	and continue...
DONE	LDA	OUTPUT	When finished, get the output
	RTS		value into acc-A and return

Figure 6.19

The major weakness of the ramp converter is that it is slow, because it has to try each value in turn until it finds the input value, and because it has to start from zero each time it carries out a conversion.

Tracking converter

If the input voltage is changing slowly compared with the sampling interval, it may be possible to save time by starting from the old value and looking for the new value by steadily increasing or decreasing the voltage until it is found. If hardware is used, the simple counter should be replaced with an up/down counter, the direction of counting being controlled by the output level from the comparator. The number in the up/down counter at any time corresponds to the current input voltage.

The method can also be programmed as a subroutine. Assuming that the interface is the same as before, the previous example can be developed to give a subroutine called TRACK as shown in Figure 6.20. This reduces the DAC output voltage if it is above the input voltage and increases it otherwise.

Successive approximation converter

The slowness of the converters described so far results from their need to try successive values when searching for the level of the input voltage. A more effective way is to carry out a **binary search** for the value desired, as Figure 6.21 shows. This technique involves first trying out a value in the middle of the range and finding out if the input voltage lies above or below this

6.4.1 ANALOG INTERFACING 261

TRACK	TST	INPUT	Test the comparator output
	BMI	LOOP	and carry on until it is less
	DEC	OUTPUT	than the input voltage...
	JMP	TRACK	
LOOP	TST	INPUT	Test the comparator output
	BPL	DONE	and exit the loop when done
	INC	OUTPUT	meanwhile try the next value
	JMP	LOOP	and continue...
DONE	LDA	OUTPUT	When finished, get the output
	RTS		value into acc-A and return

Figure 6.20

level. If it is in the lower half of the range, it proceeds to search in this range, while if it is in the upper half of the range it searches in the upper half. The process is repeated in the appropriate half of the range which is halved in turn, until the least significant bit of the code has been determined. Special hardware is available to carry out this function in the form of the **Successive Approximation Register (SAR)**. This generates the test values in turn, under the control of an input connected to the comparator output. The process is started by means of a 'start convert' input, and when it is finished the SAR indicates the fact by means of an 'end of convert' output. These two signals provide a handshake with the converter which allow it to be used with an ordinary parallel port.

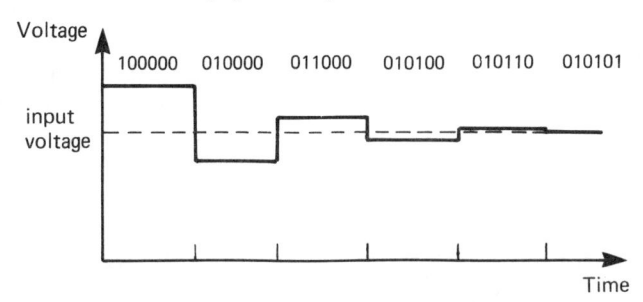

Figure 6.21 Successive approximation method

Again, a subroutine can be written to carry out the same function. First the output is set to a value in the middle of the range, that is, 1000...00, and the comparator output is tested to determine whether the input voltage is above or below this value. If it is above the test value, the bit is kept at logic '1', otherwise it is returned to logic '0'. In either case, the next less significant bit is set to '1' and the process repeated for that bit. The process ends after the state of the LSB has been determined, as shown in Figure 6.22.

```
SUCC    CLR     OUTPUT      Set the output to 0
        LDA     #$80        and start with the
        STA     ACTIVE      MSB active

LOOP    LDA     OUTPUT      Set the active bit to '1'
        ADDA    ACTIVE
        STA     OUTPUT

        TST     INPUT       ... and test the input
        BMI     HIGH

        SUBA    ACTIVE      If it is lower, set the
        STA     OUTPUT      active bit back to '0'

HIGH    LSR     ACTIVE      Move on to next bit until
        BNE     LOOP        all the bits have been done

        RTS
```

Figure 6.22

6.4.2 High speed data capture

One way to reduce this time overhead when hardware is used to generate the sequence of digital inputs to the DAC is to assume that the conversion will have been completed after a certain amount of time has elapsed. However, this requires the clock of the ADC logic to be synchronised with that of the microcomputer, and even if this is the case, the logic may take varying amounts of time to complete its sequence of operations, depending upon the input voltage.

Some time can be saved by arranging for the microcomputer to store each sample that has been read in memory after the ADC has started converting its next sample, so that the two systems operate at the same time without waiting for each other. If the time at which each sample is taken is determined by software, the timing resolution can be related to the basic cycle time of the microcomputer.

Interrupts
One problem with using the microcomputer to time the interval between samples, is that the microcomputer is not available for other tasks while in the waiting loop. If an interrupt is used, the timing will depend upon the

time at which the microcomputer responded to the interrupt, which will depend in turn upon the timing of the interrupt relative to the instruction currently being executed. This irregularity in timing can be removed if the timing of the sampling of the input waveform is controlled directly by the timer using some external logic.

Direct memory access
At sampling rates above about 100 kHz it is no longer possible to use direct program control to transfer data to memory. However, **Direct Memory Access (DMA)** can be used to transfer data via the data bus at a maximum rate limited by the capacity of the bus being used. This figure determines the maximum sampling rate of the system and is in the order of 1 MHz for most microcomputers.

FIFO buffer
Even faster sampling rates can be attained if the output of the ADC is buffered. Data from the ADC must 'queue' before it can be transferred to memory; it is put into the queue at the sampling rate and removed from the queue at the maximum rate which the microcomputer can handle. The key property of a queue is that the first item to enter it is also the first one to leave from the other end, and thus what is required is a **First In, First Out (FIFO)** system. The maximum length of the FIFO depends upon the number of data points to be sampled in a record. The maximum speed of a hardware FIFO arrangement is limited by the speed of the RAM which is used to implement it. In this application there is no need to read the FIFO until a complete record has been 'captured' in it, and speeds of up to 40 MHz should be possible with careful design.

Charge-coupled devices
Charge-Coupled Devices (CCDs) are electronic devices which can store sequences of analog charges. During each clock cycle, a sample of the input voltage is stored and each previous sample is moved to the next position, rather like an analog shift register. The samples leave the output of the CCD after a fixed number of clock cycles. CCDs are available which can be operated at tens of megahertz, and they allow a waveform to be sampled at high speed and stored directly in analog form. These samples can then be recovered at a lower speed by reducing the clock frequency, allowing slower — and cheaper — ADCs to be used for interfacing. One problem with CCDs is that, because they are analog devices, the output signal may be smaller than the input signal by an amount which depends upon the clock rate and which differs between different CCDs.

6.4.3 Averaging analog-to-digital converters

Averaging ADCs produce an output number which is proportional to the average input voltage over a defined interval, and because of this averaging process they are less sensitive to noise than their 'instantaneous' counterparts. The amount to which the noise is reduced depends upon the type of noise. If the noise is random, then it can be shown that it is reduced in the averaging process by an amount proportional to the square root of the averaging time.

A much more substantial reduction in the effects of noise is possible when the noise source is repetitive. It can be shown mathematically that the effect of averaging a repetitive signal over its period of repetition is to produce a zero result. This fact is invaluable in allowing the effects of interference from the electrical mains supply to be removed. The nominal frequency of the supply differs from country to country, but is normally either 50 Hz or 60 Hz. Thus if the averaging period is made equal to one cycle, or any other whole number of mains cycles, the effects of mains interference will be eliminated from the measurement.

In practice the exact frequency of the mains may vary from the nominal value and it is necessary to adjust the averaging interval to match the period of the mains waveform if the full reduction in noise sensitivity is to be obtained. This can be done automatically in some cases, but this can be difficult to do when integrated circuit converters are used.

Voltage controlled oscillators

It is possible to design voltage controlled oscillators which have a very linear relationship between their controlling voltage and their output frequency, and the use of an integrated circuit VCO with suitable components to set its frequency range provides a simple conversion technique. In principle, all that is necessary is to count the number of cycles in the output waveform during the averaging interval using either a software counting loop, or a hardware counter.

The limitation in accuracy using this method arises from the number of cycles that can be counted during the counting interval, the linearity of the VCO's voltage to frequency characteristic, and the effects of changes in temperature upon the characteristics of the converter. However, accuracies of up to about twelve bits are possible.

Single ramp averaging

Another way to carry out an averaging analog-to-digital conversion is to simply average the input voltage over the prescribed period using a suitable circuit. The process of averaging over a time interval T can be written mathematically as

$$\text{average} = \frac{1}{T} \int_0^T V_{in} \, dT \tag{6.20}$$

This can be performed using an integrator based upon an operational amplifier. The scaling factor can be set by a suitable choice of time constant in the integrator, but unfortunately the accuracy of the component values used in ordinary electronics is not usually better than about 5%, and components with accurate values are expensive. Even so, the tolerance on the component values can affect the accuracy of a conversion; for example an accuracy of 0.1% will limit the accuracy to about ten bits, not especially high by modern standards.

Dual ramp conversion

Although the exact values of the components used in an integrator may be suspect, their values do not usually change much from one moment to the next, and this allows another technique to be used which is inherently insensitive to exact component values. In the dual ramp conversion method, the input of the integrator is connected to the input signal in the same way as before, and after the output voltage has been set to zero it is allowed to integrate the input voltage for the required period. After this time the input voltage is disconnected and the polarity of the integrator's output voltage is sensed using a voltage comparator. Then a reference voltage with a polarity opposite to that of the input voltage being measured is connected in place of the input voltage and the integrator allowed to carry on integrating. Because the voltage connected to the integrator's input is now of the opposite polarity to the original input voltage, the integrator output voltage now starts to fall back towards zero volts. When it reaches zero volts, the comparator's output level will change again. The time taken for the integrator voltage to return to zero when the reference voltage is connected is measured and this is compared with the time for which the input voltage was connected. During the second part of the sequence, when the reference voltage is connected, the amount of charge removed from the capacitor in the integrator circuit was the same as that which flowed from the input during the first part of the sequence. This means that the total charge that has flowed into the capacitor over the whole sequence is zero.

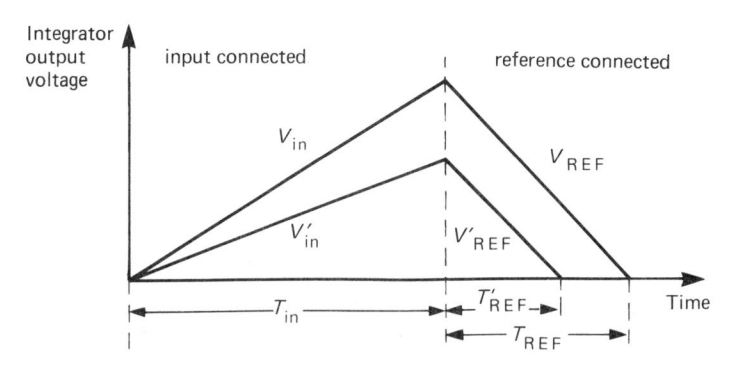

Figure 6.23 Dual ramp conversion method

The current flowing into the capacitor at any instant is proportional to the voltage at the input of the integrator. If the input voltage V_{in} was connected for a time T_{in} and the reference voltage V_{REF} for a time T_{REF} (see

Figure 6.23), then the input and reference voltages are related by the simple equation:

$$V_{REF} \cdot T_{REF} + V_{in} \cdot T_{in} = 0 \tag{6.21}$$

Now the time intervals are measured in terms of cycles of a clock waveform, so that T_{in} represents N_{in} cycles and T_{REF} represents N_{REF} cycles of the clock. If the clock period is T_c,

$$N_{in} = \frac{T_{in}}{T_c} \tag{6.21a}$$

$$N_{REF} = \frac{T_{REF}}{T_c} \tag{6.21b}$$

These relationships can be substituted into equation 6.21 and the result rearranged to give a simple expression for the input voltage in terms of the reference voltage.

$$V_{in} = - V_{REF} \cdot \frac{N_{REF}}{N_{in}} \tag{6.22}$$

In this equation the time constant cancels out, as does the frequency of the clock used to measure time. The system requires only that these quantities do not change, which can be arranged quite easily, and accuracies of up to about 16 bits are possible.

Charge balancing

The dual ramp converter depends upon the idea of balancing charge from a source controlled by the input voltage with charge controlled by the reference source over the course of a cycle of operation. Another approach is to attempt to keep the charge as constant as possible throughout the conversion process. This offers a couple of advantages: the conversion time is constant, being the same as the averaging time, and slight nonlinear effects in the dielectric of the capacitor used are minimised leading to greater potential accuracy in the system.

In a charge balancing converter the input voltage is connected all the time, and a reference voltage of opposite polarity is connected via a switch. During each clock cycle the converter monitors the output of a voltage comparator connected to the integrator to see if the voltage (and hence charge) on the capacitor is above or below a prescribed level. If the charge is higher than this level, the reference source is connected as well as the input signal for the following clock cycle, while if the charge is lower the reference source is disconnected. The number of cycles are counted to measure the time interval over which the conversion takes place, the averaging interval, and the number of cycles during which the reference source is connected are also counted separately. If the total number of cycles is N_t, and the number of reference cycles is N_{REF}, then by an argument similar to that used with the dual ramp converter the input voltage and reference voltages are related by the equation

$$V_{in} = -V_{REF} \cdot \frac{N_{REF}}{N_t} \tag{6.23}$$

6.5 Analog input circuits

The techniques which were introduced to combat noise on digital signals in Chapter 5 are largely unusable with analog signals because the signals are continuously variable. This makes them inherently susceptible to the effects of noise, and more attention must be paid to apparently minor points to reduce these effects.

6.5.1 Earthing practice

It is tempting to assume that any wire connected to the zero volt side of the power supply will actually be at zero volts, but unfortunately this attractive prospect is not borne out in practice. Any practical connection has electrical resistance, which means that if any current is flowing in the connection, a voltage drop will appear across it.

For example, the zero volt line to integrated circuits making up the logic of a microcomputer will consist of several centimetres of wire and, almost certainly, at least one pair of connectors. All the current taken by the microcomputer, which will range from a fraction of an amp to several amps, will pass via this connection. As a result, it is likely that there will be a difference in voltage from one end of this connection to the other of some tens of millivolts even during normal operation. This does not affect a logic circuit, because the noise immunity is typically (and hopefully!) several times this figure, but in an analog circuit where the difference between quantisation levels is quite small, the effect can be sufficient to 'drown' small measurements. To put this into perspective, the difference between quantisation levels will be approximately 2.5 mV for a 12-bit converter with an input range from 0 to 10 V.

Similar problems arise when two units are connected together, since power supply currents may flow in the signal leads, giving rise to voltage drops again. Interference between wires may also result from capacitative and inductive coupling between them. Care must be taken to avoid **earth loops** which are formed when careless earthing practice produces a set of earth connections which are joined in two or more places to form a continuous loop. Currents can be induced in this loop because of stray electromagnetic pickup and by the proximity of the earth and 'live' wires in the power cables to equipment. Unwanted voltages can arise at several places within a circuit as a result of forming an earth loop in this way.

6.5.2 Differential mode

The mechanisms which cause noise signals to appear in connections will always tend to produce similar voltages in adjacent wires which means that the difference in noise voltage between the two wires will be very small. This provides a method for reducing the effect of noise when sending analog signals. If the analog voltage is transmitted as the difference in voltage between the two wires, the amount of noise which will be added to this differential voltage will be very small. The voltages on any pair of wires can be expressed in terms of a **Common Mode (CM)** component which is the same on each wire and a **Differential Mode (DM)** component which is the difference in voltage on the two wires. Wherever noise is a problem, its offset can be reduced by transmitting the signal in differential mode and using differential input circuits which are as insensitive as possible to any common mode voltage. In practice any circuit has a limited range of common mode voltage over which it will operate, but this is comparable with the power supply voltage for most modern integrated circuits.

The two wires must be kept as close as possible in order to minimise the amount of DM noise pickup that will occur. This is conventionally done by twisting them together to form a **twisted pair**. Gaps between the two wires should be avoided as this leads to inductive pickup.

6.5.3 Screening

Another technique which is widely used is to cover the connections with an earthed metallic screen. The effect of the screen is to prevent any electric field from passing it, so that electrostatic pickup is prevented. It also reduces the effect of electromagnetic pickup, because the changing magnetic field induces currents in the screen which in turn create magnetic fields which tend to cancel the original field. The screen can be a sheet of metal when it is used to protect circuits within a piece of equipment, but cables are screened with a sheath of woven metal braid which can flex with the wire. This sheath must be earthed at one end of the connection only; if it were earthed at both ends it could give rise to an earth loop.

The effects of noise upon analog signals can usually be reduced to acceptable limits without difficulty by means of a combination of careful earth design, difference mode, and screening.

6.5.4 Analog multiplexing

When a large number of inputs are to be monitored, it may prove preferable to use a single ADC and an analog multiplexer rather than face the expense of having a separate ADC for each of the input channels. An analog multiplexer is similar to its digital counterpart (which was described in section 5.5), except that much greater care must be taken to prevent errors appearing in the input voltage due to the multiplexing process.

These noise and offset voltages can be produced by the various mechanisms already described. However, for most purposes they can be reduced to acceptably low levels by using MOS transistors as switches with suitable drive circuits. When the MOS transistor is conducting, the channel acts as a relatively low resistance in series with the signal, while when it is switched 'off', the series resistance is high enough to isolate the input signal. Switches of this type, with drive circuits incorporated into them, are available from a number of manufacturers, and behave well at low frequencies because the gate resistance of an MOS transistor is extremely high and negligible current leaks into the channel carrying the signal being switched. However, there is appreciable capacitance between the gate and the channel, and this means that when the transistor is being switched on and off, charge does flow between the gate and channel, appearing as pulses in the signal being switched. The effect of this coupling can be minimised by **neutralising** it. To achieve this an equal and opposite signal is coupled to the channel by means of a capacitor of value equal to the capacitance between the gate and channel of the MOS transistor. In fact this never completely eliminates the coupling because the gate-channel capacitance varies with voltage, and thus depends upon the signal being switched, but the effect can usually be reduced to an acceptable level.

6.6 Design example

The analog interfacing requirements of the two example systems are quite different: the controller needs relatively slow, low cost interfacing, while the waveform capturing instrument requires a high speed analog input in order to be able to cope with large bandwidth signals.

6.6.1 Analog input for controller

The number of bits required from the analog-to-digital converter depends upon the system being controlled and the accuracy of conversion required. The speed of conversion is not the major factor here unless very high accuracy is required as the rate at which the microcomputer can carry out the control calculations is limited to a few tens per second.

A low-cost design

Figure 6.24 illustrates how a low-cost analog-to-digital converter, the TL507C, can be used with a microcomputer. This ADC contains a 7-bit counter, a digital-to-analog converter, and a comparator. The counter can be reset by means of an external connection, and counts down in response to pulses on its clock input. The outputs of this counter are connected to the inputs of the DAC, and the analog output of the DAC is compared with the analog input by the comparator, the output of which is externally accessible.

In operation, the TL507C is reset by the microcomputer which then generates and counts clock pulses while monitoring the comparator output, until the output changes state. As the internal counter of the TL507C decrements, the DAC generates a voltage which decreases from a maximum value of $0.75V_{CC}$ towards a minimum of $0.25V_{CC}$. When this voltage passes below the input voltage, the comparator output changes and the microcomputer can now determine the input voltage from the number of clock pulses which were generated.

The clock pulses can be generated and counted using software without

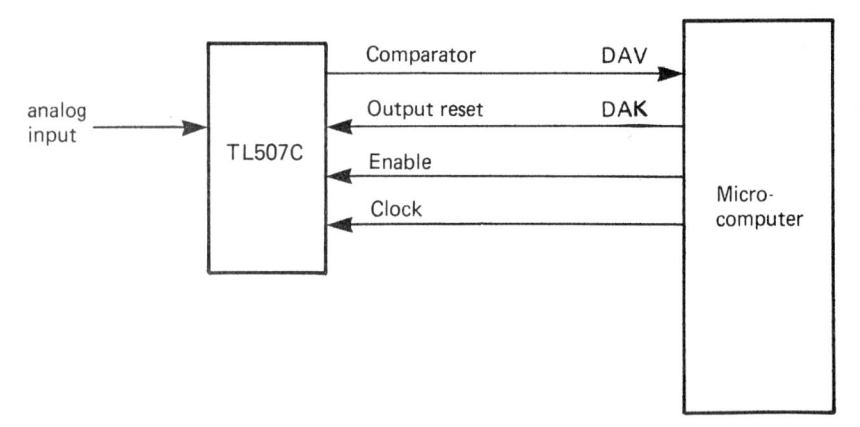

Figure 6.24 Connecting the TL507C to an 8-bit microcomputer

taking too much processor time. The internal counter of the TL507C can count at speeds up to 125 kHz, and because the counter has only seven bits, no more than 128 clock pulses are needed to complete a conversion. The interface requires only three lines — clock, reset, and comparator output — and the reset line could be dispensed with at the cost of an increase in conversion time. In the example to be discussed, the comparator output is connected to the most significant bit of a VIA port and the clock input to the least significant bit of the same port. This allows some programming tricks to be used to speed up the conversion. The connection of the reset line is less important, but for the purposes of this example it is connected to bit 1. A sample and hold circuit could be added if the analog input voltage is likely to change appreciably during the conversion period.

The initialisation of the port is carried out quite easily; the appropriate bits of the corresponding data direction register must be set to their appropriate values. In the first instruction below, the dots represent bits of the DDR which are not required here but which in practice would be assigned for other purposes. In the actual program these must each be replaced with a '0' or a '1' as appropriate:

```
LDA    #%0.....11    (Replace dots with 0 or 1!)
STA    DDR           Put this code into DDR
```

Before carrying out a conversion the counter in the TL507C should be reset to zero by outputting a logic '1' to its reset line. This line must be held at logic '0' while the conversion proceeds, so the conversion will commence with a sequence which generates a pulse to logic '1' on the reset line:

6.6.1

```
ADCONV  LDA   PORT           Read current port contents
        ORA   #%00000010     and set
        STA   PORT           bit to logic '1'
        ANDA  #%11111100     and then back to '0'
        STA   PORT           without affecting other bits
```

In the fourth line of this section both the reset line and the clock line are set to logic '0'. This will allow the next section to be programmed more compactly. Having set the internal counter of the ADC to zero, the register or memory location which is to be used as the corresponding counter in the microcomputer must also be zeroed

```
        CLRA
```

Here the accumulator has been used, as it will allow the loop which follows to execute faster.

The actual conversion is carried out by generating and counting clock pulses while monitoring the comparator output. Five lines of assembly code are needed to generate a pulse on the reset line, and a similar five lines could be used to generate a pulse on the 'clock' line without affecting the other outputs. A neater solution is to exploit the fact that the clock is controlled by the LSB of the port and simply increment and then decrement the contents of register PORT by one. (This assumes that the output register of PORT is readable by the microcomputer, which is not always the case.) Now only two lines are needed; moreover, because the comparator output is connected to the MSB of the same port, it will be treated as a sign bit and will affect the contents of flag bit N. This allows a conditional branch instruction to be used directly, and the loop needs only four instructions.

```
ADLOOP  DECA                 Decrement count by one
        INC   PORT           and pulse 'count' pin
        DEC   PORT           for same effect in TL507C
        BMI   ADLOOP         continue until MSB = 0, then
        RTS                  return with result in accumulator
```

If the last three pieces of program are placed end to end they will form a complete conversion subroutine called ADCONV. At the end of this subroutine the accumulator will contain a number in the range 128 to 255 corresponding to voltages in the range $0.25V_{CC}$ to $0.75V_{CC}$. Usually this number would need to be processed in some way by the calling program.

This use of two separate counters — one in the ADC and one in the microcomputer — allows a very simple interface to be used, and because the need to provide connections to the counter outputs is avoided the ADC can be packaged in a low-cost 8-pin integrated circuit package. The seven bit accuracy of this converter is adequate for many applications.

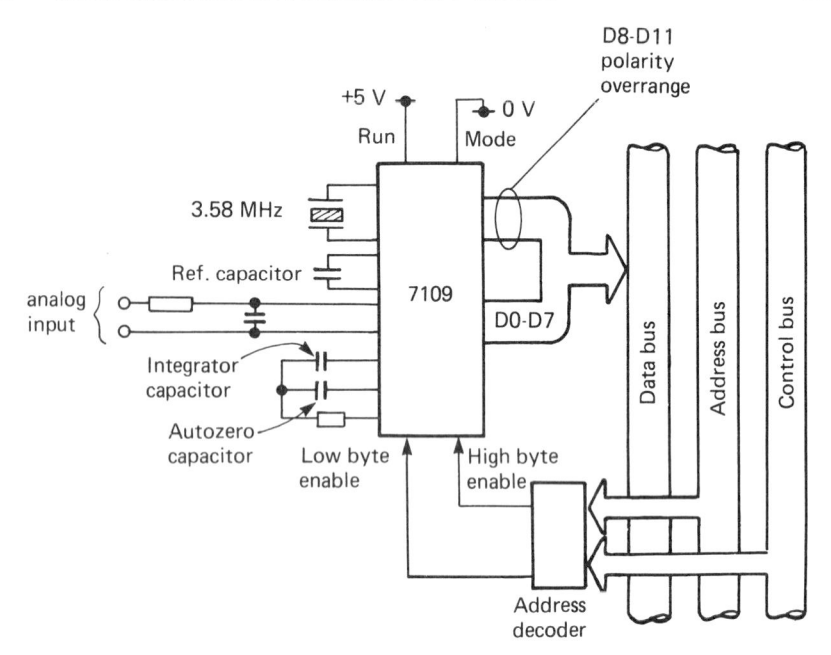

Figure 6.25 ADC interface using 7109

A more accurate design

By way of a contrast, a circuit using a 12-bit converter, the 7109, is shown in Figure 6.25. The 7109 is a dual slope integrating ADC with a 12-bit output, in addition to which there is indication of input polarity and overrange. The analog input of this device is connected via a set of analog switches to a voltage buffer, the output of which is connected to the integrator and comparator which is used in the dual ramp technique. The switches provide for connection to a reference voltage for the second phase of the conversion and to earth potential to allow voltage offsets to be compensated for. The operation of all these switches is under the control of the sequencing logic which sequences the system through three phases:

1. The 'autozero' phase, during which the voltage offsets of the analog section are measured and compensated for automatically by charging a capacitor with a voltage equal and opposite to the composite voltage offset error.

2. The 'signal integrate' phase, in which the input signal is integrated for 2048 clock cycles.

3. The 'deintegrate phase', during which the reference signal is connected and the time taken for the integrator output voltage to return to zero is measured in terms of clock cycles.

6.6.1

Interfacing the 7109 to the bus of an 8-bit microcomputer is quite simple, as the output lines are divided into two groups with separate enable lines. Alternatively two ports of an interface device, such as the VIA discussed in Chapter 4, could be used but this would be wasteful since the VIA is not necessary.

Naturally, the time taken for the 7109 to carry out a conversion depends upon the clock frequency at which it is operated, and it is designed so that a clock frequency of 3.58 MHz will give a conversion rate of 7.5 samples per second. At this frequency 60 Hz main interference will be rejected because the signal integrate phase will last an integral number of cycles. The clock signal can be generated by the 7109 itself inexpensively using a quartz crystal: 3.58 MHz crystals are manufactured in large quantities for television receivers for the NTSC system used in North America and Japan, so that crystals for this frequency are widely available.

6.6.2 Analog output for controller

The output signal from the controller must be converted to analog form and then amplified to provide enough power to operate the load. In this case the load is a resistive heating element which will be required to dissipate appreciable power, which suggests the use of a switching DAC.

Switching techniques become attractive when the load requires a power of more than a few watts, because the drive circuits can be made very much more efficient. The time taken for a heater to change temperature is usually very long in electronic terms, so that the changes in instantaneous output power which result from this approach are acceptable even when the pulsing takes place quite slowly.

Using a switching DAC with a DC supply

Given that the speed at which the heater output can be switched on and off can be made quite low, the timing for the control of the heater output can be derived from the interrupts which control the keyboard scanning. An 8-bit conversion is possible by adding the controller output, scaled to a single byte SCALED in the range 0 to 255, to a memory location TOTAL in which the total accumulates modulo 256

```
DACONV  LDA   SCALED    Get scaled control value
        ADDA  TOTAL     and add to total
        STA   TOTAL     modulo 255
```

Next the state of the carry bit is tested and used to determine the value of the output bit DAC which controls the power supply to the heater. This is switched on whenever the carry bit is set to '1', and is switched off whenever the carry bit is reset to '0':

```
              LDA    PORT          Read current value of PORT
              BCS    DAC1          and test carry bit
              ANDA   #255-DAC      C=0; switch off DAC bit
              BRA    DAC2
      DAC1    ORA    #DAC          C=1; switch on DAC bit
      DAC2    STA    PORT          and update outputs
```

These two program fragments are all that is required to generate a square wave with a duty cycle proportional to the controller output. Note that in this case the output signal is unipolar because the power input to the heater is never negative. The waveform repeats after 256 clock pulses if the controller output does not change; thus if the keyboard interrupt occurs every 20 ms, the waveform will repeat in approximately 5 seconds, a short enough period to prevent detectable temperature change in most applications.

Controlling AC power to a load
The control of the AC power supplied to a load is more difficult because of the way in which the instantaneous power changes with time. However, the switching DAC technique described in the last section lends itself directly to the control of AC power provided that the updating of the output control bit is synchronised with the mains. If 50 Hz mains are used, the repetition interval of the output waveform will again be about 5 seconds, while it will be proportionally shorter if the mains supply used has a frequency of 60 Hz. The switch can be a triac or SSR used in the same way as described in Chapter 5, because the speed of operation of the electronics used is much higher than the mains supply frequency.

6.6.3 Analog input for instrument

The highest frequency component of the input signal which can be captured without distortion is limited by Nyquist's law (Nyquist, 1928) to one half the sampling rate, and for this reason a high sampling rate is required.

The maximum speed of a successive approximation ADC is limited by the speed of the logic and the settling time of the DAC used. The 574, for example, is capable of performing a 12-bit analog-to-digital conversion in a nominal time of 25 µs, and can be interfaced quite easily as shown in Figure 6.26. The 'start convert' and 'end of convert' handshake signals can be handled quite easily by means of a parallel interface device such as the 6522 Versatile Interface Adapter. The conversion can be started by the reading of the second byte of the result if the handshake is programmed correctly. Then the microcomputer must monitor the flag register of the VIA until the end of the conversion causes the handshake flag to be set. This means that a waiting loop must be written, and as it takes several clock cycles of the microcomputer to execute the loop there will in general be an 'overhead' of some wasted clock cycles. When the microcomputer detects that the loop has been completed, it must read the twelve bits of output. The time taken

6.6.3

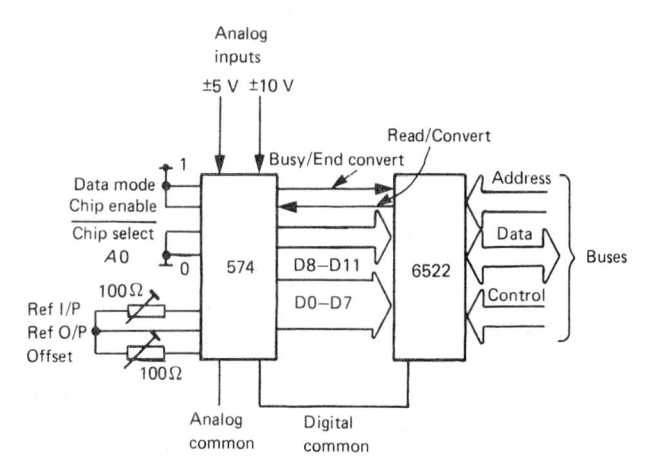

Figure 6.26 ADC interface using 574

to read these two bytes of data must be added to the overall conversion time.

High speed data capture
One way to reduce this overhead is to assume that the conversion will have been completed after a certain amount of time has elapsed. Unfortunately, the 574 does not base its timing upon that of the microcomputer but uses its own internal clock generator. The speed of this clock varies from sample to sample of the 574, so that the actual conversion time will lie in the range somewhere between 15 and 35 μs. Thus any design which expects the conversion to be completed after a certain time must be based upon the worst-case timing of 35 μs; in fact the interval between samples would be defined in any practical application, so that any design would need to be based upon the worst case figures.

Some time can be saved by arranging for the microcomputer to store each sample that has been read in memory after the ADC has started converting its next sample, so that the two systems operate at the same time without waiting for each other. If the time at which each sample is taken is determined by software, the timing resolution can be determined to the basic cycle time of the microcomputer.

6.7 Questions

1. A 'trip computer' for a car uses sensors to detect rotation of the drive wheels and rate of fuel consumption. Each of these quantities is represented by means of a sequence of pulses which must be counted. Based upon this information, the rate of fuel consumption, the total amount of fuel used, speed, distance travelled, elapsed time, estimated

time of arrival, and fuel consumption per unit distance can be calculated. Design a simple trip computer to carry out these operations and display whichever quantity is chosen in digital form.

2. A storage oscilloscope which is fast enough for many purposes can be designed as a 'waveform capturing' instrument which stores the output of an analog-to-digital converter in RAM. It can then use this information to regenerate a steady trace on an oscilloscope screen by means of a digital-to-analog converter. Devise a system which 'captures' a waveform in this way and which then allows the user to measure time intervals and voltages in the stored waveform.

3. A phase meter compares the phases of two sine (or square) wave signals by measuring the time difference between one waveform passing through zero and the zero-crossing time of the second waveform. This is then divided by the duration of a complete cycle and the result is displayed as a phase angle expressed in degrees. Design a microcomputer-based phase-meter.

4. The relative amplitude of two voltages V_1 and V_2 can be expressed in terms of decibels (dB) according to the formula

$$dB = 20\log_{10}\left(\frac{V_1}{V_2}\right)$$

A multiplying DAC can be used as a digitally controlled attenuator to attenuate an audio signal of voltage V_{in} to a value V_{out}; design a system which can attenuate a signal by a prescribed number of decibels, accurate to within 0.5 dB.

5. Design a brightness control which can vary the intensity of a 12 V, 60 W lamp from zero to full brightness. Why should a duty-cycle DAC be used rather than a resistive type?

6. A low-frequency sine-wave generator is to use a look-up table of sines to obtain samples of their output waveform. Design a unit which can generate sine waves with digitally controllable amplitude and frequencies up to 500 Hz.

Chapter 7 TRANSDUCERS: SENSORS AND ACTUATORS

A microcomputer would be useless if it could not communicate with the outside world. Sensors and actuators are the 'senses' and 'muscles' of a microcomputer-based system which allow it to interact with its surroundings. The term 'transducer' really refers to any device which can convert a signal from one form to another so that in a tape recorder, for example, both the microphone and the loudspeaker are transducers. Very often, however, the word is used simply to describe a sensor, an input transducer in which signal energy in some other form is converted to electrical output, while the word 'actuator' is used for output transducers which convert signals from the electrical form used by the microcomputer.

The signal outside the microcomputer may be a pressure, speed, or in any other one of a wide range of forms. Often there is a choice of physical principles which may be exploited to convert the signal. Consequently a vast number of types of sensors and actuators exist, and only the commonest of them can be described in the space of a single chapter. However, a number of books have been written specifically on this subject of transducers, some of which are listed in the bibliography (Sydenham, 1983; Beckwith *et al.*, 1982; Neubert, 1963).

7.1 Resistive transducers

Many sensors respond to external physical inputs with corresponding changes in resistance. These changes in resistance must in turn be converted into voltage or current signals in order to be useful. One way to obtain a voltage signal from the transducer is to wire a fixed resistor in series with it and connect a fixed voltage across the series combination to produce a potentiometer. If the values of the fixed voltage and resistor are known, the resistance of the transducer can be calculated from the output voltage from the potentiometer. Note that the actual voltages used in this resistance measurement are irrelevant provided that the working voltage ranges of the components are not exceeded. The conversion is a **ratiometric conversion**; that is, only the ratio between the input and reference voltages matters.

If the fixed voltage applied across the potentiometer is made the same as the reference voltage used by the ADC, the voltage applied to the ADC input divided by the reference voltage will represent the transducer resistance as a fraction of the whole resistance.

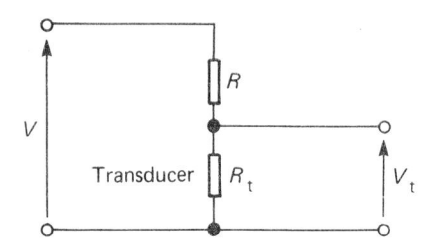

Figure 7.1 Transducer wired as potentiometer

Unfortunately this method of converting resistance to voltage has a number of disadvantages, the most important of them being that it is non-linear. Suppose that a transducer with resistance R_t is wired in series with a resistor R and that a voltage V is connected across them, as may be seen in Figure 7.1. The voltage across the transducer (and hence the voltage signal to be converted by the ADC) will be

$$V_t = \frac{R_t \cdot V}{R + R_t} \tag{7.1}$$

This will give a severely nonlinear relationship between V_t and R_t unless the resistance of the transducer is very much less than that of the series resistor, in which case the signal will be very small.

One solution to this problem is to replace the resistor with a source of current which does not vary as the voltage across the transducer changes. This can be achieved quite easily by making use of the fact that the collector current of a bipolar transistor is almost independent of the collector emitter voltage, as shown in Figure 7.2.

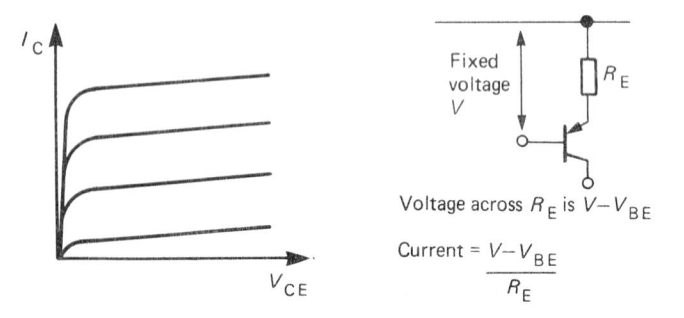

Figure 7.2 Constant current source

Another way to produce a voltage proportional to the resistance of a transducer is to put it in the feedback path of an operational amplifier and connect a constant voltage to the input resistor. Equation 6.6 shows that the output voltage will then be proportional to the resistance of the transducer.

Many resistive transducers are inherently nonlinear and these require a linearisation stage after signal conditioning. In these cases it may be more convenient to accept the nonlinearity resulting from the use of a potentiometer and correct for it when correcting for the transducer.

A problem which is sometimes encountered with resistive transducers is heating. The transducer will dissipate electrical power as heat and the resulting rise in temperature may itself change the resistance of the transducer, causing an error. This is known as **self heating** and means that it is important to operate the transducer at as low a voltage as possible, and frequently a voltage amplifier is needed to produce a signal which is large enough for the analog-to-digital converter. Often heating effects caused by the environment are more significant than self heating. It is important to calibrate the transducer at the temperature at which it is to be used, and this can include a large self heating term provided that the system is stable. Self heating can give rise to errors in systems where multiplexers are used, since the temperature of the sensor will change if the sense current is removed without replacing it with a similar current from another source.

7.1.1 Bridges

In many important cases the fractional change in resistance of a transducer is quite small, so that if the output signal is derived from the total voltage dropped across the transducer there will be a large constant component in the signal. This must be removed if the full resolution of the ADC is to be achieved. One way to remove this constant component is to use a pair of potentiometers: one containing the transducer which produces the output voltage signal and a second producing a constant voltage. The voltage of interest is the difference between the output signals from the two potentiometers. These potentiometers form the pillars of a **Wheatstone bridge** and the differential output signal is found by 'bridging' these pillars, as shown in Figure 7.3. When the bridge is 'balanced' (that is, there is no difference between their output voltages) the values of the resistors are related by the well-known Wheatstone bridge formula:

$$\frac{R_1}{R_2} = \frac{R_3}{R_4} \tag{7.2}$$

The 'conventional' method of using a Wheatstone bridge is to vary one

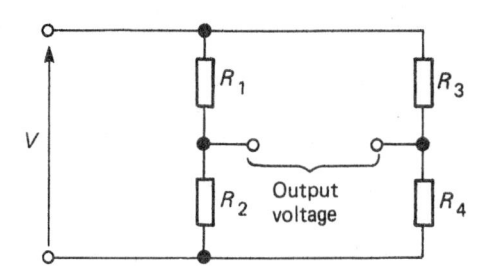

Figure 7.3 Wheatstone bridge circuit

of the resistors manually until the circuit is balanced, when no current flows in a meter connected between the outputs. In our case we wish to keep the resistors fixed and monitor the out-of-balance voltage. To do this it is necessary to convert the differential signal to a 'single-ended' one, that is, a voltage signal referenced to earth. This is discussed in section 7.1.3.

7.1.2 Relationship between transducer resistance and signal

The bridge consists of a pair of potentiometers, which means that its output voltage does not vary linearly with transducer resistance. Equation 7.1 shows that the output voltage from each potentiometer is of the form

$$V_t = \frac{R_t.V}{R + R_t} \tag{7.3}$$

The transducer resistance R_t can be written as the sum of a fixed component R_o and a relatively small variable component r which is proportional to the input quantity. The output voltage from one side of the bridge is thus

$$V_1 = \frac{(R_o + r).V}{R + (R_o + r)} \tag{7.4a}$$

while if the other side of the bridge consists of a pair of resistors with values R and R_o, the voltage at the midpoint will be

$$V_2 = \frac{R_o.V}{R + R_o} \tag{7.4b}$$

leading to a differential voltage of

$$V_1 - V_2 = \frac{(R_o + r).V}{R + (R_o + r)} - \frac{R_o.V}{R + R_o} \tag{7.4c}$$

$$= \frac{R.V}{(R + R_o)} \cdot \frac{r}{(R + R_o + r)}$$

This differential output voltage is only approximately proportional to the change in resistance r that results from the input quantity that is being transduced, the approximation depending upon the assumption that the fractional change in resistance is small.

The voltage will cause a current to flow in the input connections of the amplifier unless its resistance is very high. The current can be calculated if the output resistance of the bridge is known, and this output resistance is the sum of the output resistances of the two potentiometers which comprise the bridge. The output resistance of a potentiometer can be shown to be equal to the resistance of the two resistances in parallel, which when applied to the bridge under discussion leads to an output resistance of

$$R_{out} = \frac{R.(R_o + r)}{(R + R_o + r)} + \frac{R.R_o}{(R + R_o)} \tag{7.5}$$

The output voltage from the bridge must be modified to take account of the bridge's output resistance if it is not very small compared with the input resistance of the amplifier. The resulting expression for the differential voltage as a function of the quantity being transduced is very complicated and not particularly illuminating — which is why it will not be included here! Clearly the designer is faced with a decision: either a differential amplifier with a very high input resistance must be used, or linearisation must be incorporated. The former alternative is almost always taken, not only because it makes the mathematics much easier, but because it usually leads to a more accurate result with no need to adjust the compensating non-linearities.

7.1.3 Instrumentation amplifiers

The differential amplifier described in section 6.2.7 is useful because it allows a differential input voltage from almost any source to be converted to a single ended form, that is, a voltage referenced to the zero volt line. However, the input resistance of the differential amplifier is determined by the values of the resistors used, and typical figures for the input resistance obtained with a simple differential amplifier are not high enough to guarantee immunity from errors due to the effects of the output resistance of the bridge. One way to overcome this shortcoming is to connect a voltage follower in series with each of its inputs. This ensures that the input resistance of the composite amplifier is very high, while the differential amplifier still produces an output voltage which is the difference between the two input voltages. In fact there are many ways in which such **instrumentation amplifiers** of this kind can be designed, and considerable ingenuity has been expended in the design of amplifiers which have high input impedance and high common mode rejection. However, instrumentation amplifiers are now available as single integrated circuits, and this has made the task of the interface designer much easier.

7.2 Temperature

Temperature affects a wide range of physical processes, and a wide range of principles have been exploited in its measurement. Because of its widespread effects, temperature is measured more often than almost any other physical quantity. Scales of temperature are based upon reproducible 'fixed points' such as the melting and boiling points of pure materials which allow transducers to be calibrated. The most familiar of these scales is the Celsius or Centigrade scale which is based upon the freezing point (0 degrees) and boiling point (100 degrees) of water. The inventors of both the temperature scales in everyday use — Fahrenheit and Celsius — lived before the discovery of the absolute zero of temperature, so their scales have zero points towards the lower end of the range of temperatures ordinarily encountered in our environments. No temperature can ever fall below absolute zero, so it is used as the zero of the absolute temperature scale. Absolute temperature is measured in kelvins, one kelvin representing a temperature difference of one degree Celsius.

7.2.1 Resistance thermometers

All materials which conduct electricity change their electrical resistance with temperature. The relationship between resistance and temperature varies from material to material and it is sometimes quite complicated, reflecting the complexity of the mechanisms involved. In metals, which generally have quite low resistivities, the resistance increases approximately in proportion to absolute temperature, and for pure metals the law relating resistance to temperature is known quite accurately. Platinum wire resistance thermometers, for example, allow very accurate measurements to be made up to temperatures of several hundred degrees Celsius.

The change in resistance of a metal wire thermometer is typically only a fraction of 1% for each degree change because the resistance varies in proportion to the absolute temperature, and the zero of the absolute temperature scale is at approximately -273 degrees Celsius. Changes this small are inconvenient for many control and instrumentation applications; for example, the temperature of the human body never changes by more than a few degrees, and medical measurements require a high resolution within this narrow temperature range.

7.2.2 Thermistors

Thermistors are useful temperature transducers where high resolution is required. They are resistors made from materials which are specially formulated to show a significant change in electrical resistance with temperature, and this dependency of resistance upon temperature can be either positive or negative. **PTC (Positive Temperature Coefficient)** thermistors increase their resistance as the temperature rises, while **NTC (Negative Temperature Coefficient)** thermistors show a fall in resistance with increasing temperature. NTC thermistors are used more commonly; the resistance of a typical device might fall from 5000 Ω at 0°C to 100 Ω at 100°C. Thermistors are inexpensive and robust, but even thermistors of the same type often show marked variations in characteristics and normally they need to be calibrated individually if accuracy is important.

7.2.3 Thermocouples

Whenever two chemically dissimilar electrical conductors are placed in contact, an electrical potential exists between them, but these potentials cannot ordinarily be detected because they cancel out around a circuit. However, these potentials vary with temperature, so that if one of the junctions between dissimilar conductors is heated with respect to the others their effects will no longer cancel and a current will flow.

This is the principle of the **thermocouple**; one junction is maintained at a known temperature, while the other one is placed at the point whose temperature is to be measured. The junction between two wires made from

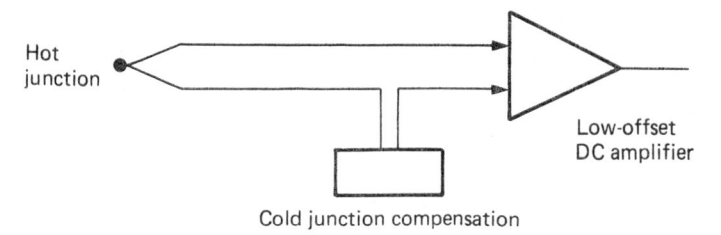

Figure 7.4 Thermocouple interface

different metals can be made very small and robust, allowing measurements to be made in confined spaces at temperatures from $-250\,°C$ to $+1500\,°C$, according to the type of thermocouple being used. In practice it is inconvenient to have to maintain the cold junction at a fixed temperature, and it is more usual to use a compensating circuit which measures the cold junction temperature and produces a correction voltage to allow for the fact that it is not at the reference temperature.

The voltages produced by thermocouples are small, typically a few tens of microvolts per degree Celsius, so that sensitive DC amplifiers with very small voltage offsets are needed to make use of them (Figure 7.4).

7.2.4 *pn* Junction

The *pn* junction lies at the heart of most electronic devices. It is simply the junction between two regions of the same crystal of semiconductor. One region is treated so that the mechanism of electrical conduction is carried out by positive charges (the '*p*' region), while the other is treated so that negative charges are used (the '*n*' region). The *pn* junction has very nonlinear electrical properties; so nonlinear, in fact, that it is used as the basis of the semiconductor diode. When the *p* region is made positive with respect to the *n* region, a current flows which rises almost exponentially with the applied voltage. If the forward voltage drop is measured at a fixed current, it will be found to vary with temperature in an almost linear fashion, the voltage changing by about 2 mV per degree Celsius. This provides the basis for a temperature sensing element, although in practice these elements are usually more complicated than single diodes because the actual voltage drop across a junction depends to a great extent upon variations in the manufacturing process used.

Transducers are available based upon this principle which have been trimmed during manufacture to give a linear dependence of output voltage upon temperature, and are very useful over a range of temperatures from about $-50\,°C$ to $+150\,°C$. This temperature range is largely determined by the properties of the semiconductor material used.

A *pn* junction can be used to measure temperature in other ways. If the *p* region is made negative with respect to the *n* region, then only a tiny 'leakage' current will flow. This current depends very strongly upon temperature, typically doubling for a temperature rise of 10 °C. Semiconductor materials also show a strong thermoelectric effect, although this is not widely used in temperature measurement.

7.3 Light

The need to detect or measure light occurs quite frequently in scientific and engineering measurements. The sensing of light falling on a detector forms the basis for a number of transducers for other quantities such as displacement. When talking about the *measurement* of light, however, we must be more specific because there are many quantities which we may wish to measure.

First of all, light can differ in colour. In fact most sources of light contain components of a wide range of different colours, as we can see when looking at a rainbow or a ray of light passed through a glass prism. The reason that each ray emerging from a prism has a different colour is that each has a different wavelength, and a measurement may need to take this wavelength into account. Our eyes analyse colour in terms of the proportions of red, blue, and green light which it contains. The human retina contains three kinds of cell, each of which has a peak sensitivity to a particular range of wavelengths, but with some sensitivity to other wavelengths as well. The same is true of electronic light sensors; each type has a peak sensitivity to a particular range of wavelengths, but this sensitivity falls off gradually as the wavelength is changed. Many such sensors are also receptive to ultra-violet and infrared light which we cannot see.

There are many specialised terms used in the measurement of light. When measuring the intensity of light, we may be interested in the overall light output of an illumination source or simply the light output in a particular direction. If the light source is part of a display system, then it is the brightness per unit area of the light source which is of interest, or it might be the intensity of the light falling on something which is to be measured. Each of these quantities has its own set of units, as shown below:

- **Luminous power** is the proportion of the total power radiated by a light source to which our eyes are sensitive, and it is measured in **lumens**.
- **Luminous intensity** is the luminous power per unit solid angle and is therefore measured in lumens per steradian, a unit which is also given the special name of **candela**. This quantity represents the overall brightness of a light source.
- **Luminance** is the luminous intensity per unit area of a light source and represents the brightness per unit area of an extended source. It is measured in candelas per square metre.
- **Illuminance** is a measure of the brightness of illumination of a surface, and it is measured in lumens per square metre, a unit which is given the special name of **lux**.

Each of these terms refers to the proportion of light which we can see, but corresponding terms can be defined for the total light including the invisible ultraviolet and infrared components.

- **Radiant power** is the total light power being radiated by a light source and is measured in **watts**.
- **Radiant intensity** is the radiant intensity per unit solid angle and is measured in watts per steradian.
- **Radiance** is the radiant intensity per unit area of a source and is measured in watts per square metre per radian.
- **Irradiance** is the radiant power falling upon a unit area of a surface and is measured in watts per square metre.

There is the added complication of having different kinds of light and different sensitivities to different wavelengths. Even more units can be defined once the wavelength of the light is taken into account, for example, the radiant power per unit wavelength. These follow quite naturally from the preceding quantities and will not all be listed here.

7.3.1 Photovoltaic devices

When light falls on a *pn* junction in a semiconductor material, it affects the distribution of electrons across the junction in such a way that a voltage appears across it, and if the diode is connected to a circuit, current will flow. This is the principle of the solar cell used for generating electrical power in specialised applications such as spacecraft; they are *pn* junctions designed with a large surface area so that they intercept a large amount of incident light. A photodiode is smaller, but it otherwise behaves in the same way. Its output signal is too small for power generation, but nevertheless it can be used for measurements of light intensity.

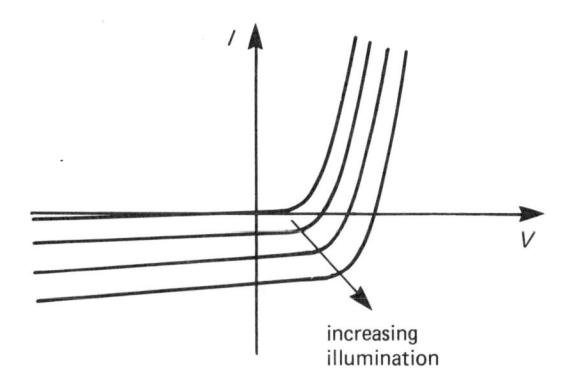

Figure 7.5 Characteristics of a photodiode

A *pn* junction made in any semiconductor material will produce a voltage output when illuminated, the amount of voltage for a given irradiance depending upon the wavelength of the incident light and the type of semiconductor material used. Most photodiodes are made of silicon, which is most sensitive to light at the red end of the spectrum, and even more sensitive to some infrared radiation. This is useful when light from a light emitting diode is being detected, since the most efficient types of LED operate in these parts of the spectrum. However, it means that the readings of light intensity obtained using a photodiode may not accurately reflect the apparent intensity of the light as seen by our eyes, which can be problematical in photographic applications. This sensitivity is essentially a property of the semiconductor material being used, and other types of photodetector made from the same material will show substantially the same effect.

7.3.2 Photoconductive devices

A major problem with the photodiode when it is used as a photovoltaic device is that the relationship between the irradiance of the photodiode and the voltage which appears across has the nonlinear form shown in Figure 7.5. This is due to the physical principles governing the operation of the device and it cannot be overcome easily. However, other physical phenomena are available which can be exploited to give output signals which depend in a linear fashion upon the irradiance of the detector.

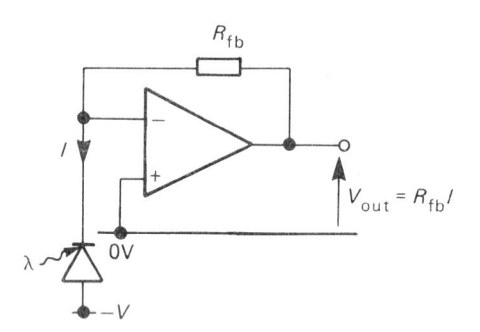

Figure 7.6 Photodiode in conductive mode

The photodiode in conductive mode
Photovoltaic transducers actually produce electricity from light; the effect of the light falling upon the photodiode is to forward bias the diode, and this forward bias voltage forms the output signal. If the same diode were reverse biased by means of an externally applied DC voltage, a negligible current would flow provided that no light fell upon it. If it were illuminated, however, the incident photons ('particles' of light) would cause a proportional number of electrons to flow. The current flowing through the reverse biased diode is therefore closely proportional to the irradiance of the diode provided that the light is of constant wavelength. This current can be converted

7.3.2

to a voltage signal using a transresistance amplifier; Figure 7.6 shows the basic form of the circuit used.

The phototransistor

Unfortunately the current produced by a photodiode in conductive mode is very small, and appreciable amplification is often necessary. In a bipolar transistor the collector current is controlled by a smaller base current, and the base and emitter form a *pn* junction. If a transistor is illuminated and no base current is provided by an external bias circuit, an internal base current can still be made to flow that is proportional to the irradiance, simply because the base-emitter junction behaves as a photodiode (Figure 7.7). This current can be amplified as if it were a normal base current to provide a larger collector current. It is because all semiconductor devices are light sensitive that transistors and integrated circuits are normally packaged in black plastic or other opaque material, since if light were able to fall upon them it would affect their operation.

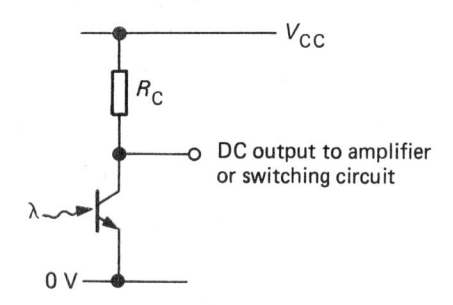

Figure 7.7 Using a phototransistor

The pin photodiode

The maximum speed of response of ordinary photodiodes and phototransistors to a change in irradiance is usually in the order of a microsecond, which is quite adequate for the great majority of applications. If faster responses are needed, for example in optical communications, a modified version of the photodiode, the *pin* **photodiode** can be used. In this type of diode there is an extra 'intrinsic' (hence the letter '*i*') layer of semiconductor material between the *p* and *n* regions of the diode structure which reduces some of the effects that limit the speed of the ordinary photodiode.

Light dependent resistors

It has long been known that some materials show a decrease in electrical resistance when they are illuminated. This change is very marked with modern designs of light dependent resistor (LDR), which have resistances in sunlight that are often several orders of magnitude less than in darkness. Moreover, LDRs can be made from materials other than silicon, and a commonly used semiconductor is cadmium sulphide (CdS) which responds to different colours of light with a variation in sensitivity which is quite like that of our own eyes. The main disadvantage of LDRs is that they are much slower to respond to changes in irradiance than the types of device which have already been described, typically requiring a tenth of a second or longer to reach a steady output again.

7.4 Position, displacement, and velocity

The measurement of position is perhaps the most basic and most widely used of all types of measurement in science and engineering. The idea of position appears under many guises because of the number of different words which we use for expressing sizes and positions. For example the word 'depth' relates the position of the top of something to its bottom; some other words which carry the meaning of 'position' are 'length', 'breadth', 'diameter', and 'distance'. A 'displacement' is a change of position, and the rate of change of position with time is called 'velocity'. 'Acceleration' is the rate of change of velocity with time, and there is even a special word — 'jerk' — which expresses the rate of change of acceleration. Physical objects are capable of **rotation** as well as linear motion, and many of the measurement techniques used to measure position along a straight line can be adapted to allow measurement of angular position.

Not surprisingly, there is an enormous range of techniques available for measuring position in its various forms. The first thing to note about position is that it must be measured with respect to a **datum point**. For example, the position of a cutting tool in a computer-controlled milling machine, can be expressed as a set of co-ordinates with respect to such a datum point.

7.4.1 Absolute position encoders

The absolute position of a detector could be determined simply by reading an electronic 'ruler' to measure the distance from the datum point. Although it would be possible to use an ordinary ruler and a large amount of electronic image processing, it is obviously much simpler to use markings that can be more easily read by a machine.

The 'obvious' code to use is a binary one. If the units were, say, millimetres, then the transducer would be required simply to read off a binary number from the 'ruler' and present it at its output. The binary numbers 0 and 1 could be coded as light and dark regions on the measuring scale if reflected light were used or as transparent and opaque regions if light were shone through the scale. However, there is a fundamental problem connected with the use of ordinary binary code in this way.

To appreciate how this type of transducer can produce an incorrect output, consider what happens as the detector moves from position 7 to position 8 on the scale. The binary output of a 4-bit detector would change from 0111 to 1000. The problem arises from the fact that the individual bits will not change at precisely the same time, no matter how carefully the transducer is designed, because the logic levels on the outputs are derived from estimates of the amount of light falling on each detector. Thus the instantaneous output from the transducer may take almost any binary value as it moves. Fortunately the solution is to choose a binary code in which only one bit changes at a time as the output changes from one value to an adjacent one.

Such codes are called 'Gray codes' and they can be produced very easily. When counting in Gray code, the trick is always to change the least significant bit that can be changed without producing a pattern that has already been used.

Table 7.1 Gray codes

decimal	binary	decimal	binary
0	0000	8	1100
1	0001	9	1101
2	0011	10	1111
3	0010	11	1110
4	0110	12	1010
5	0111	13	1011
6	0101	14	1001
7	0100	15	1000

The Gray code numbers in Table 7.1 show how only one bit changes at a time as we count from one number to the next. In the position encoder this means that only one output bit changes as the detector changes from one number to the next, thereby avoiding the chance of producing incorrect codes. The problem now, of course, is that the digital output from the position transducer is in a binary code which is different from the 'natural' binary used by the microcomputer. Fortunately the conversion from Gray to natural binary code can be accomplished quite easily using logic or a few lines of program in the interface subroutine for the transducer. All that is necessary is to exclusive-OR each bit with all the bits to its left in the Gray-coded form, as shown in Figure 7.8. In practice this can be carried out more easily using a combination of shifts and exclusive-OR operations, as may be seen in the second part of the figure. The combinations of shifts and exclusive-OR operations can be programmed very compactly using an assembly-language routine of the form shown in Figure 7.9. This is called with a Gray code pattern in the accumulator and returns with the accumulator holding the corresponding code in natural binary.

The same principle can be used to measure angular displacement. Note that the patterns which make up the Gray code repeat cyclically (as indeed do all binary numbering systems) and if the list of codes in Table 7.1 were extended we would see that the entry following the one for 15 would be the same as that for 0. In fact the codes for 0 and 15 differ in only one digit. Angular coders (often called 'shaft encoders') are designed so that one cycle of the Gray code is completed as the shaft is rotated once.

Gray code transducers allow the absolute position of a transducer to be measured very accurately but they require very precise engineering for their

Gray code pattern: 1101

first digit: 1 = 1
second digit: 1 ⊕ 1 = 0
third digit: 1 ⊕ 1 ⊕ 0 = 0
fourth digit: 1 ⊕ 1 ⊕ 0 ⊕ 1 = 1

Binary code: 1001 = decimal 9

```
Gray code:    1 1 0 1
      ⊕         1 1 0 1
      ⊕           1 1 0 1
      ⊕             1 1 0 1
Binary code:  1 0 0 1
```

Figure 7.8 Conversion from Gray code to binary

```
GTOB    STA     GRAY            Save Gray code pattern

LOOP    LSR     GRAY            Slide it one place right
        BEQ     DONE            until no ones are left.
        EORA    GRAY            Exclusive-OR with pattern
        JMP     LOOP            repeatedly until finished

DONE    RTS                     Then return...
```

Figure 7.9 Gray code to binary subroutine

manufacture and tend to be rather expensive. A related, and simpler, technique is discussed in the next section.

7.4.2 Counting methods

In this case there is only one detector and all the marks on the 'ruler' are identical. As the transducer moves along the 'ruler' the marks are counted to produce an output signal. These are **incremental encoders** which can measure position only with respect to a starting point; they cannot produce an absolute output unless they are set to a datum point before use.

Perhaps the simplest way to detect distance marks is to use light. This can be done with a light source and a photodiode separated by a sheet of opaque material in which holes are cut at regular intervals. If this sheet is allowed to move with respect to the lamp and the photodiode a flash of light will appear each time a hole passes between them. Each flash can be counted electronically to produce a measurement of the distance that the plate has travelled.

Unfortunately this simple transducer does not allow the direction of motion to be determined. For this, another detector is required, separated from the first one by a fraction of the spacing between the holes. This allows the direction of motion to be determined by noting which detector detects the hole first. Now the count can be incremented or decremented by one each time a hole is detected according to which direction the sheet is moving, and at any time this count represents the position of the sheet.

The same principle can be adapted to measuring angular position by using a wheel with holes drilled around its edge or slots cut in its edge rather like a gear wheel. If fine resolution is required the drilling of a large number of tiny holes becomes very difficult and expensive and it is preferable to print the pattern of transparent 'windows' on an opaque background using photographic methods. A piece of photographic film can be printed with much finer patterns than could be cut into a piece of metal by conventional techniques.

7.4.3 Geometric interference methods

The problem with using very fine patterns is that it becomes very difficult to resolve them using conventional light sources and detectors, and some kind of magnification is necessary. A very elegant method of doing this lies in the principle of the **Moiré Effect**. Suppose that two sheets of transparent material were each printed with a pattern consisting of opaque lines at regular intervals. If they were then superimposed, light would be able to pass through the two sheets only where both of them were transparent. If the two sheets were positioned in registration so that the transparent parts lined up with one another light would be able to pass through. However, if either of them were moved by half the distance between the lines, the transparent part of one sheet would align with the opaque part of the other one and no light would be able to pass.

This provides a very convenient mechanism for detecting very small movements. If one of the sheets is allowed to move with respect to the other in a direction at rightangles to the lines, the combined sheets will change from transparent to opaque to transparent again each time it completes a distance equal to the spacing between the lines. The fringes thus produced are known as **Moiré fringes**. The distance between the lines is limited largely by the resolution of the photographic material used and ultimately by the wavelength of the light used. Thus quite small movements can be measured.

The same principle can be used to produce an incremental transducer for angular position. In this case the lines on each sheet are radial from the centre of rotation.

Although the method as described so far allows changes in position to be sensed, it does not allow the direction of the change to be determined.

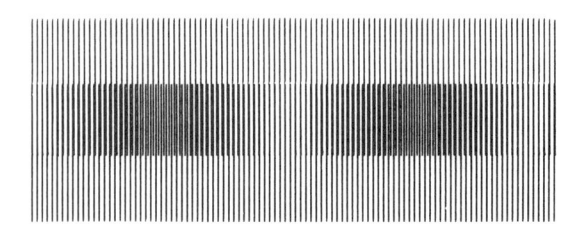

Figure 7.10 Vernier fringes

This can be done by modifying the system so that the interval between the lines on one sheet is slightly different from that on the other one. Suppose that the spacing between the lines on the two sheets is such that the distance covered by N lines on sheet 1 is the same as that covered by $N+1$ lines on sheet 2. At intervals equal to N lines on sheet 1 the lines in the two sheets will align to produce a transparent 'fringe' while midway between these light fringes a dark fringe will occur because the transparent lines on one sheet will line up with opaque lines on the other to prevent any light from passing. The overall effect is that of magnifying the pattern of lines on the sheets by a factor of about $N:1$. This is the principle of the **Vernier fringe** shown in Figure 7.10.

If one sheet moves slightly these fringes will be seen to move N times as fast because of the magnification resulting from the Moiré effect. The presence of the fringes makes it possible to detect the direction of motion by using a pair of light detectors close together and noting which detector picks up a fringe first.

Geometrical interference transducers are rather expensive but offer very high resolution. Typical resolutions before costs become prohibitive are 1 µm, or 1 arcsecond over the full resolution. In fact, in most linear systems lower resolution transducers are employed with grating pitches of 10 µm to 20 µm, and interpolation is used to obtain resolutions down to 1 µm.

Both Moiré and Vernier fringes are examples of geometric interference. The fringes are defined by the overlap of dark areas and their pitch is a function only of the grating pitches and their angle of alignment; these are all settable by the manufacturer. Transducers based upon these fringe effects are widely used for measuring changes in angular position. They are also used to measure changes in linear position but here care has to be taken to protect the very fine patterns from contamination by dirt.

7.4.4 Magnetic transducers

Many optical transducers suffer from being susceptible to errors from the presence of dust or oil in many of the places where they would otherwise be very attractive, and an alternative is to use a magnetic method of implementing the electronic 'ruler' instead. One method is to make use of a magnetic plate or disc with holes or slots cut in it. The change in position can be calculated by counting pulses as with the optical method, but in this case it is the change in magnetic flux passing through the 'ruler' rather than changes in its transparency to light which is used.

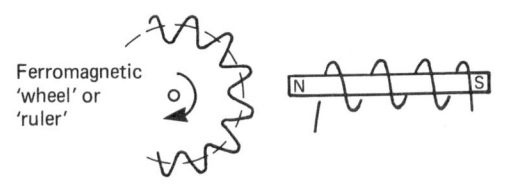

Figure 7.11 Magnetic sensor

The sensor can consist simply of a narrow magnet, around which is wound a coil. When a piece of ferromagnetic material such as iron or steel comes near the tip of the magnet the flux through the magnet (and hence through the coil) increases and a pulse of current appears in the coil. The pulses from the coil are amplified and counted. This method is very robust and can be used to monitor the rotation of a gear wheel, illustrated in Figure 7.11, even in the presence of oil and dirt. Unfortunately it cannot detect the direction of motion which severely limits its usefulness for accurate measurement. The solution is again to use a pair of detectors separated by a fraction of the distance between the 'marks' to resolve the direction of motion.

The ability of this technique to operate even in dirty environments makes it attractive for automotive applications such as distance measurement (odometry) and anti-skid systems in which the speed of rotation of the wheels is measured.

One very elegant magnetic transducer is the **inductosyn** which consists of a printed 'meander' in which alternate conductors carry the same current but in the opposite direction. The result is a magnetic field which reverses sense at each conductor, and this effect is used to form the marks for the increment counter. By using alternating current it is possible to produce an alternating magnetic field which reverses phase at regular intervals as the detector, which consists of a small coil to sense the field, moves. The direction of motion can be determined by using two closely spaced detectors and the distance travelled by counting the number of reversals. This technique can be used to resolve distances smaller than the distance between the conductors by comparing the signals from the two detectors and it is sufficiently robust and tolerant of contamination to be usable in applications such as metal cutting machines. The inductosyn is not as good as fringe methods when high precision is required, but for most purposes it is perfectly acceptable.

7.4.5 Potentiometric methods

The potentiometer provides a very simple method of converting position to a voltage signal which can be converted to numerical form using an analog-to-digital converter. Angular position can be measured using an ordinary rotary potentiometer similar to the ones used for volume controls on radio sets. These contain a resistive track across which a voltage is connected and along which a sliding contact moves as the knob is turned. The fraction of the input voltage that appears on the slider varies with the position of the slider.

In the case of a volume control the resistance per unit length of the track is not constant. Instead, a deliberate nonlinearity is introduced between the knob position and output voltage which produces the correct relationship between position and volume. However, potentiometers used for position measurement must be as linear as possible (that is, the resistance per unit length must be as constant as possible) to minimise errors. The materials used in ordinary volume controls are unsuitable for good displacement transducers intended for continuous use. As well as the problem of nonlinearity, they suffer from noise, poor resolution, and a short lifespan before degradations occur. For many years, 'slide-wire' and 'wire wound' potentiometers have been used as transducers, but although linear, they suffer from the other problems outlined above. Modern conductive plastics give smooth, long-life, low noise surfaces with 'infinitesimal' resolution, but poor resolution unless expensive techniques are used. One fairly recent approach is to use wire wound potentiometers with a covering layer of conductive plastic. With this technique the wire wound device maintains the linearity whilst the plastic material smooths out the coarse variations between the wires, and the life, noise, and resolution are improved.

If the resistance per unit length is constant, the voltage drop per unit length will be constant provided that negligible current flows in the lead connected to the slider. This means that the ratio of the output voltage to the voltage connected to the potentiometer is the same as the ratio of the distance of the potentiometer 'slider' (measured from one end of the track) to the overall length of the track. Note that in this case the potentiometer provides a linear variation of output voltage with resistance because the total resistance of the potentiometer remains constant. Consequentially it is ideally suited for ratiometric conversion.

7.4.6 Differential transformer

A fundamental problem with the potentiometer method is that there will always be wear on the track, although modern designs are very good. An alternative method is to use inductances instead of resistances and to vary the inductance, by moving a piece of ferromagnetic material into and out of the coil of wire. One way to do this is to use two coils and a single piece of ferromagnetic material which can move from one coil to another according to the position being measured. If these two coils are wired in series they form an inductive potentiometer, and the (alternating) voltage that appears at the output will vary with position. This voltage must then be converted to a direct voltage before conversion using an ADC.

This is the principle of the **linear variable differential transformer (LVDT)** shown in Figure 7.12.

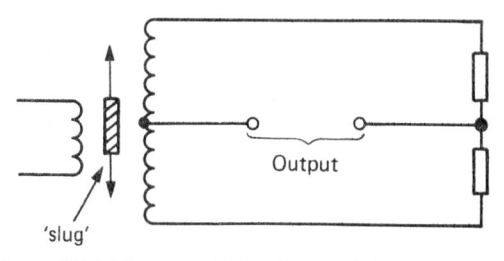

Figure 7.12 Linear variable differential transformer

The primary of this transformer consists of a single coil and the secondary consists of a pair of identical coils positioned symmetrically with respect to the primary. The two secondary coils are wired in series in such a way that their output voltages subtract. Inside the coil is a core of soft iron or similar ferromagnetic material which is free to move. When this core is in mid-position the outputs of the coils will exactly cancel out and the overall output signal will be zero. However, if the core is moved to either side of the mid-point the signal will become progressively larger. The phase of the signal will depend upon the direction in which the coil has been moved.

This signal can be converted to a corresponding DC signal of suitable polarity simply by sampling it at the peak of each cycle of the input waveform. If a high-speed ADC is used the peak value of each cycle of the output signal can be sampled and converted directly to digital form without the need for a separate circuit to convert from alternating to direct voltage.

Synchro resolver
A similar principle can be applied to angle measurement by using a rotatable sensing coil and a set of identical fixed coils wired at different angles. The amount of coupling between a pair of coils varies as the cosine of the angle between them. If sets of coils at right-angles are used the sensitivity of the sensing coil to the field produced by the two fixed sets of coils will vary according to the sine and cosine of the angle. Several techniques are available for converting these signals to provide a digital representation of the angle.

7.4.7 Velocity measurement

The techniques for measuring position can be adapted to the measurement of velocity by differentiating position with respect to time. For example the number of pulses produced per unit time from an incremental encoder can be measured and used to provide a velocity signal. If the interval between pulses is large, so that the frequency is low, a more accurate measurement may be possible by timing the interval between two pulses. A more direct method is to use a transducer which produces a signal which is directly proportional to velocity, such as a **tachometer**. This is essentially a small DC generator which produces an output voltage which varies in proportion to the rate of cutting lines of force in a magnetic field within it, and hence the speed of rotation.

When a signal is reflected from a moving object, its frequency is changed by an amount which depends linearly upon the speed of the object. This is the familiar **Doppler Effect** which can be used with sound or electromagnetic waves to measure speed. If the speed and frequency of the signal are known, the speed may be calculated directly; usually it is the difference in frequency between the transmitted signal and that reflected which is measured by the transducer. This is the principle of Doppler radar which normally uses a solid-state microwave source, and of Doppler sonar which uses sound at ultrasonic frequencies.

7.4.8 Controlling electromagnetic actuators

Almost all methods for controlling position rely upon the use of electromagnetic devices such as solenoids and motors. These present a number of problems to the system designer, mainly due to the presence of **back e.m.f.'s**.

When the current flowing in a coil changes, a voltage appears across it. This is not usually too much of a problem when switching on the current, and the only effect of the inductance of the coil is that the current takes some time to reach its final value when a voltage is applied. When the voltage source is disconnected, however, the effect of the inductance is to attempt to maintain the same current flowing in the coil.

Another way in which a voltage of this sort can appear is due to the windings in an electric motor acting as a generator as they rotate. Again, the danger arises when the source of power is disconnected, since the motor then attempts to supply power to the interface. In each case the source of the voltage is the magnetic energy stored in the actuator when it is operating. The solution to these problems lies in the provision of 'flywheel' diodes as described in section 5.6.4.

Solenoids

Solenoids provide linear motion, due to the magnetic attraction of a coil, for a piece of ferromagnetic material such as soft iron when the coil is energised. When the energising current is removed, the solenoid does not provide any force to return this ferromagnetic 'slug' to its original position, which is usually done by means of a mechanical spring, as shown in Figure 7.13. These actuators do not provide positional control in themselves; they simply create a force when a current passes through their coil. However, the fact that a spring is used means that a technique is available for translating this force

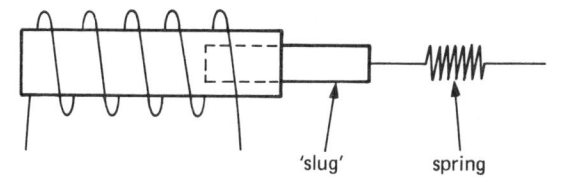

Figure 7.13 Solenoid

into a linear motion; the greater the current passing through the coil, the greater the force and the greater the motion of the slug. The presence of friction and the variations in the spring stiffness and the magnetic properties of the slug mean that this is a very inaccurate method of positioning an object unless a position measuring transducer is attached to the slug and the microcomputer is used to adjust the current in the coil.

DC motors
Many types of motors are available, and the factors which govern their speed of rotation differ widely. In the case of DC motors, the speed of rotation is governed by the current which flows in their windings and the load which is connected to their shafts. Because the load affects the speed of rotation, feedback must be supplied from a transducer if the speed must be accurately controlled. The control of DC motors requires specialist design techniques for all but the smallest types because of the high powers used and the danger of damaging the motor windings or drive electronics.

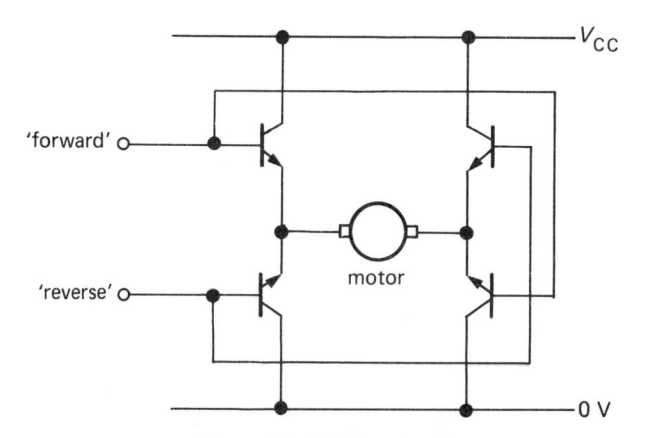

Figure 7.14 Bridge circuit

Very small motors with a power consumption of a few watts can be operated without difficulty using a DC power amplifier. Many applications for motors of this type require reversible motors, which in turn requires that the current in the motor windings must not only be controllable, but reversible as well. This can be done by connecting the motor between the outputs of two amplifiers in a 'bridge' arrangement, as shown in Figure 7.14. The voltage across the motor can be varied almost up to the power supply voltage, with either polarity, as the input voltage (for example from a DAC) is varied.

At higher ratings the power dissipated by the amplifier electronics becomes excessive and switching outputs must be used instead of continuously variable analog signals. However, the switching signal can be produced by an averaging DAC as was described in Chapter 6. The motor can be connected in a transistor bridge in which diagonally opposite pairs of transistors are switched on at the same time. The direction of the current flow in the motor depends upon the pair of transistors which are switched on and the average current depends upon the proportion of time for which they are conducting.

Stepper motors

One shortcoming of the DC motor in control applications is the fact that the position of its shaft must be measured by an external transducer. The **stepper motor** is a variation of the DC motor principle in which the motor has been modified to rotate in a 'stepwise' rather than the smooth manner associated with a well designed DC motor. As shown in Figure 7.15a, stepper motors have a number of separate windings, each of which can be energised independently. The motor can be made to rotate by energising each coil in turn in a cyclic fashion (Figure 7.15b); each time the next coil is energised the motor rotates by a fixed amount depending upon the design of the motor. If the sequence is monitored by the microcomputer, a count can be kept of the cumulative angle through which the motor shaft has turned without the need for a shaft encoder. The direction of the motor's rotation can be reversed simply by reversing the sequence of signals to the motor.

Stepper motors allow the angular speed and position of a shaft to be controlled directly by a microcomputer. The coils of small stepper motors can be controlled by series transistors as described in Chapter 5; flywheel diodes must be used because of the inductive nature of the motor coils. Larger stepper motors require more specialised drive circuitry in order to obtain the best output torque and speed from the motor. The sequence of switching on and off each coil in turn can be controlled directly by the microcomputer; normally this would be done by a subroutine which generated the next output signals in the sequence each time it was called, the direction of rotation being passed to the subroutine as a parameter. Alternatively, external circuitry could be used to carry out the equivalent operation, producing output signals which would cause the motor to step to its next position each time a pulse was received on its clock input; a second input signal would be used to determine the direction of rotation.

The inertia of the load being rotated by the motor limits the maximum rate of angular acceleration which can be used in an application, because of

7.4.8

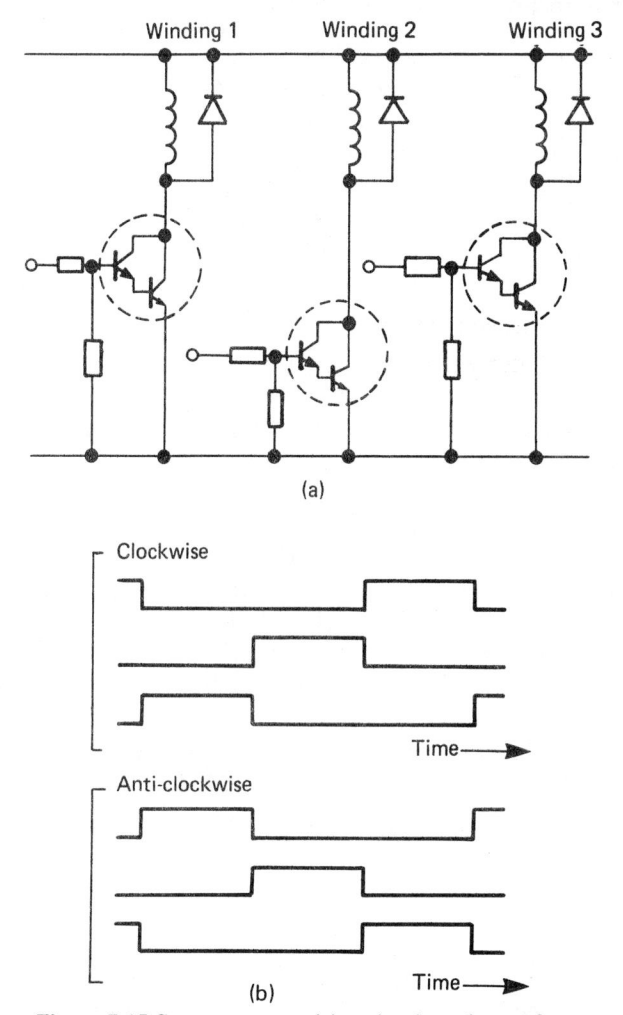

Figure 7.15 Stepper motor drive circuit and waveforms

the limited torque which can be exerted by the motor. This is true of any kind of motor, but it is more important in the case of the stepper motor, since in most cases the design assumes that the motor turns through a fixed angular increment each time the output signal to the motor is changed. The solution, of course, is to limit the rate at which the drive signals to the motor are accelerated, which directly affects the way in which the drive software is designed.

The easiest way to define a time interval to determine the speed of rotation of the motor is to divide the system clock frequency by an appropriate amount, but this approach makes the calculation of accelerations more difficult.

7.5 Acceleration and force

It is possible to make use of physical principles to measure force directly, for example by exploiting the piezoelectric effect described below. Another method is to use a coil with some control electronics to produce a force that exactly cancels out the applied force; the counteracting force can then be measured in terms of the current flowing in the coil if the geometry of the coil is known, this principle being known as a **current balance**. However, most methods of measuring force do so indirectly, by converting it into a signal of another form. The measurement of acceleration and force are closely related, and will be discussed together.

7.5.1 Piezoelectric transducers

Some materials have the property that a voltage difference appears across a crystal of the material when it is stressed mechanically. This **piezoelectric effect** is used in a number of types of transducer for measuring force and related quantities such as acceleration.

The voltage arises from the movement of electrical charge within the crystal as a result of its being deformed by the force. It is the charge which carries the signal, and if it is allowed to leak away the voltage will disappear. For this reason piezoelectric transducers are best suited to the measurement of transient effects such as vibration, but need protection from dirt and oil. If environmental protection and high quality cables and connectors are used they can be employed effectively in mechanical applications, and they are finding application in the automotive industry. A voltage amplifier with a high input impedance can be used for signal conditioning, but by reconsidering first principles a more suitable amplifier can be designed.

Charge amplifiers
In this case we require an amplifier which can produce an output voltage proportional to the input charge. In principle this can be done simply by replacing the feedback resistor in the shunt feedback current-to-voltage converter with a capacitor (Figure 7.16). The voltage across the terminals of a capacitor is related to the charge which it contains by the formula

$$V = \frac{Q}{C} \tag{7.6}$$

where Q is the charge in the capacitor, V is the voltage across the capacitor, and C is its capacitance.

Because current is a flow of electrical charge, it follows from the discussion of the current-to-voltage converter that all the charge entering the input will flow to the capacitor and produce a voltage at the output of the amplifier. This voltage will be related to the charge by equation 7.6 provided that no current flows in the input terminal of the amplifier.

It is this proviso that causes difficulties in practice because the amount of charge produced by a piezoelectric transducer is often very small. Even

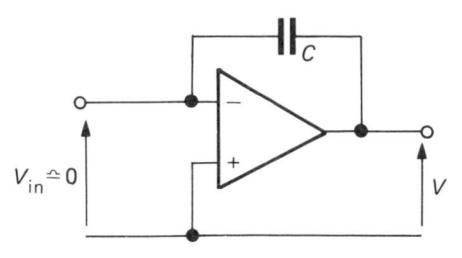

Figure 7.16 Charge amplifier

the tiny input currents associated with modern operational amplifiers will cause the charge in the capacitor to change with time so that the output voltage of a charge amplifier connected to a piezoelectric force transducer will tend to 'drift' even when the input force is steady. The design of practical charge amplifiers requires very careful attention to be paid to this problem of current leakage, and piezoelectric transducers, despite offering very great sensitivity, are best suited to applications where changes in the input signal are being measured rather than the actual signal itself.

7.5.2 Accelerometers

In principle, acceleration could be measured using one of the techniques already described and differentiating the velocity data with respect to time. However, when differentiation is carried out in electronic systems it tends to emphasise the noise and errors in the original data, so that the resulting acceleration data would be inaccurate. Another method which is less susceptible to noise is to use the well-known relationship

Force = Mass × Acceleration

which shows how a mass can convert the acceleration to a force that can be converted to a displacement using a spring. This is the principle of the seismometer (Figure 7.17) used to detect the vibrations arising from earthquakes, but so-called 'seismic mass' accelerometers using this principle find applications in many other fields.

7.5.3 Conversion to displacement

Force can be converted to a change in position by making use of the everyday phenomenon of **elasticity**: when a force is applied to a solid, it changes shape. The change of shape is proportional to the applied force provided that it does not become so large that the properties of the material are affected.

These statements can be made more precise. If a solid is compressed, it becomes shorter by a certain amount, while if it is stretched it becomes longer. The fractional change in length, obtained by dividing the change in length by the overall length when no force is applied, is called the **strain**.

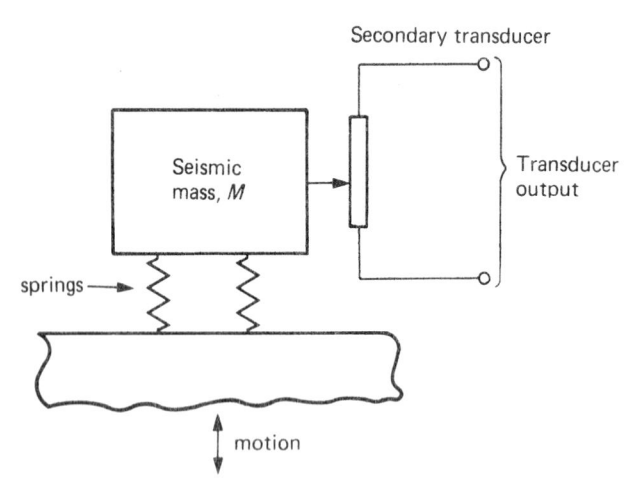

Figure 7.17 Seismic mass accelerometer

The force applied to the material will always be applied across a finite area, and the force per unit area is called the **stress** applied to the material (Hetenyi, 1950). The ratio of stress to the strain which results from it is a property of the particular material known as **Young's modulus**. The values are tabulated for most materials used in engineering.

Elasticity can be used to convert forces into displacements which can be measured in turn using the techniques already described. For example, a spring will change length if the force is applied to one end of it, and the change in length of the spring or the displacement of one end of it can be measured. Strain can be measured directly by means of strain gauges.

7.5.4 Strain gauges

The resistance of a piece of electrically conducting material depends upon its shape; for a rectangular solid it is directly proportional to the length and inversely proportional to the cross-sectional area. Whenever a force is applied to a solid it changes shape, and thus it is not surprising that the resistance of an electrical conductor changes when a force is applied to it. However, this change in resistance arises only partly from the change in the shape of the material and the main cause is a change in the electrical resistivity of the material itself.

In practice, the origin of the effect is not so important to the engineer as the fact that there is a fractional change in resistance which is proportional to the deformation of the material. This is the effect used in **strain gauges** which are resistive transducers that can be attached to surfaces to monitor the tiny changes in dimensions that arise from applied forces. Strain gauges consist of thin layers of resistive material which are usually designed with a 'meander' pattern formed using wire or by etching a sheet of metal foil. When the sheet is deformed in the direction shown in Figure 7.18 its resistance changes, and if it is attached to a surface with a suitable adhesive it can be used to measure the strain of that surface in a certain direction. If the strain is at right angles to the direction of the wires the sensitivity to the applied strain is very much lower, ideally but never quite zero.

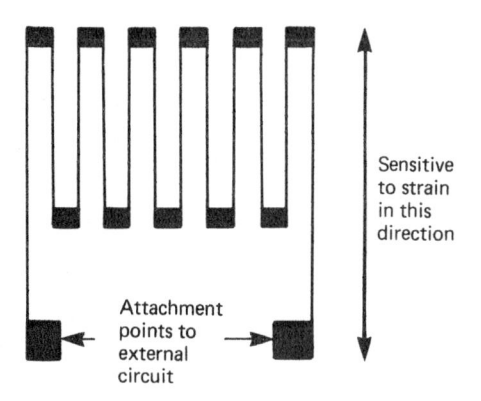

Figure 7.18 Strain gauge pattern

The fractional change of resistance, the change in resistance resulting from the applied strain divided by the resistance of the gauge when no strain is applied, is proportional to the applied strain. The ratio between them is called the **gauge factor** and it depends upon the material from which the strain gauge is made. For metal gauges it is typically about two or three, but for strain gauges made from semiconductor materials it can be much higher because these materials show a much stronger piezoresistive effect.

A strain gauge can be used in series with a fixed resistor to form a potentiometer, but as the resistance of a strain gauge varies with temperature as well as with strain, this method gives rise to considerable errors. One 'fix' to this problem is to use another strain gauge for the other resistor and to leave it unconnected to the surface under strain so that its resistance changes only with temperature. In practice strain gauges are often used in bridge circuits where each of the four resistors is a strain gauge, so that the effect of their variation in resistance with temperature is minimised. There are many ways in which the four strain gauges can be connected. They can be 'dummies' not connected to anything, or they can be attached to different points in a mechanical system to provide an output which varies linearly with an applied force.

In **load cells** the gauges are connected to surfaces of a piece of metal which has been machined to show a large strain at which the gauges are attached, as Figure 7.19a shows. Here two of the gauges are active — they measure the strain as the metal is compressed — while the other two are attached at the top and bottom of the hole where there is scarcely any strain. In the cantilever arrangement of Figure 7.19b the upper surface of the cantilever is under tension (and hence stretches) while the lower surface is under compression. This allows all four gauges to be attached in such a way that they are active.

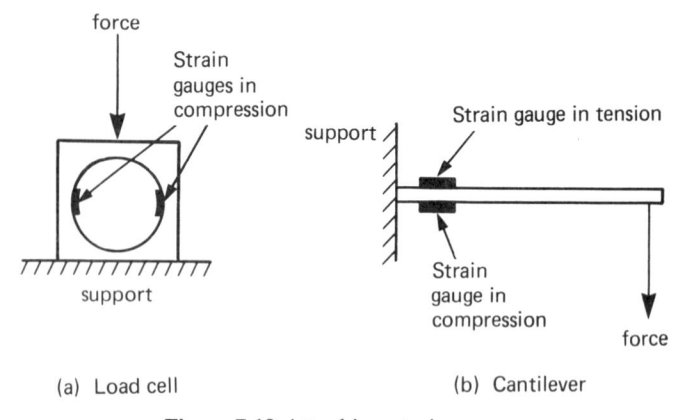

Figure 7.19 Attaching strain gauges

Strain gauges can be used directly to measure strain or if the mechanical properties of the system to which it is connected are known, the force applied to the system. This makes it suitable for measurements of torque and for weighing as well as for simple force measurements.

7.5.5 Indirect measurement of force

Sometimes the force being applied by an actuator can be measured indirectly by measuring the current taken by the actuator itself. For example, because the current which flows in a DC electric motor depends upon its speed of rotation and hence upon the load which it is attempting to drive, the force can be measured indirectly by measuring the current which it takes when its voltage is being controlled. This technique is not as accurate as some of the other ones which have been described in this chapter but it is often accurate enough; for example in measuring the load on a joint in a robot arm, or for estimating the wetness of a car windscreen by measuring the current taken by the windscreen wiper motor.

7.6 Pressure and flow

The everyday terms 'pressure' and 'flow' need to be defined rather more precisely when transducers to measure them are being specified. Pressure can be measured in two ways:

- **Absolute pressure** is the pressure measured with respect to a vacuum, that is, zero pressure. It is the absolute pressure that determines things such as the boiling points of liquids.
- **Relative pressure** is the difference between two absolute pressures, for example the differential pressure between the inside and the outside of a

pressurised container. The relative pressure which is most often measured is that with respect to atmospheric pressure, and because it is this pressure that most pressure gauges indicate, it is called **gauge pressure**. This is the pressure that a tyre gauge would indicate.

Whenever a pressure transducer is to be used, the designer must ensure that the appropriate type of measurement will be made!

The same confusion of terms is possible when flow is being measured; the speed of flow in the centre of the pipe and the overall volume flowing per second need not be related if the flow is turbulent. The situation becomes even more complicated if the nature of the material flowing in a pipe can vary. Care is needed to make sure that the correct quantity is being measured.

7.6.1 Pressure transducers

Pressure is normally measured by means of a diaphragm which deforms under the effect of the pressure. This deformation is then measured either as a displacement or by measuring the strain in the diaphragm. One type of transducer which is becoming quite widely used is a semiconductor pressure gauge in which the diaphragm and a strain gauge bridge with temperature compensation are incorporated on to a silicon chip, and housed in a package with a connection for a pressure hose. It can be interfaced in the same way as an ordinary strain gauge bridge, and some transducers even contain some of the signal processing circuitry.

7.6.2 Flow transducers

The most direct way to measure flow is to make use of a small turbine or rotor wheel placed inside the pipe which is made to turn by the flow. The bearings of the rotor must be made as free-running as possible in order to minimise the errors in measurement and to reduce the obstruction to the flow, but even with very low friction the flow will still be impeded to some extent. The rotation can be picked up very easily by means of a magnetic transducer if the vanes of the rotor contain ferromagnetic material. The output signal from the transducer is converted to pulses which are counted; the number of pulses per second provides a measurement of the linear flow rate.

Unfortunately this technique has a number of disadvantages. The momentum of the rotor will limit its ability to respond to sudden changes in flow rate, so that if the flow comes in spurts, for example due to the action of a pump causing the flow, the reading may not be accurate. At low flow rates the effect of the bearing friction will become more apparent and may give rise to excessive errors. If the fluid contains solids, these may stick in the mechanism or stick to the surface of the rotor and affect its operation, and if corrosive fluids are used they may attack the rotor or its bearings.

The result of this catalogue of despair has been a considerable amount of ingenuity in devising alternative measurement techniques. One long-established method is to use an obstruction in the pipe to cause a pressure drop as the fluid passes through it, and to measure the relative pressure across the obstruction. The relationship between this drop in pressure and the flow rate causing it can be calculated using the laws of fluid mechanics, and hence the flow rate can be deduced. **Venturi meters** and **orifice plates** are examples of transducers based upon this principle. Again, however, the effects of solids or other materials in the fluid can be to further constrict the gap causing the pressure drop, and this will affect the calibration of the transducer. This technique is also susceptible to corrosion.

Another method uses an electrical heater to raise the temperature of the fluid passing down the pipe. The slower the rate of movement of fluid, the greater will be the time that any part of the fluid will spend near the heater, and the greater will be the temperature rise. Thus if the temperature is measured both upstream and downstream from the heater, the temperature rise can be obtained, and after a calculation involving the specific heat of the fluid, the flow rate can be deduced.

If the fluid is known to be heterogeneous, that is, a mixture of different types of materials, this fact can also be used to provide the basis of measurements. The sort of mixture does not matter greatly as long as it is likely to remain there; it could be bubbles in a thick oil, or blood cells in a vein. One elegant type of measurement involves reflecting ultrasonic waves off the discontinuities in the fluid and measuring the Doppler shift — the change in frequency — of the echo, which is proportional to the linear flow rate. This is attractive in medical applications because it is non-invasive; it does not involve putting anything into the flow in order to measure it.

In some cases it is possible to use **correlation** of measurements at two nearby points in a pipe to determine how long it takes the fluid to travel from one point in the pipe to the other. The measurement can be of its colour, temperature, opacity or any other easily measurable physical property which changes because the fluid is heterogeneous. Sometimes small vortices may occur naturally due to the flow, or they may be induced deliberately, and these may be detected to provide information about flow rate. The signals from the two transducers are taken to a computer where they are stored and compared. One signal should resemble the other one shifted in time; this being the amount of time taken for the fluid to pass from one transducer to the other, and knowing the distance between the transducers allows the average linear flow rate to be calculated.

7.7 Peripheral equipment

The peripherals — the input/output equipment of the computer — contain different types of sensors and actuators, and it seems appropriate to mention them in this chapter.

7.7.1 Permanent storage media

When information is not in the computer's main memory, it must be stored in some form which can be read back when necessary. Many types of storage mechanism are available based upon a range of physical phenomena. However, the methods which are currently in widespread use are all based upon magnetisation. Most make use of magnetic media to store the information in the form of patterns of magnetisation on a surface that moves past a 'head' which can both write the patterns and read them again.

When a permanent storage device is used, the information appears to the user as if it is stored in the form of **files**. A file is conceptually like a file in an ordinary office filing system. It could contain such items as a document that was being written using an editor, or a source program written in a computer language or a set of readings from some instrumentation. The storage unit is controlled by software which looks after the organisation of the files enabling them to be stored on the medium being used.

Magnetic tapes

One type of storage uses a magnetic tape, either on reels or in the form of a cassette. The information to be stored is broken up into 'blocks' of standard length which are recorded end to end with inter-block gaps along the tape, either as a set of parallel tracks of information, or in simpler systems, as a single track. The individual bits are typically represented by changes in the level of magnetisation of the magnetic coating on the tape which are 'recorded' and 'replayed' in just the same way as in an ordinary domestic recorder, using a 'tape head' containing a small coil to cause or sense magnetisation of the storage medium.

This type of storage is suitable for keeping records of information which need to be loaded into memory relatively infrequently, the problem being the need to wind along the tape to find information whenever it is required.

Hard and floppy discs

The essential problem with magnetic tape is that it is a sequential-access storage medium. The second type of magnetic storage provides us with a 'random-access' medium in the form of rotating **discs** in which data is stored in the form of concentric circular **tracks**. The information can be located by moving a magnetic sensor, similar to the recording head of a tape unit to the radius appropriate to a particular track and waiting until the desired **sector** appears as the disc rotates.

There are two main types of disc in current use; 'hard' and 'floppy' discs. **Hard discs** are rigid, consisting of a disc of nonmagnetic metal alloy covered with a magnetic material. In some systems the hard disc is removable, but more commonly it is fixed within a sealed chamber where it is not vulnerable to the effects of contamination by dust. Disc drives of this type often incorporate **Winchester** technology; they use low mass lightly loaded heads that start and stop in contact with a lubricated disc. When the disc moves it generates a moving film of air, or air bearing, which supports the recording head. With Winchester discs the air film allows the head to hover very close to the disc surface, and this results in the detection of a larger signal than with some older disc technologies. A consequence of this is large storage density and capacities with discs of a few inches in diameter. One problem with Winchester discs is the need to 'back up' the information which they carry, that is, to make copies of the information in case it is lost either through inadvertent erasure or due to a fault. The contents of a fixed disc can be backed up using a tape system or by means of a removable disc.

The most common type of removable disc is the **floppy disc** which consists of a disc of flexible plastic sheet covered with a magnetic coating, rather like the material from which recording tape is made. This disc is held within an envelope of plastic with a hole at the centre for connecting the drive mechanism and a radial slot to allow the magnetic 'head' to contact the disc itself during operation. 'Floppy' discs have smaller capacities than their fixed counterparts but are used rather differently. In the case of fixed discs all information is available all the time, while in a system using 'floppy' discs only the information on the discs currently in the disc drives is accessible. Floppy discs are widely used in 8-bit microcomputer systems where relatively small amounts of information, typically up to a megabyte, are handled. The more sophisticated software found in 16-bit systems demands greater storage capacities and 'hard' discs are more widely used. 'Floppy' discs are commonly used for backing up information stored on non-removable Winchester discs.

Magnetic bubbles
The disadvantage of tapes and discs is that they both need mechanical moving parts, and much development work has been carried out in the area of magnetic 'bubbles' where the patterns are moved electromagnetically through special materials without the need for moving parts. The result is a storage mechanism that is much more compact and robust, but at present the storage capacities of bubble storage devices are less than those obtainable with moving media systems.

7.7.2 Printers

Just as there is a need for permanent storage of information in machine-readable form, software development requires the production of permanent records that can be read by humans. Computer output printed on paper is often called 'hard copy' to distinguish it from 'soft copy', the corresponding output presented in electronic form on a screen. The ability to produce 'hard copy' is important during the development of programs because the screen of a video display terminal normally shows only 24 rows of output and it is much easier to read material of more than one page from a printed version (with about sixty rows per page) than from a screen. This is especially true when documenting programs where printed material such as users' manuals is required. Documentation looks better when printed in a neat typeface with a high-quality printer, while the production of program listings during development really needs a high-speed printer which can operate as quickly as possible. These two sorts of application are best served by different types of printer, although a compromise may be possible.

Printers can be divided into two main groups: **line printers**, which produce a whole line of type at once, and **character printers** which type characters one at a time. Because line printers print a whole line at a time, they are usually much faster than character printers, and typically they can print several hundred lines per minute. Line printers are more useful for producing output from programs and for listings; they are not so suited to documentation purposes because the quality of their print is usually somewhat uneven, although designs are improving.

The speed of character printers is normally quoted in terms of characters per second, and typical speeds lie in the range of tens to hundreds of characters per second. Character printers are available operating on a number of different principles such as electrochemical and thermal effects which cause colour changes in specially treated paper. However, the overwhelming majority of those in general use are **impact printers** which produce the marks on the paper by striking an inked ribbon. Some printers have typefaces defined by shapes cast into a metal or plastic surface which are selected by the electronics of the printer and struck by a solenoid-operated hammer to form the shape of the letter on the page in much the same way as an electric typewriter. As might be expected, the result is comparable with that produced by an electric typewriter and quite adequate for 'correspondence quality' printing and documentation. Two main versions exist: the **golfball** in which the print head resembles a metal golfball covered with the shapes of the letters that can be printed, and the **daisywheel** which has a plastic or metal wheel resembling a flower with many petals, each petal carrying a different letter. In each case the typeface can be changed by changing the print head, which although it allows a choice of typefaces is inconvenient when several typefaces are to be used in a single document. The maximum speed of operation of both types of printer is limited by the inertia of the print head itself. The moment of inertia of the daisywheel is rather less than that of the golfball which leads to a somewhat higher speed for this type of printer of about fifty characters per second.

The second kind of impact printer in widespread use is the **matrix printer** in which the letters are formed by a pattern of dots. These dots are made by a group of fine wires contained within the print head which are operated by solenoids controlled by complicated drive circuitry. The logic which generates the pattern of dots corresponding to each character is provided by a microcomputer within the printer which translates sequences of ASCII characters into the appropriate pattern of dots on the page. Although each wire has to move several times during the printing of each character, the inertia of the wires is very low and higher print speeds are possible than with the daisywheel printer, up to two or three hundred characters per second. Unfortunately the print quality produced is not as high as that of the daisywheel because the individual dots can be seen if the page is examined closely. Some of the more recent designs of matrix printer overcome this shortcoming by optionally allowing the printer to operate more slowly so that the characters can be produced by means of overlapping dots. In this way much better print can be obtained. Another advantage of the matrix printer is that the typeface can be changed much more easily because it is generated in software, so that with the use of special control characters it is possible to have mixed typefaces within the same computer-printed document without having to change the print head.

Laser printers also generate images on paper in the form of dots, but in this case the dots are very close together and are generated using a controlled light source and an electrostatic technique to define the patterns of ink on the paper, rather like a photocopier. They are capable of high quality printing in a wide range of fonts.

7.7.3 CRT display units

The commonest form of visual display is still based upon the **Cathode Ray Tube (CRT)** of the type used in television sets for decades. This projects a beam of electrons which forms a point of light when it strikes the layer of phosphor material which forms the screen. The brightness of the dot can be controlled by a voltage applied to a control electrode, and the position of the dot may be controlled by deflecting the electron beam either magnetically or electrostatically.

The extremely low mass of the electron allows it to be scanned very rapidly across the screen. The conventional scanning format causes the dot to be moved at a steady speed from left to right across the screen, starting at the very top. When the point of light created by the beam reaches the right-hand side, it is returned to the left-hand side once more and the scan is repeated slightly lower down the screen. This scan from left to right is repeated until the whole of the screen has been covered by what appears to the viewer as a finely spaced grid of lines, after which the electron beam is returned to the top of the screen to repeat the scanning procedure. This grid of lines, known as a **raster**, must be repeated sufficiently rapidly that the human eye is not aware of the motion of the beam, due to 'persistence of vision'.

In practice there are two widely used scanning standards for television. In general, the screen is scanned from top to bottom 50 times per second in countries where 50 Hz electrical mains are used, while in those countries where 60 Hz mains are used, the screen is scanned 60 times per second. In both standards, **interlacing** is used: even and odd numbered rows are scanned during alternate traverses from top to bottom of the picture. This reduces visible 'flicker' while maintaining the same signal bandwidth. However, the standards differ in the number of lines making up the raster; in the 50 Hz system there are 625 lines in all, while in the 60 Hz system there are only 525. This television format is a useful basis for information displays.

7.7.3

Instead of using the output from a camera, a signal is generated by a special purpose interface which causes written or graphical information to appear on the screen.

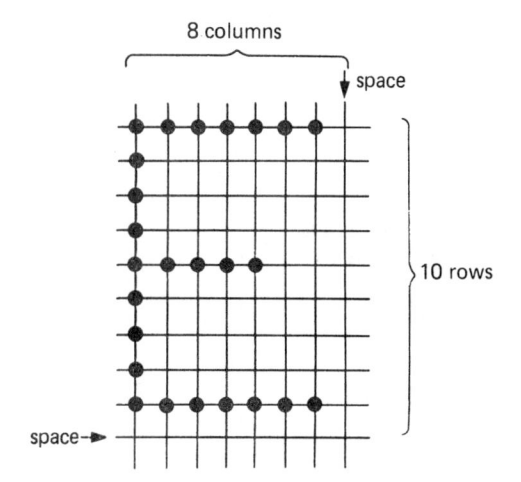

Figure 7.20 Forming characters from dots

Character displays
Information may be displayed on the raster by varying the brightness of the point of light as it moves, and in this way characters can be drawn on the screen. Usually, each character occupies a fixed amount of space on the screen and is formed as a pattern of dots which merge on the screen to form the outline of a character. Most video display systems allow up to 80 characters to be displayed in each of 24 rows, although many other arrangements are available. Similarly, different systems use different numbers of dots to display characters, although an array of dots occupying 8 columns and 10 rows is commonly employed, as shown in Figure 7.20.

The **video** signal used to control the brightness of the dot on the screen must carry information at a very high rate. If a character is represented by $8 \times 10 = 80$ dots and there are 24 rows of 80 characters in a display, $24 \times 80 \times 80 = 153\,600$ bits are needed to 'draw' the contents of a single 'frame' on the screen. If a non-interlaced display is scanned at 50 Hz, 50 frames must be drawn per second, implying an information rate of almost 8 million bits/s, and this rate can be much higher in the case of higher resolution displays. This very high information rate means that the video signal cannot be generated directly in software, and the special purpose hardware needed for this task is provided in the form of a **Cathode Ray Tube Controller (CRTC)**. CRTCs are available for most microprocessor families, and although they differ in their detailed design, all CRT interface circuits contain the same major blocks, as shown in Figure 7.21. The main sections are:

- timing controller generates synchronising signals for display scanning and the remainder of the interface.

- refresh RAM contains the ASCII codes of the characters to be displayed.
- character generator ROM contains the patterns of dots which are used to represent each of the characters. The address supplied to this ROM consists of the ASCII code of the character to be displayed and the number of the row in the character pattern required.
- shift register is loaded in parallel from the output of the ROM and the resulting data is shifted to produce a serial signal at its output.

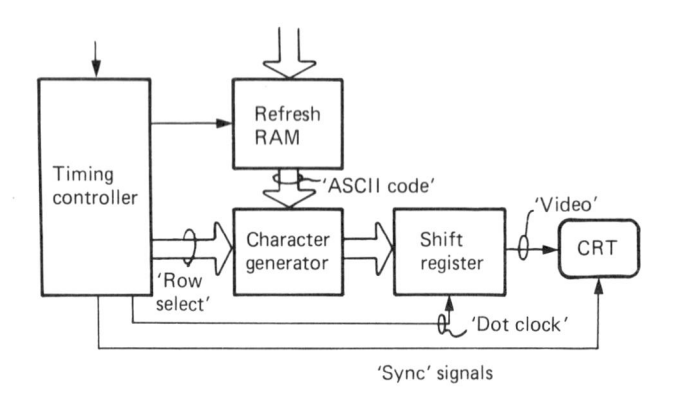

Figure 7.21 Character display interface for CRT

Most CRTCs allow a **cursor** to be displayed on the screen to indicate a character position and this can be moved by special commands to the CRTC. This cursor consists of a distinctive feature such as a flashing rectangle which moves one position to the right each time that a character is 'written' on the screen. When the screen is full, the information displayed on the screen can be changed either by 'wiping' the screen to start a new 'page' or by moving the contents of the screen up by an amount equal to one row of characters to remove one line at the top of the display and reveal a new blank line at the bottom, a process called **scrolling**.

Although characters are written on the screen at a scan rate which is too fast for the human eye to detect, any character on the screen is in fact flashing at 50 Hz or 60 Hz, the exact time of each flash depending upon the position of the character on the screen. This flashing can be detected by a sensor such as a phototransistor, and with suitable circuitry the video interface can determine the position of the sensor if it is pressed against the screen. This is the principle of the **light pen** which allows a user to interact directly with a video display. It is especially useful in 'menu-driven' systems when the user chooses from a list of options displayed on the screen rather than typing them in.

7.7.3

Graphics displays

Diagrams may be drawn on a character display by using special symbols to form commonly used shapes such as straight lines and corners. If a display is intended primarily to draw graphics, however, a special-purpose graphical display controller should be used. The main difference between this type of display and a character display is that each individual 'dot' of the picture is controllable individually instead of being treated as part of a character.

Each 'dot' forms a 'picture element', normally abbreviated to **pixel** or **pel**. The information describing each pixel is stored in a special-purpose RAM known as a **bit plane**. The pixels can be accessed using registers in the controller which contain the x and y co-ordinates of the pixel currently being 'pointed' to by the controller. Alternatively, the bit plane can be memory-mapped, so that the microprocessor can address the elements of the display directly without the need to use pointers in an interface circuit. The controller circuit usually provides facilities for drawing lines and other shapes on the display without the need for the microprocessor to calculate the co-ordinates of the pixels making up those shapes.

Displays of this type are called 'raster graphics displays' because they make use of a raster to draw images on the screen. 'Vector graphics displays' control the motion of the electron beam to draw shapes on the display as a set of lines or 'vectors'. Raster displays are superseding their vector counterparts in most applications because of the availability of low-cost RAM for the bit plane. The spatial resolution of the display depends upon the application for which it is intended. Low resolutions of 64×64 or 128×128 are adequate for tasks such as monitoring object identification in automatic and robotic applications. Medium resolutions of 256×256 and especially 512×512 allow images approaching those of broadcast television to be created. Some applications such as flight simulators and computer aided design require higher definition than this. Such equipment needs special scanning circuits and higher resolution than can be achieved using standard television components and it tends to be considerably more expensive as a result.

A bit plane with a single bit defining the state of each pixel allows black and white diagrams to be drawn on a raster display. Increasing the number of bits allows a variable voltage to be produced using a DAC, allowing the brightness of each point to be varied to create a monochrome picture with half-tones. Colour pictures can be drawn by using three DACs, one for each colour component (red, green and blue) of the picture. Simple colour displays, such as those used on some personal computers, use only one bit to control each colour component to produce eight different 'composite' colours, as shown in Table 7.2.

The use of several bits to define each of three colour components allows much more subtle shading to be achieved. Unfortunately, doing this for all the pixels of a display requires a very large amount of RAM compared with that used for storing the program in most systems, and many more shades and hues can be produced than is necessary in almost any display. For

Table 7.2 8-Colour display

BGR	Colour
000	Black
001	Red
010	Green
100	Blue
011	Yellow
101	Magenta
110	Cyan
111	White

example, if eight bits were used for each of the three colour components, 2^{24} or approximately 16 million different colours and shades could be represented, more than the total number of pixels! A common solution is to use a **palette** which is a look-up table which maps the bits corresponding to each pixel to the appropriate signals to drive each of the three DACs to produce the exact colour and brightness desired. In this way complicated shading effects can be achieved with only a few bits per pixel. The generation of complex images by computer is a complex and specialised field, and the reader is referred to texts such as Foley and van Dam (1982, pp. 95-135) for a more detailed coverage of graphics display hardware.

Interactive graphics peripherals

The light pen has already been mentioned during the discussion of raster displays. It can be used as a 'pen' to write on the screen by programming the microcomputer to leave a trail of white pixels behind the light pen as it moves across the screen. This is rather like drawing freehand on a drawing board, and does not give particularly accurate results, and programming 'tricks' are normally used to make drawing easier. For example, the trail of brightened pixels can be made to act like a taut rubber band attached to the end of the light pen, so that the line drawn is always straight. The ends of the lines are sometimes made to move automatically to join the end of other nearby lines, a technique known as 'gravity'. Alternatively, the light pen may be used as a 'pointing device', for indicating different regions of the screen, often to give commands by means of a menu. The same pen may be used in both types of application during drawing, for example to indicate that the next line to be drawn is to be dotted rather than solid.

Many other techniques are available to allow the user to interact with computer-generated images. Some can be used for drawing, while others are more suited for pointing, but the techniques can be categorised into digitising devices, direct pointing devices and indirect pointing devices.

Direct pointing: *touch sensitive screens*

These allow the operator to point directly at objects drawn on the screen of the display. The four main methods described below are in current use, the choice of method application depending to a great extent upon the applications envisaged. All methods suffer from the fact that continued touching of the screen leads to the accumulation of dirt, so that periodic cleaning is needed.

- **Resistive membranes** use a pair of translucent sheets placed over the display. One sheet has an almost invisible pattern of wires etched into it in the x direction, while the other has a similar array of conductors running in the y direction. When pressure is applied to the sheets by the finger or a stylus, some of the wires touch, and the position of contact may be determined by methods similar to those used for scanning keyboards. This technique can achieve high resolution and is not susceptible to accidental triggering because some force is needed to push the two sheets together. However, the sheets are prone to move with time, and are susceptible to puncturing.
- The **capacitance sensing** approach uses a set of transparent electrodes which are directly bonded on to the screen or on to a separate sheet. If one of these electrodes is touched with a finger, its capacitance changes, and this change in capacitance may be detected electronically. The disadvantage of this method is that it will not necessarily work if touched with anything other than a bare fingertip.
- **Acoustic sensing** makes use of acoustic surface waves (a type of sound wave) which are generated by piezoelectric transducers located at the edges of the screen. Normally these waves propagate across the screen from one side to the other, but if the screen is touched by any object some of the sound energy will be reflected from the point of contact back to the transducers. Although this approach offers the advantage that no electrodes or wires are present to obscure the view of the screen, it detects the edge of an object rather then its centre, which limits the accuracy of pointing. Also, the screen must be kept clean to avoid spurious reflections.
- The **optical touch screen** uses rows of LEDs and photodetectors around the edge of the screen to form a grid of infrared beams; any object such as a finger which is placed against the screen will interrupt some of these beams, allowing the position of the object to be detected. In practice the LEDs can be multiplexed and the photodetectors scanned to provide a matrix arrangement; more sophisticated processing of the signals from the photodetectors can allow the centre of the object to be located quite accurately.

Indirect pointing devices

In this case the operator does not touch the screen directly, but causes a cursor to move on the screen to point to the desired position or object. Three main types of pointer are in use.

- The **mouse** is a handheld device used to control the position of a cursor on the screen, and it earns its name from the fact that it is a small rounded object with a 'tail' consisting of a wire leading to the computer. A mouse is used to indicate relative movement, and if it is picked up, moved, and replaced on the surface, the cursor will not move on the screen. It usually operates by means of a spherical roller on its base which operates potentiometers or incremental shaft encoders that output signals corresponding to motion in the x and y directions as the mouse is moved across a flat surface such as that of a desk.

- A **trackerball** is a spherical roller which can be moved with the fingertips to control the position of a cursor. It is similar to a mouse in principle, but because it is mounted in a panel, large and rapid changes in position are more difficult.

- Alternatively, the motion may be applied to a small **joystick** (like a miniature gear lever) which can be moved forward, back, left, or right. Changes in position are sensed by potentiometers and springs are often used to return these potentiometers to their centre positions when force is removed from the joystick. The motions of the operator's hands are smaller when using a joystick than with a mouse or trackerball, and fine control of position is difficult because of the rather jerky motion that results. In practice, joysticks are normally used to control speed of motion across the screen rather than directly to control position. In this case the speed of motion is related to the force applied to the joystick, and 'isometric' joysticks are available which directly sense the force applied by means of strain gauges instead of using potentiometers and springs. Joysticks can also be used as three-dimensional orientating devices by adding a third sensor to detect rotation.

Digitising tablets

Digitiser pads or tablets are indirect pointing devices with high absolute accuracy. They are designed to have high resolution, linearity and repeatability because they are used for converting accurate graphical information such as maps and engineering drawings into electronic form. A digitiser pad has a two dimensional surface analogous to the display screen, and uses an accurate pointing device such as a pointed 'pen' or a 'mouse' carrying a lens with cross-hairs, as shown in Figure 7.22. Large pads of 1 m or more across are used in engineering draughting and in applications such as computer aided design (CAD). A number of techniques are in use.

- **Electromagnetic grids** consist of two superimposed pairs of grids of wires, one arranged in the x direction and one in the y direction. Two

Figure 7.22 Interactive graphics terminal

types are in use, 'active probe' and 'passive probe' systems. Active probe systems use a 'pen' which radiates a high frequency signal which is picked up by the wires of the grid; the position of the probe can be determined by scanning the wires of the matrix. In passive probe systems the wires of the grid are scanned to cause them to radiate high frequency signals which are picked up by the probe. In each case the position of the probe can be calculated to a resolution better than the distance between the wires by interpolation.

- **Magnetostrictive pads** rely upon the fact that when some ferromagnetic materials are magnetised, they change size. This principle of 'magnetostriction' can be used to generate a mechanical pulse at one end of a ferromagnetic wire which then propagates along the wire. This pulse can be detected electromagnetically by means of a probe, allowing the distance of the probe along the wire to be calculated from the time delay of the mechanical pulse and the (known) speed of sound in the wires. Magnetostrictive pads use two superimposed grids of ferromagnetic wires which allow the x and y co-ordinates of the probe to be determined.

- **Acoustic pads** also use sound waves, but in this case the sound is generated within the probe, usually by means of a spark which produces a sharp pulse of ultrasound. Long ultrasonic transducers are attached along two adjacent edges of the pad to pick up this sound, and the time delays thus measured can be processed to yield the x and y co-ordinates of the probe.

7.8 Machine vision and speech systems

Visual information in the form of text or graphics has long been an accepted form of computer output, but it is much more difficult to input such information directly to a machine. The problem here is the large amount of information which makes up the image, and the difficulty of interpreting this information.

7.8.1 Computer vision systems

Before a computer can begin to extract useful information from an image, it must 'acquire' the image in an electronic form which it can use. Many types of image sensor are available; most of them output a time-varying analog signal as the image is scanned, having been developed for television applications. Usually the scanning is electronic, although mechanical methods dating back to the earliest days of television are still used.

Vidicon camera
This is a vacuum tube in which the optical image is projected on to a photoconductive target within the tube. Where no light falls on the photoconductor, no current can flow from one face to the other, but incident light allows a current to flow, discharging any voltage across the target at that point. The rear face of the target is scanned by an electron beam which charges it up to a fixed potential; during the time that the electron beam is scanning the rest of the target this potential discharges at a rate which depends upon the amount of light. When the electron beam again scans across this point on the target, it re-establishes the original potential, and in doing so it deposits a charge which depends upon the change in voltage. As the beam continues its scan across the target, it deposits charge at a rate which depends upon the amount of light falling on that point, thereby producing a current with a waveform which depends upon the image being scanned.

Image orthicon camera
The vidicon has a number of shortcomings, notably poor sensitivity, resolution and tolerance to excessive illumination. In a camera based upon the image orthicon the image is focussed on to a 'photocathode' which emits electrons at a rate which depends upon the illumination at that point. These are focussed on to a screen where a pattern of charge accumulates with the same spatial distribution as the image. The resulting 'charge image' is scanned by an electron beam, and as a result of a phenomenon known as 'secondary emission' electrons are emitted at a rate which depends upon the charge at the point in the image currently being scanned. These secondary electrons are passed through a device known as an 'electron multiplier' which produces an increased electron flow with the same spatial pattern as the original electrons derived from the image. The current signal produced in this way has a waveform similar to that produced by the vidicon, but with an improved signal to noise ratio.

Solid state arrays
More recently, solid state imaging devices have become available. The most common types that exist are **photodiode arrays**, in which arrays of photodiodes are scanned by multiplexers, and **charge coupled arrays** in which the signals from an array of photodiode-like elements are shifted out by a CCD analog shift register. Two-dimensional arrays allow an image to be scanned without any moving parts, and they are available in arrays of up to several hundred pixels on a side. One dimensional **linear arrays** are available with up to several thousand elements. These are used in robotic applications such as position detection, and in facsimile applications where the image on a sheet of paper is passed in front of the sensor.

Other image sensors
Other scanning systems are available which use rotating optics to scan a laser beam. Examples of their use are to be found in facsimile transmission, and in bar code sensing of products at supermarket checkouts.

Light is not the only way by which images can be obtained. Sound is used in applications such as medical imaging, where reflections from internal organs are processed to form an image. X-rays and Nuclear Magnetic Resonance are also used in medical imaging. Here tomographic techniques are used: beams are passed through the patient or other object to be scanned in many different directions and the information obtained from an array of detectors is processed to yield an image. Images may also be produced using radar (reflections of radio waves) and sonar (reflections of sound waves). Again, mathematical techniques are available to improve the resolution of the images thus obtained.

Vision processing
The signal from the image sensor must be converted into digital form using an analog-to-digital converter, after which it is usually stored in RAM. This is simply the reverse of the graphical display method already described, and many integrated circuits intended for raster-scan displays are also suitable for capturing images. Usually the information processing rates needed to 'recognise' objects within the field of view are very high, and only rudimentary processing is possible with even the most powerful microprocessors. The best approach to the problem appears to lie in the use of multiple processors, but the hardware and software techniques used lie outside the scope of this book. For the interested reader, two useful texts in this area are Ballard and Brown (1982) and Faugeras (1983).

7.8.2 Speech synthesis

Speech is the most natural form of human communication, and much research effort has been expended on devising systems for the computer input and output of speech. The use of speech obviates the need for the operator to use a keyboard and to look at a visual display, allowing him or her to move around.

Human speech arises from various parts of the vocal tract. Air from the lungs is expelled through the vocal cords which vibrate, chopping the flow of air to form the basis of 'voiced' sounds. The mouth, tongue, nose, and lips are used to modify the sounds from the vocal cords and to create other, 'unvoiced' sounds. Voiced sounds are characterised by being quasi-periodic and rich in harmonics; vowel sounds and some consonant sounds are voiced. Unvoiced sounds are more 'random' in nature, being produced by constricting air flows (e.g., 's') or by suddenly releasing air (e.g., 'b', 'p'). All the sounds in spoken English are made up from a few dozen basic sound elements called **phonemes**. In order to produce a pleasing result, however, the sounds must have natural emphasis and intonation.

Computer speech can be generated either by replaying stored pieces of human speech or by reconstructing speech from coded data. Speech can be stored either in analog form, using magnetic media such as tape, or in digital form, but in either case a large amount of data must be stored — telephone quality speech requires about 8 kilobytes per second. Despite the large amount of computer data needed to store speech signals, they do not contain more than a few dozen bits of 'human' information per second. Thus many techniques for coding speech into forms requiring fewer bits per second have been devised.

Two main approaches are used. The first is to code the speech waveforms directly, storing changes in waveform and taking advantage of the almost periodic nature of voiced sounds. The second approach is to attempt to model the waveform generating system of the vocal tract by electronic means. This requires a large amount of processing, more than can be tackled by most 8-bit microprocessors, but many special purpose chip sets are now available. One difficulty with these chips is the need to code the phrases required, which involves the use of special equipment; usually this must be done by the manufacturer. Alternatively words may be constructed from phonemes, with extra information being supplied to attempt to produce a more human intonation. Speech coding techniques allow intelligible speech to be reconstructed from rates as low as 1000 bits per second or less.

7.8.3 Speech recognition

The automatic recognition of speech is much more difficult and research is continuing in this field. The transducer is simple enough — a microphone, but processing the waveform obtained in this way to extract the intelligence is difficult. Problems arise because of the differences between the voices of different speakers, the need to reject background noise, and the difficulty of identifying the point where one word stops and the next one starts. Most systems attempt to analyse the sound spectrogram, the variation of the spectrum of the sound wave with time, to detect a sequence of phonemes. Current technology allows a small set of words to be recognised when spoken by any speaker, or some hundreds of words from a single speaker when the machine has been 'trained' to that voice. Other potential applications of speech recognition are being discovered as technology improves. For example, the difficulty encountered in the design of speaker-independent systems suggests that security systems which work by identifying the voice of the speaker should be possible. Wolf (1980) offers a more thorough coverage of this subject than is possible here.

7.9 Design example

In this section a few examples of transducers and their interfacing will be examined. Transducers tend to be very specific to their applications, and for this reason it is difficult to choose examples which illustrate a range of general points. Three examples are given here: a thermistor used for temperature measurement with a controller, a position control system using a DC motor, and a floppy disc drive system.

7.9.1 Thermistor temperature measurement

As has already been indicated, thermistors have large temperature coefficients, which allow relatively high sensitivities to be attained in comparison with other types of temperature transducer.

This temperature coefficient is often large enough to allow the thermistor to be used in a simple potentiometer arrangement without recourse to the complexity of a bridge circuit. Figure 7.23 shows a thermistor used in conjunction with the 507 analog-to-digital converter encountered in one of the examples at the end of the last chapter.

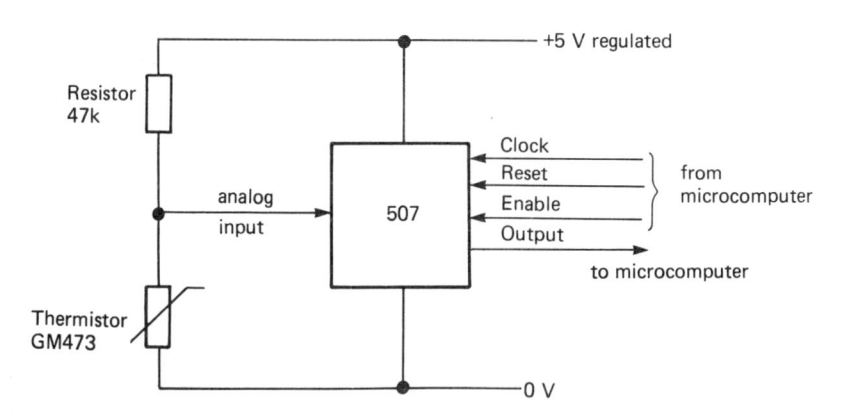

Figure 7.23 Thermistor interface

A high speed interface is not needed here because the temperature of a body cannot change rapidly in most applications. The thermistor will require linearisation, but this can be done in software by means of a look-up table once the characteristics of the thermistor are known. This can be calculated from manufacturer's data or carried out empirically using a sample of the thermistor type to be used. The linearised reading may then be scaled by multiplying by a constant, or the linearising table could contain entries showing the temperature directly in degrees.

The driver subroutine for the thermistor will thus consist of two sections, e.g.

function *temperature*(*channel*: **integer**);
begin
    volts: = *adc*(*channel*);
    temperature: = *table*(*volts*)
end;

This Pascal function calls two further functions; the device driver for the analog-to-digital converter and the lineariser.

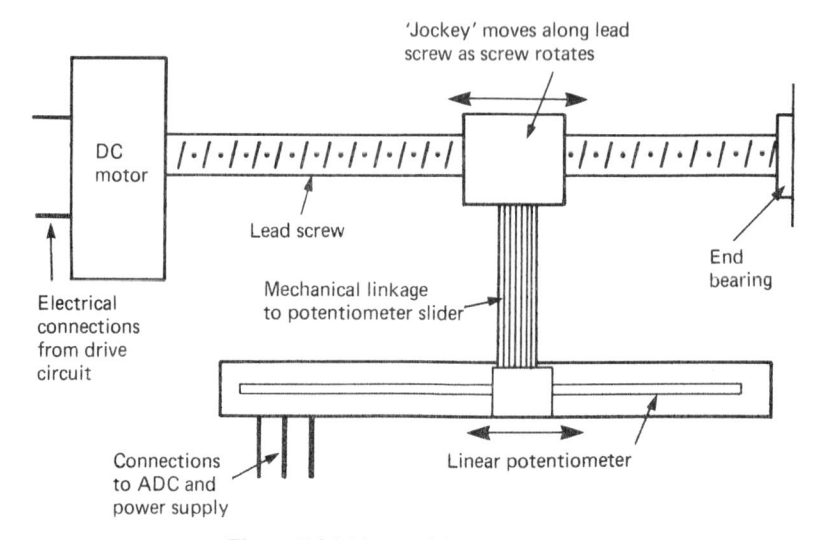

Figure 7.24 Linear drive mechanism

7.9.2 Driving nonlinear systems

DC electric motors offer a low-cost means of producing movement, but they present problems when precise motion is required. If they are driven with less than their normal operating voltage their output power is markedly reduced, and frictional losses mean that below a certain voltage there will not be enough force available to move the output shaft at all.

If positional control is required when using a DC motor, some form of feedback is necessary. Figure 7.24 shows a reversible DC motor connected to a lead screw which converts rotation of the motor shaft into linear motion. In this example the positional feedback is provided by means of a linear potentiometer. The position control system must compare the current position of the 'jockey' moving along the leadscrew with the demanded position and compute a suitable voltage for the motor from the positional error.

7.9.2

The position servo requires an ADC to convert the potentiometer output to a number position for use in the program and a DAC to convert a number in the program to a voltage for the motor. This voltage must be bipolar because the motor is reversible by reversing the voltage across its terminals, and a power amplifier is needed to provide sufficient current for the motor.

In the circuit in Figure 7.25 the ADC is interfaced to one port of a VIA with the handshake lines carrying the 'start convert' and 'end of convert' signals for the ADC. The DAC does not require a handshake; in this circuit a unipolar type is used, and its output voltage is shifted electronically to provide a bipolar output. This is because a single-polarity supply is used; two amplifiers are used in a bridge circuit to provide the reversible voltage across the motor. Their inputs are wired in opposite directions so that an increase in output voltage from the DAC results in an increase in voltage at the output of one amplifier and a decrease in voltage at the output of the other.

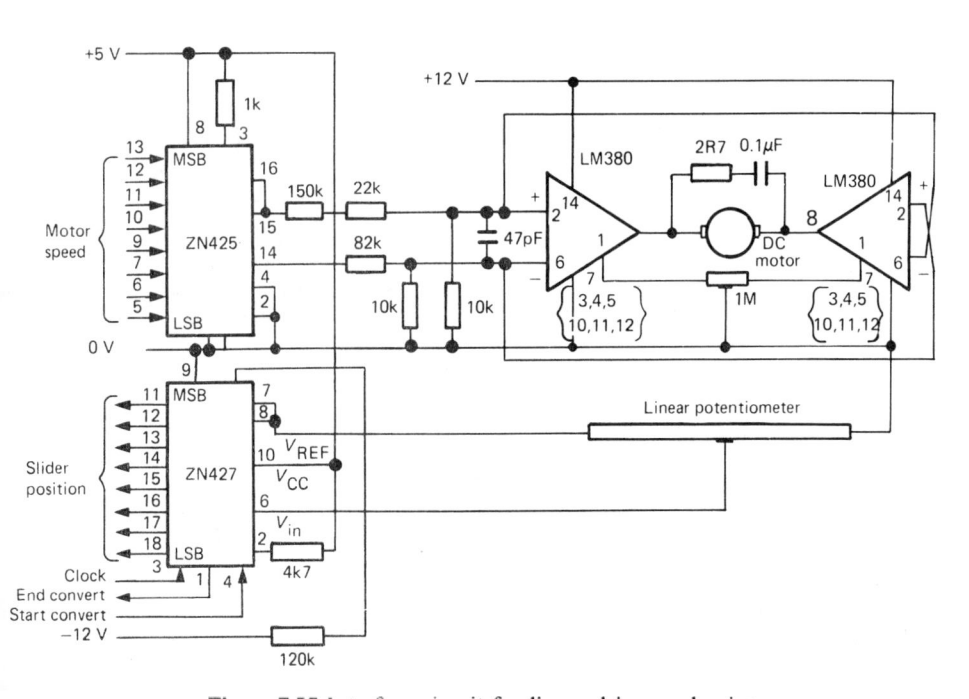

Figure 7.25 Interface circuit for linear drive mechanism

The design of optimal control software for positional control requires a considerable knowledge of control theory as well as a set of characteristics of the motor drive system. The speed of rotation of the motor will depend nonlinearly upon the applied voltage, and the linear change in position of the 'jockey' will be proportional to the integral of the motor speed with respect to time. The system will be unable to respond instantaneously to changes in drive voltage because of the mechanical inertia of the system, which means that if the drive voltage is maintained at the full value and only the polarity of the motor is reversed to cause it to move the jockey towards the desired position, it will tend to 'hunt' backwards and forwards about the desired position. Conversely, if the drive voltage is made proportional to the size of the positional error it will fall to a value where the motor ceases turning before reaching the demanded position.

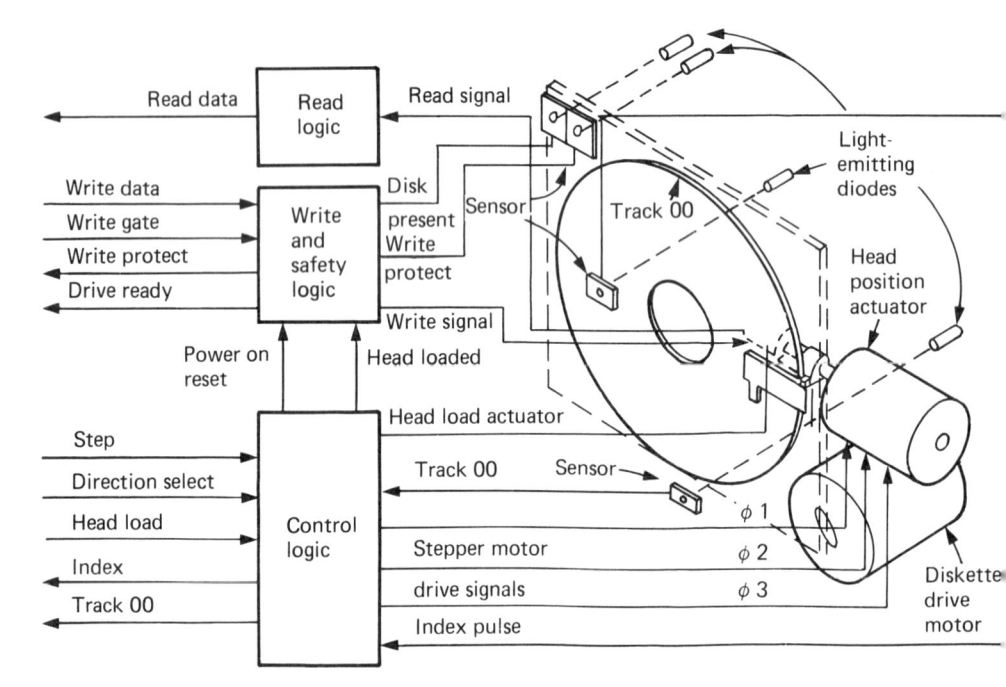

Figure 7.26 Floppy disc drive unit

7.9.3 Floppy disc drive

One common use of linear actuators is within floppy disc drives. Information is coded on the disc as small regions of magnetisation which are 'written' as a series of concentric tracks. The magnetic transducer, or 'head', which writes and reads these magnetic patterns must be driven radially inwards or outwards to the correct radius for the track to be used, which requires a close degree of positional control. In practice stepper motors are used rather than DC motors, and although a lead screw arrangement can be used, other mechanical arrangements are usually preferred because they are faster in operation.

The mechanism of a floppy drive system (Figure 7.26) provides other examples of transducers. Because a stepper motor is an open loop device with no built-in datum point, the drive system requires a separate sensor to detect the limit of travel which may be either a mechanical switch or an optical detector. Another detector is necessary for determining the angular position of the disc as it rotates, and this is done by sensing a hole (sometimes there are more than one), punched in the disc. An optical sensor is ideal for this purpose. Another sensor is used to detect a slot in the side of the floppy disc which indicates whether the disc is 'write protected' to prevent inadvertent erasure of information on the disc. This can be either mechanical or optical. Similar sensors determine whether a floppy disc drive has been placed within the drive unit and whether the door of the unit has been closed. In addition to the radial movement of the head provided by the stepper motor, provision must be made for a 'head load' mechanism to press the head against the surface of the floppy disc when it is to be used for reading or writing. The easiest method of doing this is by means of a solenoid.

Although there are many ways of providing the transducers needed by the floppy disc drive, the electronic interfaces to these transducers are included within the drive itself. This allows a standard interface to be used, so that (in theory) drives made by different manufacturers can be interchanged.

7.10 Questions

1. Colour photography requires very close control of the temperatures of the chemicals used. Design an interface for a control circuit which uses an electric heater and a thermistor to measure temperature.
2. Suggest transducers for sensing the following quantities within a car: engine speed, road speed, air speed, and rate of fuel flow.
3. Design a weighing machine which displays weight directly as a digital quantity expressed in pounds or kilograms. The maximum weight displayable should be about 5 kg, making the machine suitable for vegetables, etc. How can the cost of a purchase be calculated by the machine if the price per pound or kilogram is entered using a keyboard?
4. Many small identical objects, such as pills or electronic components, are more easily counted by weighing rather than by counting directly. If the weight of a single item is known, the weight of any required number can be calculated by the system. All the user need do is to add items until the correct number appears on the display. Modify your design for the previous question to allow this function to be carried out.
5. A sonar ranging system emits a brief pulse of sound at an ultrasonic frequency (typically about 40 kHz) and times the delay before an echo is received. This time delay is converted to a corresponding distance and displayed. Design a microcomputer-based sonar using a hardware timer counting at a frequency of 1 MHz and write a program to generate a brief pulse of approximately 1 ms duration and measure the time interval before a return pulse is detected; there is no need to design a power amplifier for the transmitter. Remember to allow for the fact that no echo may be detected, and that if an echo is received, the sound wave

will have traversed the intervening distance twice. The speed of sound in air is approximately 330 m s^{-1}.

6. A light pen contains a phototransistor or similar light-sensitive device which can detect when the bright 'dot' of an oscilloscope trace is near when the pen is pressed against the screen. Devise a system which senses the output of the light pen and causes the position of the dot (or other shape drawn by the oscilloscope), to follow movements of the light pen. You may assume that the position of the dot can be moved in the x and y directions by means of suitable output voltages.

7. Design an interface for a 'mouse' which employs a pair of incremental encoders which pick up the motion of a spherical roller on the under surface of the 'mouse'. Two signals are produced, corresponding to movements in the x and y directions. Each of these signals consists of a pair of pulse sequences produced by sensors which are 90° out of phase. The relative timing of these pulses indicates the direction of motion as well as the distance travelled. The interface should count pulses to determine the distance moved in the x and y directions and it should also be able to detect the pressing of a button contained within the 'mouse'.

8. Severely disabled people often retain control over limited parts of their bodies; for example, they may be able to suck and blow, and this ability can be used to communicate with a machine that can carry out a range of services for them, such as switching room lighting on and off, or opening and closing curtains. Design a command interpreter which recognises signals coded as changes in output from a pressure transducer of different durations and calls corresponding subroutines.

Chapter 8 MACHINE COMMUNICATIONS

The idea of inputting and outputting information is fundamental to the use of microcomputers and it has already been discussed in considerable detail in Chapters 4 to 6. When the output of one computer is connected to the input of another computer, a special case of input/output arises. The same principles apply, but because the transactions can become relatively complicated, a separate chapter has been devoted to this topic of communication between machines.

The word 'communication' covers not only inter-computer communication, but also that between a computer and a remote input/output device. The use of inter-computer communications includes many remote control and instrumentation applications where the control panel, displays, transducers or actuators of a system may be situated at a distance from the computer to which they are connected. Communications techniques are also needed in distributed control systems where a number of microcomputers are interconnected to distribute the system's computing tasks, and in hierarchical control systems in which microcomputer-based equipment is supervised by a larger computer. In such cases the data being transferred between one machine and another usually needs to be translated into a specially designed form to suit the properties of the communication link. The function of the transmitter is to convert the data into this form, while that of the receiver is to return it back to its original form.

If the data appearing at the output of the receiver is to be the same as that which was presented to the transmitter, it must carry out exactly the reverse operations to those which take place in the transmitter. This means that they must conform to the same **communication standard**. As technology has advanced, a huge range of standards has appeared covering all aspects of data communication, and only the underlying principles and a few relevant examples can be given in a single chapter such as this.

8.1 Error detection and correction

The first thing to note about communication links is that, because they link together parts of a system which may be situated some distance apart, they are vulnerable to the effects of electrical interference. There are many sources of such electrical **noise** in a communication system. It can be due to other signals in adjacent wires in a cable system or from interfering signals on a radio link, and other common sources are badly suppressed electrical equipment and natural lightning occurring near to the communication link. Even if all these sources of noise are eliminated, noise will still occur because of the natural electrical activity of the materials from which the circuits are made. Normally this last source is very small, but if the signals in a link become attenuated by distance it provides a fundamental limit in communication.

The signal being received at any instant contains two components: the desired component from the transmitter, and a random noise component from external sources. Provided that the noise is less than a certain amplitude, its effects in a digital communication link can be completely removed by the receiver. This is because a digital receiver 'decides' at any time whether the incoming signal corresponds more closely to a logic '0' or a logic '1'. Provided that the noise is not large enough to change the incoming signal so that it resembles an incorrect logic level it will not affect this decision, and no error will occur.

With the types of noise which are encountered in practice, errors are almost unavoidable, and the communication system must be designed to cope with them. This is done by including some **redundant information** in the digital signal. The extra information allows the received data to be checked in the receiver in an attempt to determine whether an error has occurred.

Two courses of action are open if an error is detected. Either the message can simply be recognised as incorrect and a retransmission requested, or the receiver can attempt to correct the errors in the message as received. In the latter case the receiver must use the check information to localise and correct the bits which were wrong. The technique of generating and using check information is called **error-control coding** and it has been the source of considerable research of a rather mathematical nature since it was first applied by Hamming (1950). Unfortunately it is possible only to design codes with a maximum error-detecting or error-correcting 'power', and it must be accepted that there is always a possibility of erroneous data being received without the receiver being 'aware' of the fact.

8.1.1 Parity

The simplest example of error-control coding is also the most widely used. It allows the presence of an incorrect bit appearing in a sequence of data to be detected by the use of an extra bit called the **parity** bit. In the transmitter the total number of binary '1's is counted, but instead of sending this total, the transmitter sends only the least significant bit of the binary total. This is the parity bit, which simply indicates whether the total is odd or even. When the message is received, the data is checked in the same way. If the parity bit calculated locally is different from the one transmitted, an error has occurred, whereas if they are the same, no error has been detected. This does not mean, of course, that no error has occurred; if two bits in the data were changed during transmission, the parity bits would still match. The simple parity check is suitable only for detecting the occurrence of single errors, but it is often used in applications where the effects of errors are not usually critical, such as in communication with computer terminals. When errors arise independently, for example as the result of a pulse caused by something being switched on or off, the probability of two errors arising in the same group of bits is extremely small, so that a single parity bit is sufficient to detect essentially all errors.

The result of adding the parity bit in this way is that the transmitted words each have an even number of logic ones, and for this reason the system is often called 'even parity'. It is no more difficult in practice to arrange for the transmitted words always to have an odd number of logic ones, in which case the system is, of course, called 'odd parity'. Odd parity is more commonly used in practice. When systems are connected together, it is important to ensure that they are both using the same parity convention, since standards vary as to whether odd or even parity should be used.

8.1.2 Checksum

The process of calculating the parity bit at the end of a sequence of bits can be regarded as a process of adding up the bits modulo 2, that is, discarding any carry after the first bit. The same principle can be extended to groups of bits. Because almost all microcomputers use wordlengths which are multiples of eight bits, it follows that the messages that pass between them tend to consist of sequences of eight-bit bytes. If all the bytes in a message are added up and any carries beyond the group of eight bits are discarded, the resulting eight-bit total will provide a check upon the contents of the message; this is addition modulo 256. If any byte is received incorrectly, this total will be different unless there are two or more errors, the effects of which cancel out exactly. This provides another method of error detection. The total, called the **checksum** is calculated at the transmitter and sent along with the message. At the receiver it is calculated again and compared with the checksum which was received from the transmitter. If the two checksums are different, there is an error somewhere in the message.

8.1.3 Error correction

A simple form of error correction can be attempted if the two techniques of parity and checksum calculation are combined. If the bits in a message are written in groups of eight, with each group of eight below the previous one, a rectangular pattern of bits results. At one end of each row is the parity bit, while at the bottom of the column is the checksum, as shown in Figure 8.1.

If a single bit is incorrect somewhere in the message, the parity bit of one of the words will be incorrect as will one of the bits in the checksum. This allows the row and column where the error occurs to be indentified, and since each bit can have one of only two values the error can be corrected by changing a 1 into a 0 or a 0 into a 1 at the position where the error occurred. This method allows only single errors to be corrected, but it

	parity	data
	1	1010010
	0	1101001
	1	1001100
	...	
	...	
	1	0110111
checksum	0	1101100

Figure 8.1 Checksum and parity

requires much less extra information to be included in a message than would be the case if the message without checksum and parity were transmitted three times and errors found by comparing the copies received.

8.1.4 Cyclic redundancy codes

The simple techniques of parity and checksum calculation are suitable for checking information when its structure is known, but if a message consists of a stream of individual bits whose structure is unknown to the coding system it becomes more difficult to use these methods. A method which could be used would be to append a single parity bit at the end of the entire message, but because this could only distinguish between there being an odd or an even number of ones in the overall message it does not provide a very secure mechanism for spotting errors; the correct check information could be obtained even with random data 50% of the time. However, if more check bits were used, the probability of obtaining the correct check information even when errors are present would diminish rapidly. In fact the chances of obtaining a given set of check bits is one in two raised to the number of check bits, a figure which can be made as small as desired.

The mathematical problem lies in designing a method of calculating these parity bits in such a way as to be as effective as possible in identifying errors. Many methods have been developed over the years, most of them based upon the shifting of the data along a shift register with feedback through a set of logic gates wired in a special way determined by sophisticated algebraic methods.

However, the exact set of connections does not matter so much as the result. The register in the system is zeroed before the data starts and the data emerging from its output is connected to the output of the transmitter. At the end of the data transmission the system is run for an extra number of cycles equal to the number of bits in the shift register. This process produces extra bits which provide the check information for the error detection process. In the receiver the incoming data is again passed through a shift register with feedback logic until the end of the message is reached. The contents of the register at this point indicate whether or not an error has occurred. This technique is known as **cyclic redundancy coding** (Martin, 1970) and it is usually performed by special logic incorporated in the integrated circuits used for data communications. Obviously the coding scheme used in the transmitter must correspond to the decoding scheme used in the receiver, but provided that both the transmitter and the receiver conform to the same standard the user need not be concerned with the detailed theory of operation. There is, however, a CCITT standard for the generation of cyclic redundancy checks, details of which are included in McNamara (1977).

8.2 Serial communications

In serial communications the information is sent one bit at a time. The signal emerging from the transmitter's output consists of a voltage or current that is switched between logic '1' and logic '0' levels in some controlled fashion. When this signal arrives at the receiver, not only must the signal be strong enough to allow the logic levels to be discriminated easily, but the receiver must be able to identify when each bit occurs in the waveform, as shown diagrammatically in Figure 8.2. This process is called 'bit clock recovery' or 'bit timing extraction' and is normally performed by logic in the special integrated circuits used for transmitting and receiving serial data.

The sequence of bits leaving the transmitter usually form some sort of data pattern; successive groups of bits may represent characters, successive groups of characters, a line, a group of lines, a page, and so on. Once the receiver has determined the time at which each bit appears, it must find out when each new character starts, and eventually the structure of the original message must be pieced together.

Communication systems can be divided up into two groups: **synchronous systems** in which the timing of all the bits in an entire message is controlled by a master clock, and **asynchronous systems** in which successive characters appear at arbitrary times and the timing of the bits within successive characters is unrelated. Each system has its advantages; synchronous systems are more effective when large amounts of data are to be passed between two machines, but asynchronous systems are more suited to the irregular data rates associated with human interfaces such as visual display units.

Despite the fact that there is no need to know exactly how the circuits operate, it is useful to have an understanding of the principles involved and the sorts of signals used by the different kinds of communication link.

8.2.1 Baud rate

It is important for the receiver to 'know' the time interval between each bit in the sequence of bits as they arrive in order to be able to 'make sense' of the signal being presented to its input terminals. The rate at which successive bits appear is called the **baud rate** and although any choice of baud rate could be used in principle, practical systems always use standard rates. The

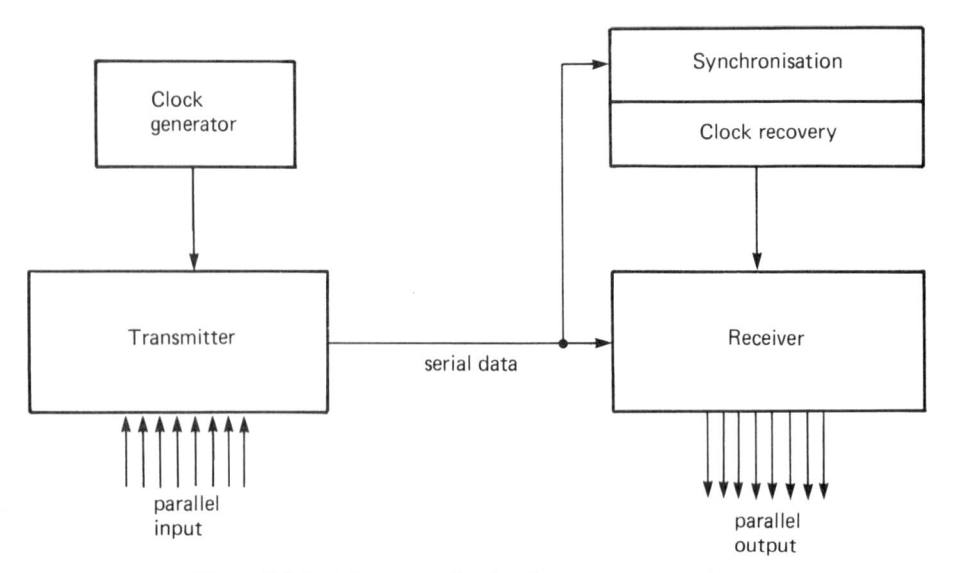

Figure 8.2 Serial communication between two machines

earliest electromechanical systems used quite low rates of a few tens of bits per second, but with few exceptions the links between electronic machines and connections to more modern electromechanical printers all use speeds which are related to a basic 75 bits per second. This speed itself is just about adequate for some long distance keyboard input circuits (75 bits per second corresponds to quite a respectable typing speed) but it is too slow to allow data to be displayed as fast as it can be read. The other speeds in common use are obtained by successive doubling of this modest rate to give values of 150, 300, 600, 1200, 2400, 4800, 9600, and 19 200 bits per second.

8.2.2 Synchronous communication

In a synchronous system, the start of a transmission is indicated by a **synchronising pattern**, which is a distinctive pattern of '1's and '0's. When the receiver recognises this pattern at the start of a message, it starts counting the bits which follow it. Each time a predetermined number of bits has been received, that group of bits is passed to an output register, from which the receiving computer or other machine can pick up successive data 'words'. These 'words' may be characters of a readable message or bytes of a program. Figure 8.3 shows how a typical message might appear with the synchronising information at the beginning of the message. Note that the communication link 'idles' at logic '1' level when no data is being sent.

The receiver must maintain synchronism with the incoming bits as they arrive throughout the message, reading each bit that comes in exactly once. It must never miss out a bit or read the same one twice. This is done by detecting the change in logic level that occurs each time that successive bits are different and adjusting its own internal timekeeping if it appears to be data into transmitter:

10110010
10101101
....

data as transmitted:

... 1 1 1 0 0 0 1 1 0 1 0 1 0 1 1 0 0 1 0 1 0 1 0 ...
... idle | sync. pattern | data word 1 | data ...

data from receiver:

10110010
10101101
....

Figure 8.3 Synchronous serial communication

running fast or slow. However, unless some restrictions are placed upon the sort of data that can be transmitted, it is not possible to guarantee that there will not be times when a long sequence of '1's or of '0's will occur. This would mean that there were no changes in the signal, and it would not be possible to maintain the receiver's synchronisation with the transmitter. To prevent this happening, coding schemes are often used which put in extra bits if they are needed at the transmitter to break up a sequence of '1's or '0's and which remove them at the receiver. This bit insertion and removal is done automatically by the integrated circuits used, and no extra code is needed in the software used to operate the interface.

8.2.3 Asynchronous communication

Asynchronous methods get around the problem of synchronising the receiver to the transmitter throughout a message by not attempting to keep the receiver synchronised between one word and the next. Each time that the transmitter wishes to send a word it switches its output to logic '0'. This level is maintained for the same amount of time that is used to transmit one bit, and is called the **start bit**. The start bit is used to indicate to the receiver that data bits are to follow, and the time at which the data bits will arrive is determined from the start of the start bit, since this always appears as a change from the 'idle' level of logic '1' to logic '0'.

The first data bit will start immediately after the start bit, and if the receiver samples the incoming signal during this time, normally at the centre of the bit duration, one and a half bit times after the start of the start bit, the first bit can be read. Exactly one bit time after this, the second bit can be sampled, and so on until the end of the word. The end of the word is marked by a bit which is always at logic '1', called the **stop bit**. After the stop bit the transmitter can repeat the sequence of start bit, data, and stop bit if more data is ready; otherwise it will commence idling once more until data is presented to the transmitter.

The error control techniques used by the two types of system tend to be different. In synchronous systems a cyclic redundancy code is normally used at the end of a message, while in asynchronous systems a parity bit is often sent after the last data bit and before the stop bit. Figure 8.4 shows the waveform produced when sending the binary pattern 1001101 (which corresponds to the ASCII letter 'M') as seven data bits followed by an even parity bit and a single stop bit. Note that the least significant bit of the data is sent first.

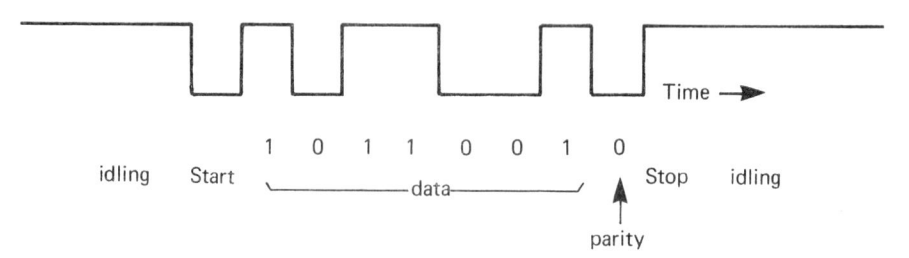

Figure 8.4 Asynchronous character waveform

8.2.4 Simplex and duplex methods

Thus far we have considered only messages travelling in one direction, from a transmitter to a receiver. This is called **simplex** transmission, and while it is sometimes used, most systems which are complicated enough to support serial communications are also sophisticated enough to require two-way communication. For example, typing information into a keyboard might enable an operator to send information to a computer, but that same operator would almost certainly want to see some response from the machine!

Communication systems in which messages can pass in either direction are said to be 'duplex' systems. In fact we can differentiate between **full-duplex** systems in which communication can take place in both directions at the same time, and **half-duplex** systems in which each unit can either send or receive, but not both at the same time. In the case of half-duplex systems, a single circuit can be used; the various units connected to the internal bus of a microcomputer, for example, can be thought of as being capable of half-duplex communication if they are 'read-write' devices. Full-duplex systems require a separate circuit in each direction unless some quite sophisticated circuit techniques are used.

8.2.5 Line buffering

At the transmitter, information is presented in the form of parallel data words which are converted by the communications interface into some sort of serial signal. At the receiver, the incoming data is converted back to parallel form once more. At each end of the link, therefore, registers are needed to hold words of data, that is, to act as data **buffers** between the computer and communication link.

In effect these buffers are similar to the input and output data registers of parallel interface devices. Their function is to hold the next word of data while the previous one is being sent by a transmitter, or to hold a word of data after it has been received. The buffer can either hold just a single word, as is the case in simpler communications devices such as those used for asynchronous communication, or a number of words arranged as a first in, first out (FIFO) queue. The use of a FIFO can reduce the overhead on the microcomputer, since the execution of a main program is disturbed less by having to wait for a message to be transmitted before continuing. The FIFO buffers used in interface devices usually contain a few bytes of storage, but the principle can be expanded to use a larger buffer in RAM controlled by software in the input/output subroutine.

Often a buffer large enough to hold an entire line of text is set up in memory. Output material from a program is passed to this buffer a line at a time, and the output subroutine takes each character and passes it to the output device in turn until the whole line has been sent. At the other end of the link, characters from the receiver are passed by the input subroutine to an input buffer, until a 'carriage return' character is received which indicates that the line is complete, and that the line can be passed to the main program. This approach allows the contents of the receiver line buffer to be edited if it is connected to a keyboard input, enabling typing errors to be corrected before the line of input is acted upon by the computer.

8.2.6 Control and status information

The buffer mechanism requires a 'handshake' system to ensure that data is passed correctly. Each time that the source of data on a communication link wishes to send a character, it must check to see if the buffer is ready to accept any more data. If it is not ready, the program must wait until it becomes ready before continuing. Each time that the transmitter is ready for more data, it checks the buffer to see if there is any data available, and if so, it takes it from the buffer, converts it to the form used in transmission, and sends it. As soon as there is space in the buffer, the transmitter will indicate that the buffer is ready to accept more input data.

Thus a handshake is needed both at the input and output of the buffer. When the buffer is within the same integrated circuit as the transmitter circuitry, the output handshake is handled by a logic circuit and the programmer need only be concerned with the handshake on the buffer input. Similarly, if the receiver buffer is in the same device as the receiver, only the output handshake need concern the programmer. However, when the buffer is implemented in memory, both handshakes must be set up in software. The handshake will consist of 'data available' and 'data acknowledge' wires if it consists of an electrical connection to the microcomputer, or of corresponding bits in a register if it is within the microcomputer itself.

In the transmitter circuit there will be a control/status register containing some read-only bits which indicate the status of the transmitter and some bits which can be written to that are used to control aspects of the transmitter's operation. One of the status bits within this register will be a read-only 'Transmitter Data Register Empty' (TDRE) bit which indicates that there is space in the buffer. This cannot be directly modified by the microprocessor because it is being used to indicate something about the interface's status. To clear this bit, it is necessary to write more data into the buffer register, and the bit will be set again by the transmitter once it has taken data from the buffer for transmission. The receiver will have a similar 'Receiver Data Register Full' (RDRF) bit which indicates when data is available.

However, the control/status register can be used for many more purposes. For example, it can be used to set the number of bits in each character being transmitted or received, the type of parity being used, and perhaps the number of stop bits being used (some systems use two stop bits instead of one) and the baud rate. It can be used to indicate when the receiver has detected a parity error, or when a character has been received with a missing stop bit, or if the output buffer has overflowed because it was not read soon enough after the RDRF bit was set and the receiver tried to put even more data into the buffer. Many communication systems have extra connections to the receiver which are used to indicate when the link is working correctly, so that no attempt is made to communicate when the link is broken. The state of the connection between the transmitter and receiver can likewise be indicated by status bits.

8.2.7 Special purpose integrated circuits

It will be apparent from the preceding sections that the circuits of the devices used on communication links are quite complicated, and indeed they are comparable with the simplest microprocessors. However, the use of large-scale integration means that integrated circuits for serial communication are in fact quite cheap. A number of different types are available, each with its own set of abbreviations.

The simplest is the **UART**, short for **Universal Asynchronous Receiver Transmitter**, which integrates all the logic necessary for an asynchronous link between two logic circuits except for the circuit for generating the clock for the bit rate generation (although this is sometimes included) and the circuits for changing the voltage levels to and from those used in the communication link itself. The transmitter section has a set of connections for the input data, together with a pair of handshake lines, while the receiver has a corresponding pair of connections for output data with a pair of handshake lines. There are also inputs for controlling wordlength, the number of stop bits and the type of parity to be used, if any, and outputs indicating the status of the transmitter and receiver sections.

The UART requires a large number of connections because each signal is carried by a different connection to the circuit, and normally they are packaged in a 40-pin integrated circuit. The advantage of the UART is its widespread applicability, because it does not make any assumptions about the microprocessor to which it is connected. In fact, it need not be connected to a microprocessor at all, and the UART provides a simple method of providing input/output at a place remote from the microcomputer. However, if an asynchronous communications link is required for a particular type of microprocessor, special purpose devices are available which replace the data, control, and status input/output connections with a set of registers which can be accessed via the buses of the microcomputer.

Integrated circuits of this type appear under a number of different names, the commonest perhaps being **ACIA**, an abbreviation for **Asynchronous Communications Interface Adapter**.

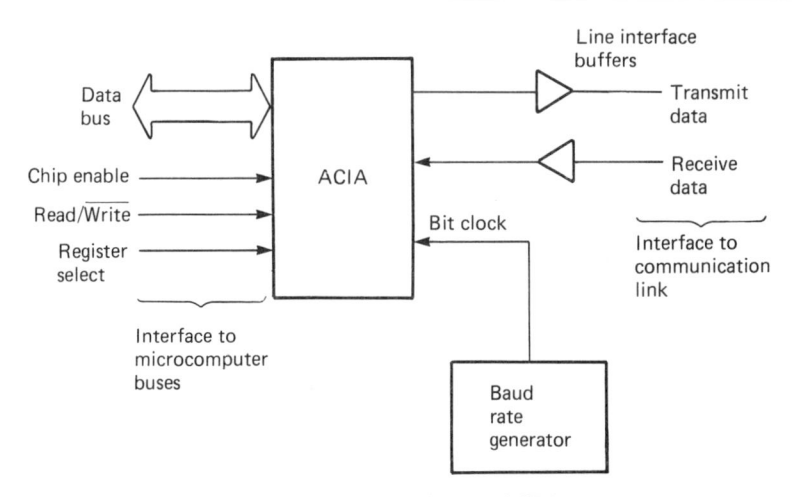

Figure 8.5 Connecting an ACIA

They can be used in exactly the same way as a UART except that interfacing to the microcomputer bus is easier. Since fewer pins are required, a smaller integrated circuit package can be used, saving money and space on printed circuit boards. Figure 8.5 shows an ACIA type 6850 connected to the buses of a microcomputer. Note that an external **baud rate generator** is required; this is an oscillator which produces the clock signal for the logic of the ACIA. Newer designs of ACIA, such as the 6551, have a baud rate generator incorporated into them, which usually avoids the need for an external oscillator for generating this clock signal, although provision is made for connecting an external oscillator if required. This integrated circuit requires more pins on its integrated circuit package than the 6850, because two connections are needed for the quartz crystal associated with the oscillator.

Integrated circuits are also available for synchronous communications interfacing. The main difference in this case is the need for the transmitter to send a synchronising pattern at the start of each message, and the need for the receiver to detect this synchronising pattern before it will start receiving data from the message. Some of the newer interfacing devices are programmable to allow them to be used for either synchronous or asynchronous communications, giving increased flexibility when they are incorporated on a computer board.

8.2.8 Communication standards

Just as input/output interfaces require some kind of signal conditioning, the serial signals produced by communications interface devices need to be converted to a more robust form before they can be connected to the transmission system. The various standards which have been used reflect the improvements in electronic technology which have taken place over the years.

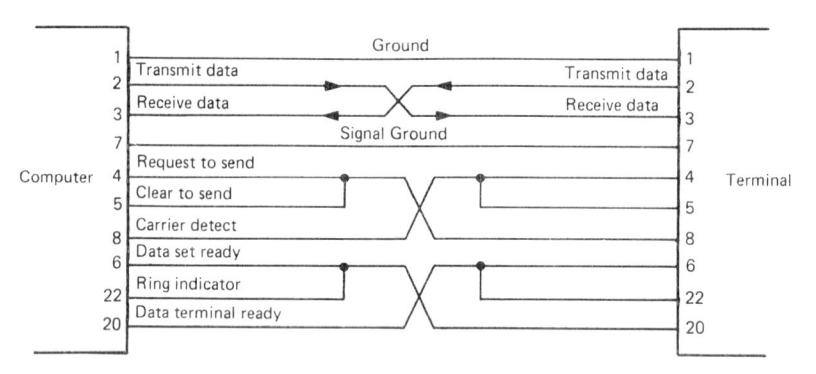

Figure 8.6 Connecting a terminal using the RS-232-C standard

RS-232-C
This standard was produced by the EIA (Electronic Industries Association) in the USA, although it is essentially identical to the V24 standard of the CCITT (International Telegraph and Telephone Consultative Committee) in Europe. This standard specifies that logic '1' is to be sent as a voltage in the range -15 to -5 V and that logic '0' is to be sent as a voltage in the range $+5$ to $+15$ V. These voltages are much larger than those handled by ordinary logic circuits, so that interfacing circuits are needed. The standard specifies that voltages of at least 3 V in amplitude will always be recognised correctly at the receiver according to their polarity, so that appreciable attenuation along the line can be tolerated. The **RS-232-C standard** specifies that the maximum length of cable between the transmitter and receiver

should not exceed 100 feet, although in practice many systems are used in which the distance between transmitter and receiver exceeds this rather low figure.

The standard also specifies the connectors to be used. The standard RS-232-C connector has 25 pins, 21 pins of which are used in the complete standard. The original standard anticipated that much of its usage would be via interface equipment to the public telephone network, and many of the connections specified relate to the monitoring and control of the **modems** used to provide this interfacing. Modems will be discussed in section 8.2.12. Many of the modem signals are not needed when a computer terminal is connected directly to a computer, and Figure 8.6 shows how some of the 'spare' pins should be linked if not needed.

The limited range of the RS-232-C standard is one of its major shortcomings compared with other standards which offer greater ranges within their specifications.

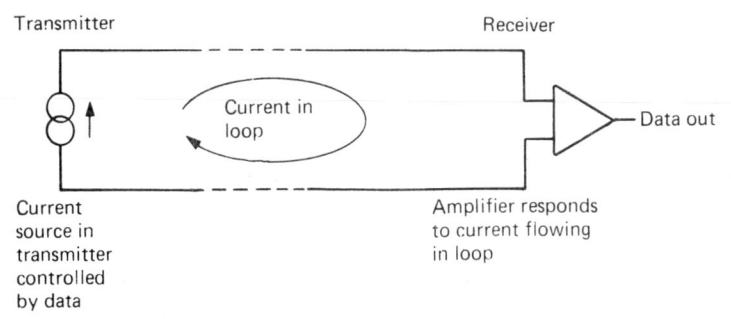

Figure 8.7 Current loop connection of two machines

Current loops

One reason why the range of the RS-232-C standard is limited is the need to charge and discharge the capacitance of the cable connecting the transmitter and receiver. This is made worse by the large difference in voltage between the two logic levels used and the fact that the maximum current available at the output of a RS-232-C driver is deliberately limited to prevent it from being damaged by short circuits.

Figure 8.7 illustrates an alternative approach in which the signal is specified in terms of current rather than voltage. In this case the receiver has as low an input resistance as possible and it detects the current flowing in its input leads. The lower the input resistance of the receiver, the lower is the voltage that appears between its input terminals, and the less the time that is taken to charge and discharge the cable capacitance.

Current loop systems have been in use for a long time in teleprinter circuits, and a number of different standards have been used. However, the commonest convention uses zero current to indicate logic '0' and a current of 20 mA to indicate logic '1'. These systems often use optoisolators to isolate the transmitter electrically from the receiver. The voltage drop across a light-emitting diode is only about 1.7 V, which means that the voltage changes appearing on the line are much less than those to be found with RS-232-C. A current loop circuit using optoisolators is shown in Figure 8.8.

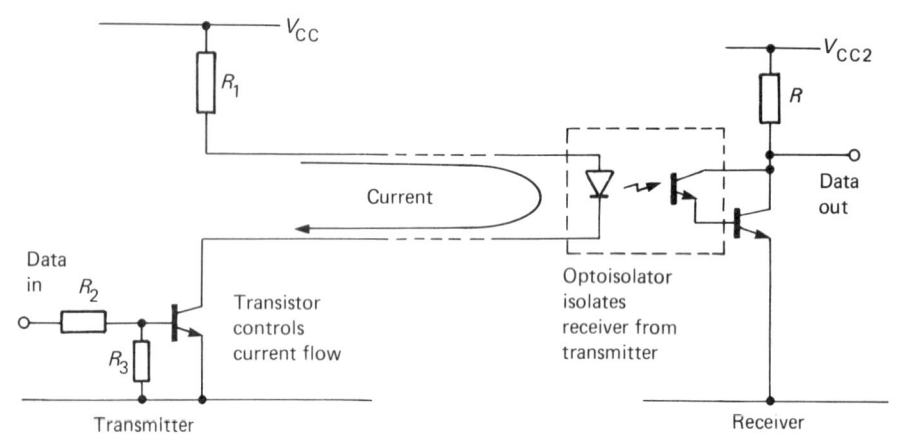

Figure 8.8 Current loop with optoisolators

RS-423

Improvements in electronic technology have allowed the **RS-423 standard** to be introduced which offers considerable improvements in speed and range over the older RS-232-C. The output resistance of the driver circuits used with the older standard was deliberately kept high so that they could tolerate being shortcircuited, while the output resistance of the newer drivers can be made low, but circuitry incorporated which prevents excessive currents flowing when a fault occurs. At lower speeds, a single voltage referenced to earth can be used. Since a sensitive receiver capable of detecting very small voltage changes allows very small signals to be received, the transmitter can be used to send voltage levels of 0 and 5 volts instead of the larger voltages used by RS-232-C. This often means that the entire microcomputer system can operate from a single five volt supply, rather than having to have two extra supply voltages simply to operate the RS-232-C interface.

RS-422

At higher speeds the attenuation of the signal due to cable resistance and capacitance can become more of a problem. The integrated circuits used for the RS-423 standard can also be used for the closely related **RS-422 standard**, which uses a pair of wires carrying a signal represented by the difference in voltage between the two wires rather than having a 'live' and a 'zero volt' line. This 'balanced connection' greatly reduces the effect of interference, which tends to affect the two wires equally, and allows quite small signals to be detected at the receiver without problems. In this case the driver circuit has two outputs, each switching between 0 and +5 V, but arranged so that when one is at 0 volts, the other is at +5 V, and vice versa. This means that the differential voltage between the two lines switches between +5 V and −5 V. At the receiver these two lines are connected to the inputs of a differential amplifier which is designed to be able to tolerate common voltages of several volts while discriminating between differential voltages of a fraction of a volt.

This allows data to be sent at speeds of up to ten million bits per second, and for distances of up to 4000 feet, although the standard does not allow the full speed to be used over the full distance. The cable used for this system is usually a twisted pair of conductors. At high frequencies, a long length of this twisted pair appears to have a characteristic impedance of about 100 Ω, and improvements in performance can be obtained by connecting a resistance of the same value across the inputs to the receiver. The effect of this is to reduce the effect of signals being reflected back from the receiver, which distorts the signal waveform when high data rates are being used.

8.2.9 Interconnection technology

The choice of the medium which is used to connect the transmitter to the receiver determines the speed at which they can communicate and the probability of errors occurring during transmission. Several techniques are possible, each with advantages and disadvantages.

Twisted pair

Perhaps the most obvious way to connect together two systems is by means of a pair of wires forming a circuit between them. Although this can often be all that is required, several precautions are worth observing. First, the pair of wires forming the circuit should be kept as close together as possible. This minimises the effects of electromagnetic coupling of undesired signals from other sources, which will appear as noise in the circuit. The effects of this pickup can be reduced even further if the pair of wires is twisted, so that the effects of pickup tend to average out. Connections of this kind are widely used for all sorts of electronic communication with bandwidths up to about 1 MHz, for example telephones.

Another source of pickup is due to electrostatic coupling from external sources, and this can be reduced only by covering the cable with an electro-static screen connected to earth. This screen usually takes the form of a woven braid of wires surrounding the twisted pair and insulated from it. Where several circuits are in use, it is common to find several twisted pairs all protected by the same braided screening.

Coaxial cable

One development of this idea is the coaxial cable, which consists of a single conductor at the centre of a tubular braided conductor. In this case the braid is kept at zero volts while the centre conductor provides the 'live' connection. Coaxial cables have relatively low attenuation at high frequencies of up to several hundred megahertz and their construction is such that very little pickup of extraneous signals occurs. This type of cable is widely used in applications where the frequencies involved are too high for twisted pairs, for example in the downleads of television aerials. The problems with coaxial cable are that it tends to be rather more expensive than twisted pair cable and that it is more difficult to install because of its larger diameter and the fact that it needs special coaxial connectors. Consequently it is mainly used in situations where very high speed data connections are in use, for example in networks of computers. Baseband data rates up to tens of megahertz are possible, the principal limitations arising from the limitations of the drive electronics.

Fibre optics

One way to circumvent the problem of interference to data signals where the communication link passes through electrically noisy environments is to avoid the use of an electrical connection altogether. This can be done by means of fibre-optic cables which consist of a special glass coated with a protective sleeve so that they look quite like 'ordinary' cables. The signal in this case consists of a beam of light produced by a light emitting diode, or a rather more sophisticated injection laser diode, connected to the input end of the cable and detected by a special high-speed light-sensitive diode at the receiving end of the fibre-optic cable.

This type of connection is essentially insensitive to electrical interference along its length, and because the glass and its plastic coating are both good electrical insulators it can be placed close to high voltage cables or immersed in water without problems. Fibre-optic cables provide a means of signal transmission that is secure, because it is relatively difficult to 'tap' an optical fibre, at data rates of up to 1 Gbps (1000 million bits per second). The difficulties of using fibre-optic cables arise from the difficulty of making connections to them and of jointing them. Another problem which limits the design of networks using fibre-optic cables is that no satisfactory way yet exists of making 'bus' systems with multiple connections to the same cable, in the way that buses can be built using electrical connections via metallic cables.

8.2.10 Signalling techniques

When data is to be sent from one point to another, a signal waveform can be produced simply by generating one voltage for a fixed period called the **bit period** if a logic '1' is to be sent and another voltage for the same time interval if the level to be sent is a logic '0'. This signalling technique is called **Non Return to Zero (NRZ)**.

Problems arise with NRZ coding when a long sequence of logic '1's or of logic '0's occurs without a break. Because the bits being transmitted are all the same, the output voltage does not change. The first consequence of this is that the circuits carrying the signal must be directly coupled, which rules out the use of systems designed to carry voice signals, such as telephone lines and radio links. The second is that if there are no changes in logic level in the signal appearing at the receiver, it will not be possible to determine the point in time where one bit finishes and the next one begins. Changes in logic level are essential for this purpose, and if they do not appear in the signal as generated, they must be added by some other means.

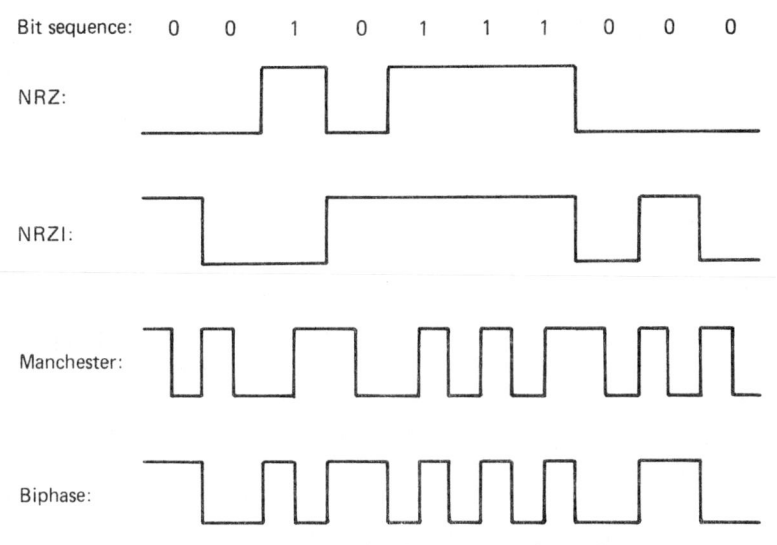

Figure 8.9 Waveforms of different bit coding schemes

One simple modification which can be made to NRZ is to arrange for the bits to be coded not as logic levels but as changes in logic level between one bit time and the next. In **NRZI (Non Return to Zero Inverted)** code a logic '0' is represented by sending the opposite logic level to that sent during the previous bit period, while a logic '1' is sent as the same level. This means that a succession of '0's at the transmitter will still produce a signal with a usable number of transitions between logic levels, but a succession of logic '1's could still cause problems. Practical systems using NRZI ensure that long sequences of logic '1's are never sent.

In the **Manchester code**, which is named after the place where it was first used, each bit is coded in such a way as to contain at least one transition. A logic '1' is represented by a waveform which is at logic '0' for the first half of the bit time and then at logic '1' (i.e., it contains a negative transition), while a logic '0' is represented by a waveform of the opposite polarity, as shown in Figure 8.9. This signal has no steady component and can be transmitted via links containing transformers because it changes level so frequently. The large number of transitions in the waveform ensures that there is no difficulty in maintaining synchronisation. Manchester coding is used in applications such as aircraft 'fly-by-wire' systems where the data signals must be sent over links which are subject to noise and which are transformer coupled.

Biphase coding is a related system in which there is always a transition at the end of a bit period. The two possible logic levels are distinguished by inserting a transition in the middle of the bit period when a logic '1' is sent but not when sending a logic '0'. Thus it is the transitions between logic levels rather than the levels themselves which convey the information, so that if the signal wires are transposed or if the signal is connected using a transformer, the communications link will still work. This simple but robust coding technique is used for applications such as communication between single-chip microcomputers. The signal may need to have extra changes in logic level included in order to assist the synchronisation process in the receiver, or it may need to be modulated on to a carrier wave in order to pass it over the communication link being used.

The coding techniques which have been discussed so far are all examples of **baseband systems**. Although in some cases the form of the logic waveform is modified to remove DC components or to improve synchronisation, in each case the result still contains essentially the same range of frequency components as the original signal. A radically different approach to generating a signal capable of passing through links which are not suited to logic signals is to use the logic waveform to control the amplitude, frequency, or phase of a 'carrier' wave. This is the same principle as used in radio transmission, except that the carrier frequencies used can extend from about 1 kHz to several hundred megahertz, depending upon the application. The lower frequencies are used where the communication path is one which was designed for audio signals, by far the most common being the ordinary telephone system. Specialised systems exist using coaxial cables where carrier frequencies of hundreds of megahertz allow very high data rates to be used. Systems of this type are known as **broadband systems**.

8.2.11 Sharing techniques

When a communication link extends over a considerable distance, the cost of making a connection to it becomes small compared with the cost of the link itself, and methods of improving the use made of the link become attractive. This technique of sending more than one signal on the same communication channel is called **multiplexing**. Several techniques for multiplexing exist. The essential principle is that the system used must not only be able to combine many signals into one communication link, but it must also be capable of unravelling the signals at the other end!

The simplest technique to arrange using modern electronic techniques is **Time-Division Multiplexing (TDM)** in which the available time is divided up into a number of consecutive 'slots'. Figure 8.10 shows how the multiplexing and demultiplexing processes can be likened to the use of rotating switches which in the multiplexer connect to each of the inputs in turn, and

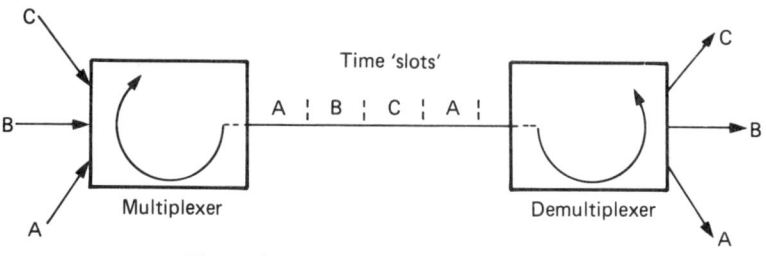

Figure 8.10 Time division multiplexing

in the demultiplexer connect to the corresponding output at the same time.

Conventional TDM uses equal time slots for each of the channels, but a development of TDM, known as **statistical TDM** devotes an amount of time to each channel which depends upon how busy it is. Most data communication takes place at varying rates, and statistical multiplexing allows each channel to use a higher data rate when it needs to communicate than would be possible with ordinary TDM.

A third technique is possible in broadband systems by making use of the fact that a carrier wave is used. If each channel is modulated on to a carrier with a different frequency, the different channels can be separated at the receiving end by simply tuning to that channel in the same way that a radio receiver can select just one of a number of radio transmissions. This is called **Frequency Division Multiplexing (FDM)**.

8.2.12 Modems

Where signals have to travel a very long distance, it is normal to use telephone lines or radio links. These are designed to carry voice signals, and usually it is not possible to connect serial data signals directly. The data must instead be modulated on to a carrier wave with a frequency to suit the communication link used, and demodulated at the receiving end of the link.

The two processes of modulation and demodulation require some supervision when they are carried out as part of the communication path between two computers. For example, the transmitter should not start transmitting until the system at the other end of the link is ready to receive signals, and the receiving end of the link must be informed if the signal carrying the data is not present. The various tasks involved in modulation, demodulation, and the monitoring of these two processes is performed by a unit somewhat unglamourously called a **modem**, the name being formed from the first parts of the words 'modulator' and 'demodulator'. The carrier can be modulated in several different ways; the mathematical representation of the carrier wave is

$$f(t) = A(t)\sin[2\pi f(t)t + \phi(t)] \qquad (8.1)$$

which shows that the instantaneous amplitude $A(t)$, frequency, $f(t)$ or phase $\phi(t)$ of the carrier wave can be varied to represent the data being communicated. There are many standards in operation for the type of modulation to be used and the frequencies of the 'carrier waves' to be used to carry the modulation. Many of these standards allow duplex operation over a single telephone circuit so that a terminal can be used with a remote computer via a telephone connection dialled in the normal way.

Modems can be used when instrumentation is installed at a remote site; the site can be dialled automatically at appropriate intervals to allow the instrumentation system to be interrogated by a computer. The same technique can be used to send control information to a computer controlling a remote process. The use of radio links for remote control also requires the modulation and demodulation of a carrier wave; this can be done directly on to the radio frequency carrier used for the radio system itself, or on to a 'subcarrier' of a frequency comparable to those occurring in voice signals, so that a communication link designed for voice communications can be used.

8.3 Communication between several machines

As the complexity of microcomputer systems increases, the need to include more than one microcomputer in a single system becomes greater. This introduces a number of additional problems, notably the need to identify the sender and recipient of messages passing through the system. Perhaps the greatest problem which appears as the complexity of systems increases is the need for standardisation not only at the level of the interconnection between the machines, but at all levels of the communication process between a program running in one machine and its counterpart in another. In a simple system it is a relatively simple matter to design all the elements of that system to conform to a standard which might only be used in that particular system, or by a single manufacturer. This is called a 'closed' system because, in effect, it is closed to all components except those which were designed for that system.

If all communication equipment were designed to conform to the same set of rules at all levels of the communication process, then a system could be made using 'off the shelf' components from a number of different sources, greatly easing the problems of the system designer. This 'open' system approach has been the subject of much work by the International Standards Organisation (ISO, 1981).

The first problem which faces a large standardisation exercise such as this is to define the function of a communications interface between two programs running on different machines. This can be divided into two layers as shown in Figure 8.11; at the lower level are the functions which relate to the actual process of communications, while the upper level contains functions which are more oriented toward the programs which are communicating. The two programs send messages to each other via the lower 'communications' layer, but they are no more 'aware' of this layer than two people using a telephone are usually aware of the intricacies of the communication system linking them.

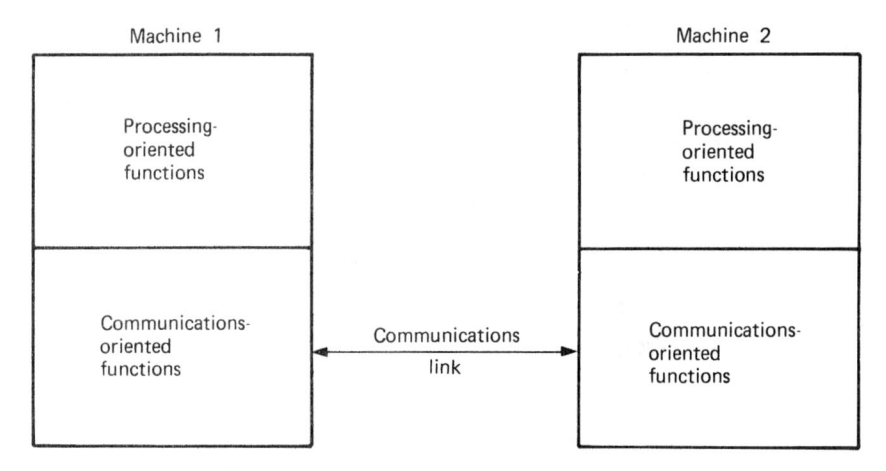

Figure 8.11 Layers in a communication system

This model can be refined further to reveal seven distinct layers which are involved in a general communications system. The function of each of these layers is listed briefly.

- The **Applications Layer** supports the applications programs which are communicating via the lower layers.
- The **Presentation Layer** performs the two-way functions of taking information from applications, converting it to a form suitable for common understanding (making it machine independent), and presenting it to the applications in a form which they can understand. It provides services which give independence of character representation, command format and machine characteristics.
- The **Session Layer** is responsible for establishing a logical connection between applications wishing to exchange information. The two applications form a liaison called a 'session', and the layer maintains this liaison and ensures that data reaching a system is routed to the correct application. Remember that a computer or microcomputer may be 'timesharing': running several programs at the same time.
- The **Transport Layer** is the uppermost of the 'communications' layers and it provides, in association with the layers below it, a universal data transport service which is independent of the physical medium in use.
- The routing of data from one machine to another is carried out in the **Network Layer**, in which a logical connection is set up between the two machines.
- The **Data Link Layer** looks after the passing of data from one end of the link to the other, carrying out tasks such as error detection and correction.

- **Physical Layer** provides the interface with the physical medium carrying the communication link and the means of controlling its use.

These layers have been identified as a basis for standardisation in the design of networks of computer-based systems. However, the ideas involved are similar regardless of whether the computers are separated by a few centimetres or by thousands of kilometres. The main effect of increasing the distance between the machines which make up a network is to reduce the 'tightness' of the coupling between them.

8.3.1 Tightly vs loosely coupled systems

If the machines making up a network are more than a few metres apart, the only satisfactory method of interconnecting them is to use serial communications because of the high cost of the transmission media. However, when the machines are closer together, parallel communication between them can be used because the cost of the cables is a less important consideration. Consequently data can pass from one machine to another very quickly; essentially each machine is connected to another via a parallel interface, and if more than two machines are involved, a parallel bus similar to the internal bus of a single microcomputer can be used. This is the approach taken in many modern instrumentation systems, and in the widely used IEEE-488 standard which will be described in section 8.4.1.

The result of using a parallel bus to link the various machines is to form a composite bus structure in which each microcomputer has a **local bus** which is used for its internal data transfers, and access to a **global bus** which is used for communication between the different microcomputers. If the addressing scheme used by each microprocessor is arranged so that some addresses refer to memory and input/output devices connected to its local bus, while others refer to devices on the global bus, the linkage between the microprocessors can be made even tighter. In this way a composite microcomputer with several microprocessors is formed. When the distance between the machines making up the system is greater, it is not possible to couple the microcomputers this closely and a serial bus must be used.

In systems of this complexity, the data passed between one part of the system and another usually consists of relatively long messages rather than simple bits, bytes, or words. Passing messages between one microcomputer and another in a tightly coupled system like this is simply a matter of one microcomputer writing the message into RAM on the global bus and the other microcomputer reading it. This mechanism is very fast, but it raises problems in coordinating the various machines making up the system.

An **arbitration** system is needed to ensure that only one of the microprocessors can access the global bus at any time. A message cannot be read by the recipient until it is ready to be read, and the originator of the message must tell the recipient that the message is there, which means that control and status information is needed. The identity of the recipient must

8.3.1   MACHINE COMMUNICATIONS

be indicated with an address, but this extra information must be invisible to the programs which are communicating with each other.

Tightly coupled microprocessor systems
Much of the increase in computing power which microcomputers have achieved since their inception has arisen from increases in their processing speed. It appears increasingly likely, however, that future improvements will accrue more from the use of **concurrency**. Although systems with more than one microprocessor offer an attractive method of increasing the processing power of a microcomputer, problems arise because of the need for communication between the microprocessors. Both data and programs must be communicated, because of the need to share the workload between the microprocessors, and these overheads detract from the potential improvement which might be expected from using several microprocessors. Figure 8.12 shows the reduction in overall processing power which results from this need to coordinate the microprocessors and communicate between them.

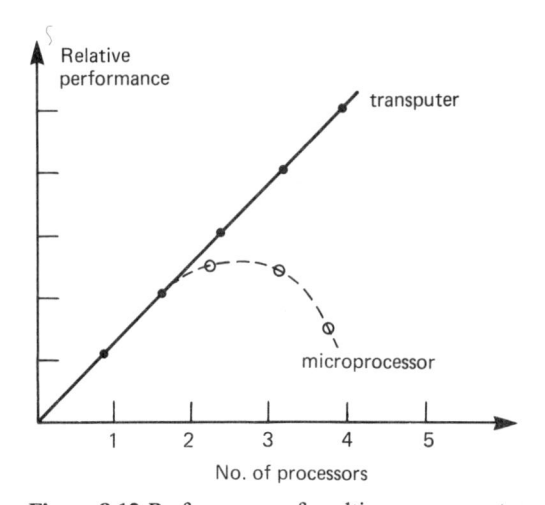

Figure 8.12 Performance of multiprocessor systems

Note that above a certain point, adding extra processors actually leads to a decrease in performance. The IMS T424 **transputer** (Figure 8.13) represents a solution to this problem in which communication with similar devices is designed into the circuit at a fundamental level. The transputer is essentially a single-chip microcomputer with a simple but powerful microprocessor, RAM, and input/output circuitry including four communication channels. The supervision of these communication channels is directly under the control of hardware within the transputer, and by using these channels a number of transputers may be interconnected to form a composite system. All the channels of a transputer can operate simultaneously, unlike a system using a bus to link together several processors in which the speed of communication is limited by the overall capacity of the connecting bus.

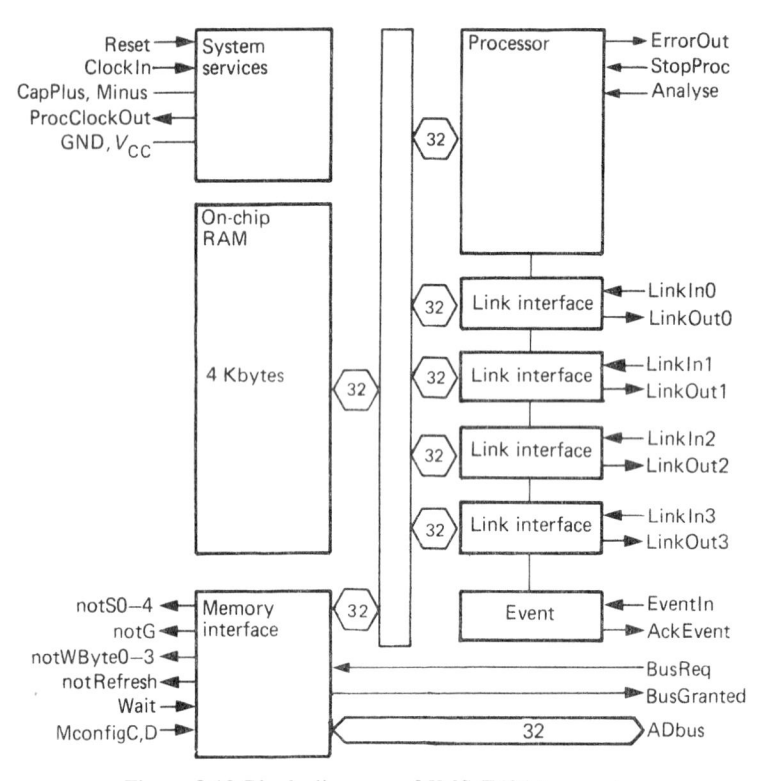

Figure 8.13 Block diagram of IMS T424 transputer

When several processors are operating concurrently in the same system, a language must be devised to allow them to be programmed as easily as possible. In the case of the transputer a language called **occam** (INMOS, 1984) has been developed, which introduces three new programming constructs. These are:

seq Processes within a block labelled **seq** are executed sequentially.

par Processes within a block labelled **par** are executed in parallel, that is, concurrently.

alt Only one of the processes within an **alt** construct is to be executed; the system determines which, if any, of the processes can execute (they may be waiting for a channel to become ready, similar to waiting for a flag with a single-microprocessor system) and waits if necessary.

Using these three constructs in conjunction with those already encountered such as **if** and **while** allows systems with several processors to be programmed, with the problems of communication between the processors being hidden from the programmer by the compiler.

8.3.2 Addressing in distributed systems

Just as a microcomputer bus system needs addresses to specify the source or destination of data passing between the microprocessor and memory, so the network which interconnects a number of microcomputers also requires addresses to identify the destination of information passing through it. Exactly how this address is included with the data depends upon the type of system. It could be carried on separate parallel wires at the same time as the data, as is usually the case within microcomputers, but because it requires a large number of connections this method is usually reserved for tightly coupled systems.

Alternatively, it can be carried on the same wire or wires as the data, but at a different time. In a parallel system an extra signal is needed to indicate whether the information on the bus at any time is data or an address, while in a serial system the address must be indicated in some other way. Some examples of systems for interconnecting microcomputer-based equipment are given later in this chapter.

Control and status

The statements made about addresses in distributed systems apply equally to the control and status information which is needed to manage the flow of information around a network. A separate bus can be used, or it can be sent as part of a 'frame' of serial information.

Transparency

Problems can arise when control and address information are sent on the same wire as the data, when there is a danger that data can be mistaken for other information. Ideally a communication link should be **transparent** to data, so that any bit pattern can be sent without fear of the data affecting operation of the link. Again, the examples which follow will show how transparency is achieved in different types of communication link.

8.3.3 Arbitration difficulties

When different machines want to 'talk' at the same time, some means of arbitration is needed to ensure that each machine will take its turn in the conversation. One way to do this is to have a network 'master' which has software to allow it to dictate which of the remaining 'slaves' is allowed to talk at any time. In tightly-coupled parallel systems, where access to the bus must be achieved quickly, the arbitration must be carried out much more rapidly, by high-speed logic circuits.

Arbitration is also needed when two machines wish to make use of the same 'resource'. For example, if two microprocessors wanted to send messages to the same terminal at the same time, their messages would become hopelessly jumbled. The solution is for each program attempting to make use of a communication path, for example to the terminal, to check first that the path is clear. Unfortunately this is not a sufficient solution, because if two programs were to try at almost exactly the same time, they would both discover that the path was free, but would still end up transmitting at the same time. The arbitration process needs an **indivisible test and set** operation which will 'claim' the resource automatically if it is free before another program tries to gain access to the same communication link.

8.4 Standards: handshakes and protocols

When information is passed between communicating machines, a handshake should be used to make sure that the data has been received correctly by the recipient before the sender sends any more data or terminates the 'conversation'. Standardisation is needed because of the necessity of connecting machines from different suppliers. The set of rules for communicating between machines is called a **protocol** because of its similarity with diplomatic practice.

In this section three examples of standards will be given: a parallel one which is useful over the sorts of distances involved in the interconnection of systems within a single room, and two serial systems which can be used over greater distances.

8.4.1 IEEE-488 instrumentation standard

IEEE standard 488 (IEEE, 1982) is intended for use in interconnecting 'intelligent' instruments, that is instruments containing electronic logic circuitry which is sophisticated enough to format their readings into a standard format suitable for communication via the parallel bus defined by this standard. IEEE-488 uses a set of eight wires that act as a parallel data and address bus and a further eight wires used for control functions, as shown in Figure 8.14. The standard also specifies how a number of separate earth lines should be connected for optimum high speed performance.

The standard allows up to a million bits per second to be carried by each line, but as the system is asynchronous, the actual data rates depend upon the properties of those items of equipment connected to the bus that actually take part in any given interaction. The equipment can be divided up into controllers, talkers, and listeners. Controllers, as their name implies, are capable of controlling the bus system. Talkers are capable of acting as sources of information, while listeners are capable of receiving information. Up to 15 separate items of equipment — 'devices' — can be connected to the bus, and the specification places limits on the total length of the bus and the maximum length of each connecting lead. The bus system is designed in such a way that more than one listener can be programmed to receive information when a talker sends it, and because the system is asynchronous, a sophisticated handshake is needed to control the passing of information.

This handshake uses three wires instead of two; in addition to the normal 'data available' and 'data accepted' signals of the two-wire handshake there is an additional signal which is used by any listener to indicate that it is not ready for data. The sequence of logic levels which appears on these

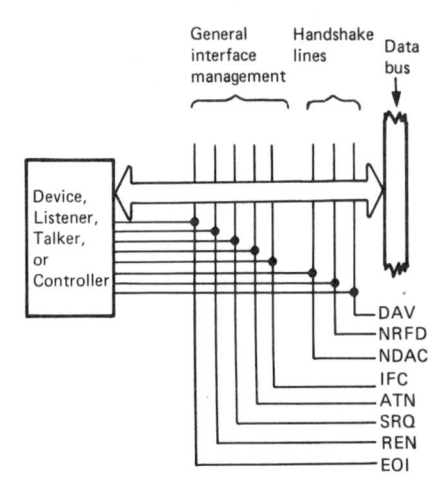

Figure 8.14 IEEE-488 bus lines

logic lines as data is transferred on the bus is shown as a set of waveforms in Figure 8.15. The functions of the three handshake signals are explained below.

NRFD   'Not Ready For Data' is returned by the listeners and remains low as long as one or more of them is not ready to accept the data from the talker. When all the listeners are ready, NRFD switches high.

NDAC   'Not Data ACcepted' also comes from the listeners and it remains low as long as one or more of the listeners has not yet accepted

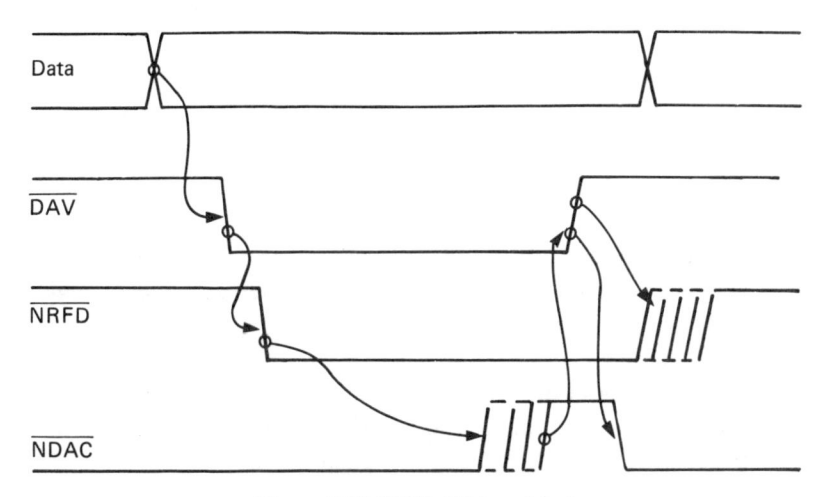

Figure 8.15 IEEE-488 handshake

the data. When all the listeners have accepted the data, NDAC switches high.

DAV 'Data AVailable' is a signal from the talker which switches it 'low' when it has data available for transmission and it senses that all the listeners are ready for data, and switches it 'high' again when it senses that the data has been accepted by all the listeners.

DAV is similar to the conventional two-wire handshake signal with both active pull-up and active pull-down, whereas the other two signals NDAC and NFRD are active-low and have open-collector outputs. This is because they are designed not to change until the slowest listener responds, with the result that the bus will allow the transferral of data to proceed as fast as the slowest listener and talker will allow.

In fact the system is much more complicated than this introduction indicates. For example, because there are usually several potential talkers in a system, the DAV outputs of the talkers must be designed to be three-state, with only the output of the active talker being in the active state.

The complete IEEE-488 specification is too complicated to allow it to be explained in detail here. However, it allows for data, addresses, and control commands to be sent along the eight bus lines. The overall result is not very different from a microcomputer bus (in fact, at least one manufacturer has used it as a microcomputer bus) and it allows the equivalent of an interrupt to take place from a device on the bus to the controller. This is called a 'service request' and two techniques are specified for polling the devices on the bus to ascertain the origin of the request: a serial poll, in which the status of each device is checked in turn, and a parallel poll in which up to eight devices can report whether they originated the service request by means of the eight bus lines.

The remaining five control lines are used to manage the flow of addresses, data, and control information on the interface bus. Their functions are as follows:

ATN 'Attention' is a signal from the controller to indicate that a message is present on the lines DIO1 to DIO8. If the level is 'low', the bus is carrying address or control information, while if it is 'high' data is being transferred.

SRQ 'Service Request' is a signal from the other devices on the bus to the controller, and it represents a sort of interrupt line; a logic 'low' indicates a request for service.

EOI 'End Or Identify' is a signal from a talker device which is used to indicate the end of a multiple byte transfer sequence, for example at the end of a multi-digit reading from an instrument. It has a second purpose during 'polling' of instruments by the controller.

REN 'Remote Enable' is a signal from the controller used to switch devices to 'local' (front panel) control or to remote control via the bus.

8.4.1

IFC   'Interface Clear' is a signal from the controller which is used to put the bus system into a known quiescent state, usually 'idle'.

8.4.2 Serial communications

As the distance between microcomputers increases, the expense of parallel systems becomes excessive and the lower cost of serial systems becomes more attractive. The same elements are needed in a 'transaction' between two devices on the bus: in addition to the data itself, address and control information is required. Also, because the greater length of a serial link makes it even more vulnerable to errors, error control mechanisms are often built into the system.

8.4.3 BISYNC

BISYNC (IBM, 1970) is a serial system, and as its name might suggest, it is both BIdirectional and SYNChronous. Although rather old-fashioned, it provides a good example of a synchronous communication system in which the information is divided up into bytes. The message is sent from one machine to another as a set of **frames**. Before communication can begin, the transmitter sends a synchronising pattern continuously and waits until a synchronising pattern is received from the other transmitter. Then simple handshake messages are sent from one to the other to confirm that the communication channel is functioning correctly. The structure of each message is as shown in Figure 8.16. The abbreviations SOH, etc., are special bit patterns included within the ASCII specification for this purpose.

| SYN | SOH | Header | STX | Data block | ETX | Check bits | SYN | • • |

Figure 8.16 BISYNC message structure

Each of these special characters is used to structure the message in some way.

SYN   This is a SYNchronising character and at the start of a message the receiver waits until it has detected this pattern before starting to pick out further bytes.

SOH   This marks the 'Start Of the Header' on this section of the message which contains addressing and control information.

STX   This marks the 'Start Of the Text' in the message and the end of the 'header'.

ETX   The 'End Of the Text' is indicated by the special ETX character. If it is the end of the current block transmission, an ETB, or 'End of Transmitted Block' character is sent instead.

BISYNC has problems because the bit patterns corresponding to the special characters above will cause the communication system to malfunction if they appear in the message being sent. To overcome this problem, another special 'Data Link Escape' character, DLE, is used to prefix any occurrence of a special character. Any character is treated as an ordinary character if it is prefixed with a DLE. Although this appears to get around the problem, if the checksum pattern at the end of a message happens to be the same as a DLE, then the SYN character following it may be lost.

8.4.4 Bit oriented protocols

The essential problem experienced with BISYNC is its lack of data **transparency**, which means that the system is inherently susceptible to certain bit patterns. Also, it insists upon the data passing over the communication link being sent as a series of bytes, even if the data itself does not naturally break down into eight-bit groups. HDLC, an abbreviation for 'High-level Data Link Control', is a development of two related earlier standards, SDLC ('Synchronous Data Link Control') and ADCCP ('Advanced Data Communications Control Procedure'). In fact these three systems are quite similar, and their main features are essentially the same. They are all examples of **Bit Oriented Protocols (BOPs)** in which data is not structured in any way; there is no division into bytes, and transparency is assured.

HDLC (ISO, 1979) uses the special pattern '01111110' to maintain synchronisation. The key feature of this pattern is that it contains a sequence of six ones. In HDLC, the transmitter ensures that this cannot happen when data or any other information is being sent by inserting an extra zero whenever a sequence of five ones is detected. In the receiver these extra zeros are removed, so that the unique pattern of six ones can only appear during a synchronising pattern. In a system of this sort, the transmissions consist of sequences of 'frames', each frame having the form shown in Figure 8.17.

Flag	Address	Control	Information	Check	Flag

Figure 8.17 HDLC frame

The size of the address and control 'fields' of this frame depend upon the system being used. In SDLC (IBM, 1975) both address and control fields are eight bits long, but in HDLC and ADCCP (ANSI, 1979) more than one byte can be used in each case. The function of each field is shown below.

- The start and finish of each frame is shown by means of a **flag** consisting of the pattern '01111110'. The flag which finishes one frame can be the same as the flag which starts the next one, so that the flags simply act as separators between frames.
- In SDLC the **address** consists of a single eight-bit pattern, but in HDLC and ADCCP it can be extended to several bytes in which all but the last

byte consist of an eight bit pattern starting with a binary '0' and the last byte consists of a pattern starting with a '1'.
- The **control** field consists of an eight-bit pattern (SDLC) or a sixteen bit pattern (HDLC and ADCCP). The detailed coding of the bits within the control field is specified by the various standards and is too complicated to be discussed here. However, the information contained is not especially complicated, consisting of sequence numbers, whether this is the final frame of a transmission, and suchlike.
- The **information** field can be of any length, and is not even limited to an integer number of bytes. This contains the data to be transmitted and is completely transparent; any binary data can be included in this field.
- Finally comes the **check** field, which contains the cyclic redundancy check information for the message. This is identified as the last sixteen bits before the closing flag at the end of the frame.

Note that these systems do not require any special control characters as BISYNC did. The lengths of the fields are all either set out in the standard or they can be determined by the receiver. If a transmitter wishes to 'abort' a frame once it has started it, either because it ran out of data or for some other reason, it can do so by sending a succession of more than six ones, which is recognised by the receiver as an 'illegal' bit pattern.

8.4.5 Storing data on disc

The storage of data on magnetic media such as tapes and discs presents similar problems to the communication of data between sites. The information must be written to the medium in serial form and read in the same way. In practice the data is divided into frames which are stored as **blocks** on the disc. Each block requires a serial number to identify it; this is similar to the address at the beginning of a communications frame and in the case of discs it usually indicates the physical position of the block on the medium. The block address is preceded by a synchronising pattern and followed by a cyclic redundancy check word. Next the data itself is stored, followed by its checksum.

Despite the fact that the basic form in which data is packaged into blocks for storage varies little from system to system, it is an unfortunate fact that very many different formats have come into use, especially in the case of floppy discs. Not only are floppy discs available in several different physical sizes, but the number of tracks and the number of sectors into which each track is divided (each sector contains a block of storage), the method of representing each bit of data, and even the number of sides used (one or both) can differ between systems. The existence of standard integrated circuits for floppy disc control is of little help as a standardising influence because these can be programmed for a wide range of different formats.

8.5 Local area networks

The ability to interconnect numbers of computer-based systems to provide communication between them leads to the idea of a **network** of such systems. Each system in the network is given an identity in the form of an address, and some means of interconnecting these systems is needed. **Local Area Networks (LANs)** are networks which, as their name suggests, interconnect systems in the same area (up to a few kilometres apart) at speeds of about 10 Mbps. They are used to join together equipment within the same room, in the same building, or on the same site. Serial techniques are used over such distances, in contrast with tightly coupled microprocessor systems which often use shared memory, or with **High Speed Local Networks (HSLNs)** in which the separation between processors is small enough to allow parallel techniques to be used.

LANs are finding application in a wide range of areas, almost all of which are more or less relevant to engineering and science. These include:

- **Remote access:** users can gain access to computers anywhere within the LAN without being aware of the structure of the network.
- **File transfer:** files of information can be transferred quickly and easily from one system to another within the LAN. These files may be data or programs.
- **Job transfer:** it will often prove more efficient to transfer a program to run on another computer within the LAN rather than on the nearest computer. Thus an automatic control system may continue functioning even when some of the equipment is faulty.
- **Resource sharing:** the ability of a LAN to move information quickly means that expensive resources, such as special printers, can be shared by several systems.
- **Database access:** it is easier to maintain a database in a single system than to attempt to keep parts of it on separate systems. LANs allow the data — for example, parts lists or stock lists — to be available anywhere in the system.
- **Electronic mail:** the LAN provides an excellent method for communicating messages between various users of the network.
- **External services:** connections from the LAN to external data services such as Videotex (Prestel in the UK).

8.5.1 Topologies

The component systems cannot be interconnected without some sort of plan. Although the process of connection may appear to be a simple one of providing circuits between them, in fact the **topology** of the connections is important. Figure 8.18 shows the topologies in common use.

8.5.1

Random interconnection

Of course, units can be connected by any number of direct links if desired. This places a requirement upon each unit in the system to 'know' the way in which the system is interconnected so that the messages can be forwarded from one unit to any other unit across the network as desired. Usually this is a nuisance, and other more systematic interconnection schemes are preferred.

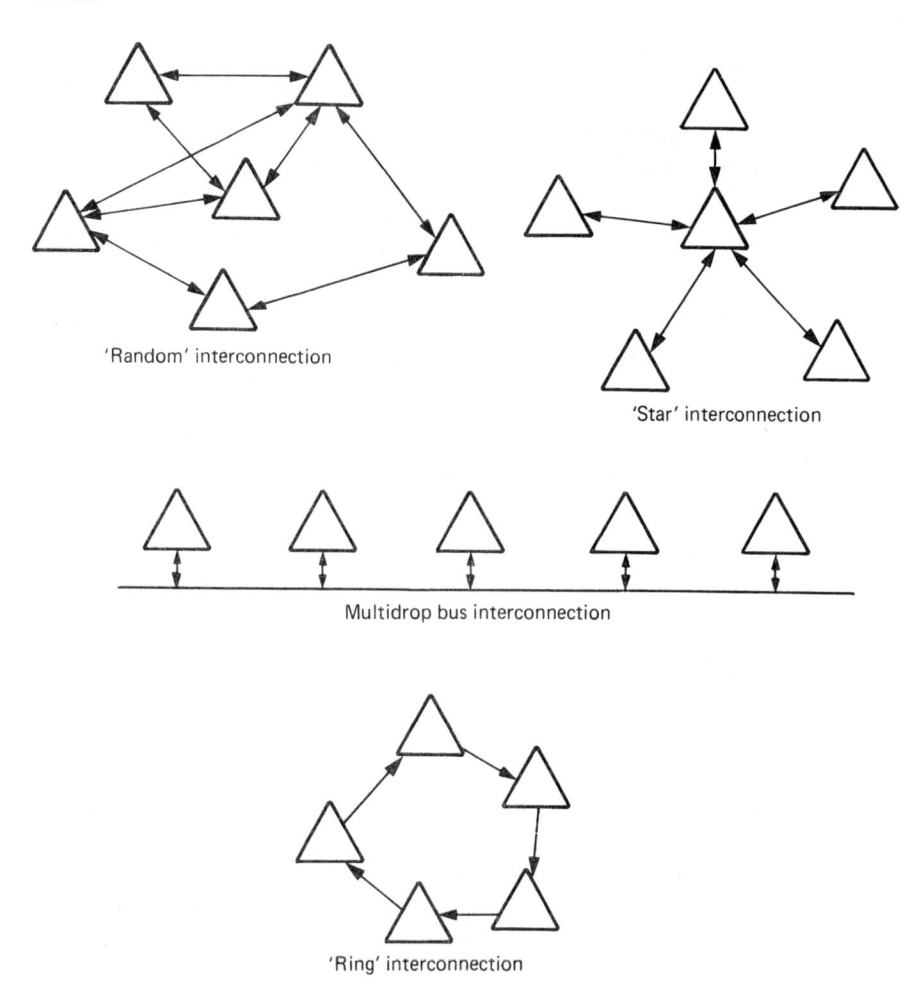

Figure 8.18 Network topologies

Star connection

The 'natural' way to connect a controlling computer to a number of 'slave' units is to use a separate connection from the controller to each slave. This means that the controller must select a slave internally by choosing which communication port to use. This corresponds to the system employed by a microcomputer when used with a number of input/output devices, and it has the advantage that the logic required in each slave unit is relatively simple. The disadvantage is that it becomes expensive in terms of the connections used if the slaves are remote from the controller.

Bus connection

The most familiar method of interconnecting a number of devices to make a composite system is to use a bus, just as in a microcomputer. In this case the address that selects a unit is sent at the start of a message frame as explained in section 8.4. An advantage of this system is that it needs only a single communication channel, which makes the wiring less expensive.

However, this approach requires each device connected to have a bidirectional connection on to the bus so that it can both receive and transmit on the same connection. This need for a bidirectional connection can be a disadvantage with some interconnection technologies, either because the relevant standard does not provide for three-state outputs or because the technology does not allow bidirectional connection with a bus, as for example, with optical fibre communication.

Rings

In a ring, each unit is connected to the next unit by a simplex communication link. Each unit has an input and an output, and the links are arranged so that messages always pass around the loop in the same direction. Because the data flow is unidirectional, simple technology can be used in the transmitter and receiver, and optical communication can be used if required. This type of system needs quite sophisticated signalling techniques of the type provided by bit orientated protocols such as HDLC. Rings do not need a controller, although some sort of 'monitor' to check system operation and restart the system if it fails for any reason, is useful.

Internetworking

As the use of LANs increases, there is increasing interest in interconnecting them to provide larger-scale communication between systems. The interconnections are provided by a **gateway unit** which is necessary to convert messages from the format used in one LAN to that used in the other. A global addressing scheme is needed to uniquely identify each system in a set of interconnected networks.

8.5.2 Network access methods

When several units have access to a common bus or other medium, their behaviour must be co-ordinated in such a way that each of them has a chance to access the medium without being disturbed by any of the others. This can be arranged by designating one of the units as a master to supervise the behaviour of the others or by using a 'protocol' which ensures that any unit can transmit only when it has permission to do so in the form of a 'token'. Alternatively, some kind of **contention technique** can be used which allows them to 'fight' for access according to a set of rules.

Contention techniques
One such contention technique is rather like that which we use ourselves when in conversation around a table — we wait until the previous speaker has stopped talking and then start ourselves. If two people start at the same time, they both stop momentarily and then one of them carries on. This is the basis of the system which rejoices in the acronym of **CSMA/CD**: Carrier Sense Multiple Access, Collision Detection.

In this system a unit monitors the line and waits until there is no signal present. The absence of signal indicates that there is no transmission taking place, and so the unit begins sending. However, there is a chance that two units will independently check at the same time to see if there is a signal present, and if none is detected, begin transmitting. In fact this is quite likely to happen because two or more units may have been waiting for the previous transmission to end. Each unit monitors the line while it is transmitting, and if it senses that two units are transmitting at the same time (a 'collision'), it stops transmitting for a short while. If each unit stopped for the same time and then started again, another collision would inevitably ensue, so instead the units each delay by a random amount of time. If another collision still happens, the units delay by a random amount once more, but this time the random time is chosen so that the average delay is somewhat longer.

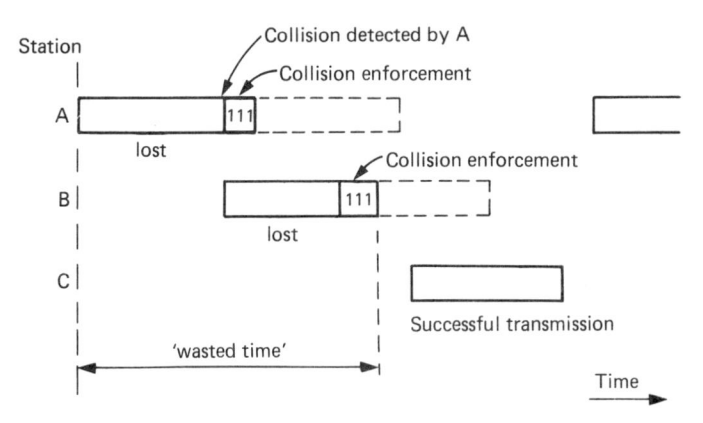

Figure 8.19 Bus contention in an Ethernet network

Eventually, after one or more attempts, one unit will start transmitting long enough before its competitors to ensure that the others will detect its signal before starting to transmit themselves. This complicated system can be implemented using special purpose integrated circuits, and forms the basis of the **Ethernet** system (Metcalfe and Boggs, 1976). This originated in Xerox's Palo Alto Research Centre during the early 1970s, but the CSMA/CD system has since been suggested as one of the IEEE-802 local area network standards which are discussed later in this chapter. Ethernet is designed to run with a data rate of ten million bits per second over distances of up to 1.5 km using low-cost coaxial cable. Although the maximum number of stations which can be accommodated on a single Ethernet network is 1024, each station has a six byte address in order to ensure that stations can be uniquely identified on a world-wide basis, so that Ethernet systems can be interconnected without problems.

Token access

An alternative approach uses a 'token' which is effectively an invitation to the holder to transmit if it wishes. Only the holder of the token is allowed to transmit, and if it does not wish to do so, it passes the token to the next unit. Rules are necessary to oversee the passing of this token; each unit is allowed to transmit for only a limited amount of time, and after finishing it must pass to the next unit according to some list. Each unit must know the number of the next unit to which it must pass control.

Further complications are introduced by the possibility that units may be connected and disconnected (or switched on or off) while the system is running. This means that a unit must confirm after passing the token that the next unit in the sequence has received it. The units must also check periodically to see if any other units have become operational while the system is running. This can be done by polling the numbers of previously non-operational units to determine if they have since become operational. For example, if unit 35 normally passes the token to unit 38, it should periodically check to see if unit 36 or unit 37 has become active.

Ring access

Token access can also be used in ring systems. In this case the sequence of passing the token is determined by the order in which the units are connected around the ring. When nothing is taking place on a ring, a token passes round and round until one of the units wants to transmit, at which point the token is replaced with the information to be sent.

The **slotted ring** is another access technique, and this is used in the

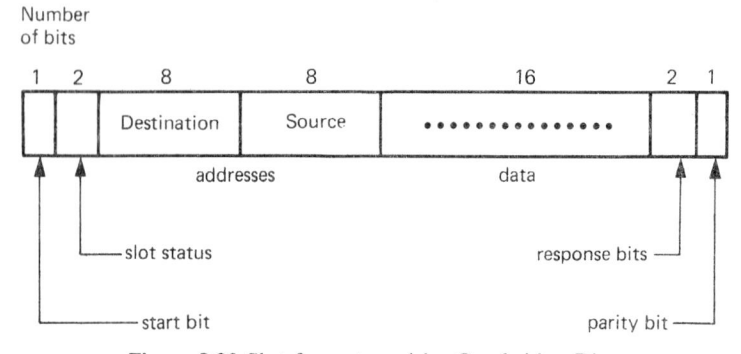

Figure 8.20 Slot format used by Cambridge Ring

Cambridge Ring (Hopper et al., 1985). This system uses relatively small packets of the form shown in Figure 8.20. A potential transmitter waits for an empty slot to appear, inserts the data and address information, and marks the packet as 'full'. When the packet returns, the originating station inspects the response bits and marks the packet as 'empty'. Each packet is carried in a time slot which contains the source and destination addresses and two bytes of data. Each address consists of eight bits, which allows up to 254 different addresses to be used (addresses 0 and 255 are reserved for special functions). The start of the packet is marked with a start bit, which is followed by two status bits which are used to monitor the status of the packet: whether it is empty or full, and whether it has passed the ring's monitor. At the end of the slot is a pair of response bits which are used to set up a handshake between the source and destination and a parity bit which allows errors to be detected. The basic slot format does not provide for error control and higher level protocols are necessary to handle the communication between the source and destination of the data.

A third type of ring access method is known as **register insertion** or **buffer insertion**. The transmitting station simply diverts incoming data into a buffer, transmits its own message, and then issues the buffered material. The main problem with this system is the unpredictable ring length and corresponding transmission delay.

Network performance
The performance of a station depends on the number of stations connected, the access method employed, and many other factors. The token passing and CSMA/CD schemes are compared in Figure 8.21, which shows how the throughput S of a network changes as the number N of stations on the network is increased (Arthurs and Stuck, 1981); in each case the lengths are measured in terms of bit times. The parameter a is a measure of the network size, defined as

$$a = \frac{\text{length of data path}}{\text{length of frame or packet}}$$

IEEE-802 standard
The CSMA/CD and token passing access methods already discussed form the basis of a draft set of standards produced by the IEEE 802 committee (IEEE802, 1982). The three layer standard of Figure 8.22 defines the functions of the lowest two layers of the OSI reference model.

- The **Logical Link Control layer (LLC)** provides for the exchange of data between one or more service access points from a single physical connection on the local area network. On transmission, data is assembled into a frame with address and CRC fields, and on reception the frame is disassembled, and address recognition and CRC validation is carried out. The protocol and frame format used closely resemble that of the HDLC system described in section 8.4.4.

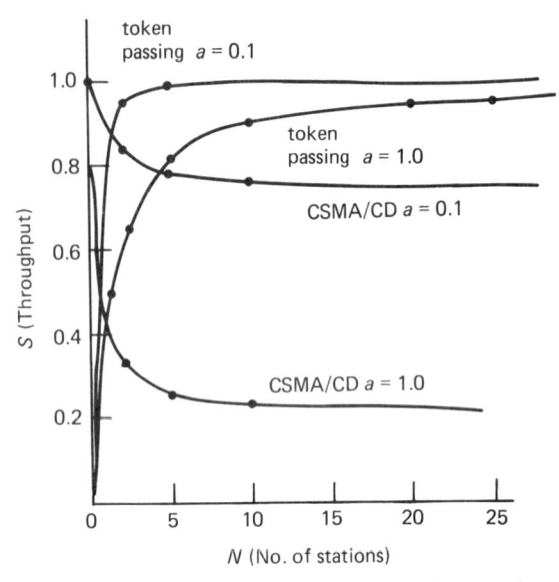

Figure 8.21 Throughput of CSMA/CD and token-passing networks

- The **Medium Access Control layer (MAC)** manages data communication over the physical link. It is included as a separate layer since in the older OSI Layer 2 Data Link no provision was made for access in multiple-source, multiple-detection systems. This MAC layer encompasses three standards: CSMA/CD, token-passing bus and token-passing ring access techniques.

- The **Physical Link layer** provides a similar function to that of the OSI model: it is concerned with the encoding and decoding of the signals, preamble generation and removal for synchronisation purposes, and bit transmission and reception. The standards included in the IEEE-802 model are summarised in Figure 8.23.

(OSI)	(IEEE 802)
Network	Network
Data Link	Logical Link Control
	Medium Access Control
Physical Link	Physical Link

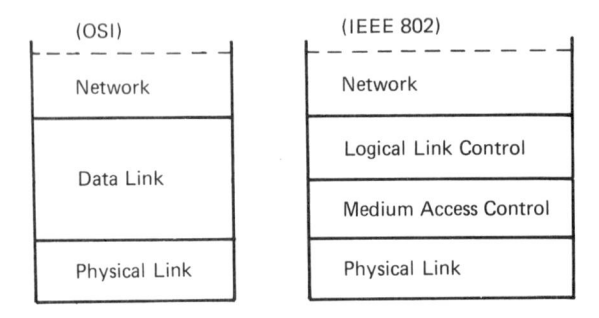

Figure 8.22 IEEE-802 three layer model

8.5.2

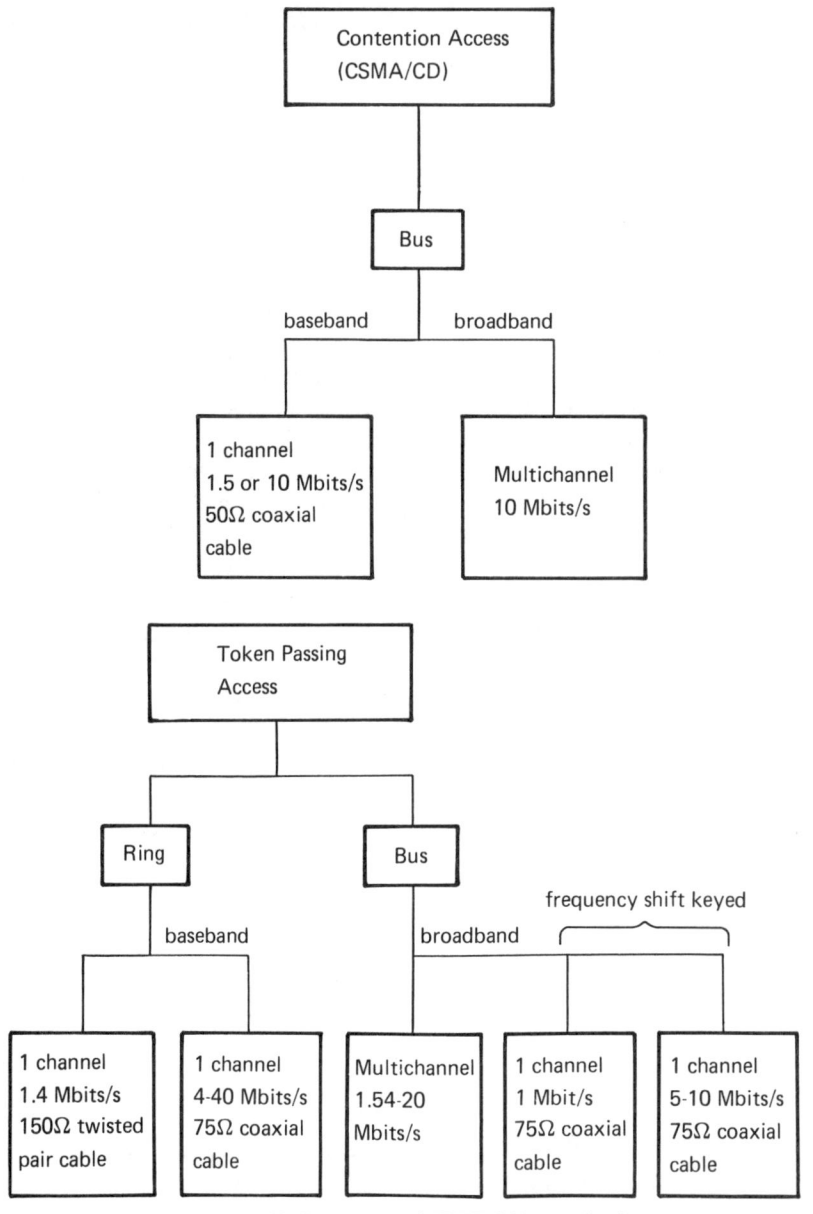

Figure 8.23 Summary of IEEE-802 standard

- Contention Access (CSMA/CD)
  - Bus
    - baseband: 1 channel, 1.5 or 10 Mbits/s, 50Ω coaxial cable
    - broadband: Multichannel, 10 Mbits/s
- Token Passing Access
  - Ring (baseband):
    - 1 channel, 1.4 Mbits/s, 150Ω twisted pair cable
    - 1 channel, 4-40 Mbits/s, 75Ω coaxial cable
  - Bus:
    - broadband: Multichannel, 1.54-20 Mbits/s
    - frequency shift keyed:
      - 1 channel, 1 Mbit/s, 75Ω coaxial cable
      - 1 channel, 5-10 Mbits/s, 75Ω coaxial cable

The introduction of a widely accepted LAN standard permits full use to be made of VLSI techniques in the design of interfaces, because the large market offered by conforming to a single standard allows the high design cost of such circuits to be shared over a large number of units. It is expected that the bulk of future LAN development work will be within the scope of the IEEE-802 standard, particularly in the USA, and that the cost of interface circuits for LAN use will fall considerably. Further discussions of this standard may be found in Myers (1982) and Graube (1982).

8.6 Network security

Information which is being communicated from one place to another is vulnerable not only to the effects of electrical noise, but to deliberate attempts to interfere with it. Such problems, irrespective of the motive, can be included together under the heading of **network security**. This may be defined as the protection of network resources against unauthorised disclosure, modification, utilisation, restriction or destruction. Of course, the controls used to ensure that a network remains secure can be physical as well as automatic; for instance, ensuring that access points to the network are in rooms which are kept locked when left unattended.

Access control can be provided automatically by requiring a prospective user of the network to 'log in' using their identity code and a secret password known only to the user and the network. Unfortunately this method is not particularly secure and can be defeated in several ways, for example by leaving a program running on a system which impersonates the network's login sequence and attempts to trick other users into revealing their passwords.

Eavesdropping is another problem to which LANs are susceptible. This can be done by physically 'tapping' the cable used or by using an electronic sensor near to the cable to pick up signals radiating from the cable.

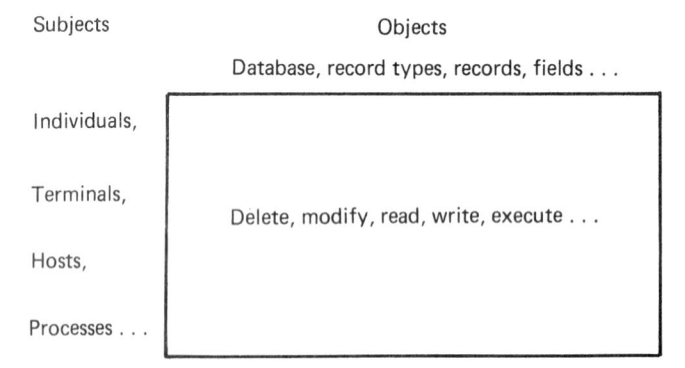

```
Subjects                    Objects
                    Database, record types, records, fields ...

Individuals,      ┌─────────────────────────────────────┐
                  │                                     │
Terminals,        │                                     │
                  │  Delete, modify, read, write, execute ... │
Hosts,            │                                     │
                  │                                     │
Processes ...     └─────────────────────────────────────┘
```

Figure 8.24 Access matrix to database

Access to data on a system, for example a database, can be controlled by a database management system, which is a program running on the same system as the database itself. This uses an **access matrix** which is a table listing who or what has access to which item of data, as shown in Figure 8.24. The access rights granted to the potential user depend upon the operation to be carried out upon the data; the five main types of operation are to read, write, delete, modify and execute data or programs in the database.

8.6.1 Encryption

As has already been observed, eavesdropping represents one of the biggest problems in LAN security. One solution is **encryption**; the information in its raw form, called **plain text**, is processed to form an encrypted version called **cipher text**. The encryption process requires a sequence of operations which is as difficult as possible for an unauthorised user to determine and which can be reversed by the intended recipient. 'Breaking' a code is a matter of much time and computational effort, and a large amount of research has been applied to the design of codes which are difficult to break (Tanenbaum, 1981). Two methods are attractive for use in LANs.

Data encryption standard
The **Data Encryption Standard (DES)** was developed by the National Bureau of Standards in the USA. This uses a 56-bit 'key' which together with the DES code transforms 64-bit blocks of the plain text into corresponding 64-bit blocks of cipher text. At the destination the cipher text can be reconverted into plain text using the same key as used at the source. The DES encryption can be performed using software or by means of special purpose integrated circuits; up to 2^{56} different codes are possible simply by changing the keyword.

One weakness of this method lies in the need to send the key to the receiving end. If this key were intercepted by a third party, messages could be intercepted and decoded, or spurious messages could be generated. The possibility of messages being received from unauthorised users of a system could have serious effects in applications such as remote control. Another weakness is that the key size is probably not sufficient to prevent access to a determined codebreaker (Hellman, 1980).

Public key encryption
An alternative encryption method, developed in 1976, uses an encrypting key E_i and a decrypting key D_i which are different, such that it is effectively impossible to derive D_i from E_i. This means that E_i need not be secret, because only the intended recipient of the message will have D_i, the means to decrypt cipher text encrypted with E_i. Thus a 'telephone directory' of encrypting keys could be published, allowing anyone to encrypt keys for an intended recipient without fear of the messages being intercepted by a third party. This is known as **public key encryption** (Diffie and Hellman, 1976). Note that although this method prevents unauthorised decryption of messages by a third party, it cannot prevent that party from sending messages. In this case a handshake with the sender using the sender's public key could request confirmation before the receiving system takes any control action. By its very nature, encryption is a complex subject, and it is impossible to give more than an overview in this section. Those readers who are interested in pursuing the subject in greater depth are referred to an excellent text by Denning (1982).

8.7 Design example

Two examples of communication between machines will be considered in the case studies for this chapter. The first is a simple serial communications system for the controller, which illustrates the main points that have been discussed in this chapter, and which would allow the controller to be controlled remotely from another computer. The second example centres on the IEEE-488 interface used for instrumentation, and shows how the waveform capturing and processing system described in the earlier chapters could be controlled by a computer.

8.7.1 Serial communications for controller

In this section a simple communications protocol will be examined for use with the three term controller. The members of the 6801 family of microcomputers each contain a serial port which can be used for this purpose. The serial format used is asynchronous, with a start bit and a single stop bit, and with a waveform that can be programmed to be either NRZ or biphase. The baud rate of the signal can be derived by dividing the system clock by a power of two, or from an external clock input.

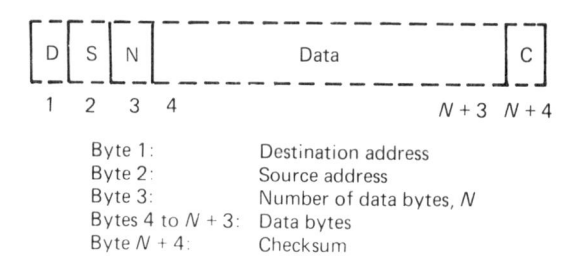

```
┌─┬─┬─┬──────────────────┬─┐
│D│S│N│       Data       │C│
└─┴─┴─┴──────────────────┴─┘
 1 2 3 4                N+3 N+4
```

Byte 1: Destination address
Byte 2: Source address
Byte 3: Number of data bytes, N
Bytes 4 to $N+3$: Data bytes
Byte $N+4$: Checksum

The checksum is calculated over all the bytes in the message and should give a zero result

Figure 8.25 Serial data format

A useful feature of the serial port is that it can be programmed to operate in a 'wakeup' mode. This means that receiver interrupts can be disabled during a message by a special mask which is cleared automatically as soon as the receiver detects that the message has finished. There should be at least one logic '0' (a start bit) during each consecutive group of ten bits while data is being transmitted, because the format of the signal is asynchronous. The receiver will assume that the message has finished as soon as ten consecutive logic '1's have been received, and clear the wakeup bit. If each message begins with a byte containing the address of the intended recipient of the message, the receiver in each of the microcomputers monitoring the signal will cause an interrupt when the first byte is received. If they are programmed to set their 'wakeup' masks if the byte received is not the same as the address of that unit, the message will be ignored by all except the microcomputer for which the message was intended.

Starting from this point a format can be developed for the messages being sent between microcomputers. In addition to indicating the intended recipient of the message, the format should indentify the sender of the message and its length, and it should make some provision for detecting errors.

The format to be used in this example is shown in Figure 8.25. After two bytes containing the destination and source addresses of the message, the third byte shows the number of bytes of data to be sent; this is a byte-oriented protocol because the design of the asynchronous port of the 6801 family makes it much easier to design protocols of this type. At the end of the message, after all the data has been sent, a single byte is used to contain the checksum. Again, the design of the hardware makes the calculation of a simple checksum much easier than a CRC check.

---

type *datarray* = **array**[0..*maxbytes* − 1] **of** *integer*;
procedure transmit(*source,destin,nbytes*: *integer data*: *datarray*);

var *i, checksum*: *integer*;

begin
  sendbyte(*destin*);
  sendbyte(*source*);
  sendbyte(*nbytes*);
  for *i* := 0 **to** *nbytes* − 1 **do**
  begin
    checksum := *checksum* + *data*[*i*];
    sendbyte(*data*[*i*])
  end;
  sendbyte(*checksum*)
end;

Figure 8.26 Pascal implementation of transmitter function

The sequence of operations required in the transmitter could be expressed using a simple Pascal procedure as shown in Figure 8.26. This calls a procedure *sendbyte* which looks after the transmission of a byte by means of the serial interface. The corresponding procedure for receiving messages would have a quite similar form. However, Pascal is an unsatisfactory language for programming this application because in its standard form it does not allow efficient access to the registers of the interfacing device. More importantly, Wirth's definition of Pascal does not allow interrupts to be programmed. Interrupts allow the microcomputer to carry on with other tasks while waiting for a message to be received; even the ability to perform other tasks during the transmission of a message may be important in control applications. Consequently, assembly language will be used in this study; the assembly language of the 6801 is similar to that of the 6809.

The communications program is designed to operate by means of an interrupt. Interrupts caused by the serial communications interface (SCI) of the 6801 family have a separate interrupt vector. However, this is at the 'high' end of the address space of the microcomputer at $FFF0, and as will be shown shortly, the interrupt handler modifies the vector to be used during the reception and transmission of messages. To avoid the hardware complication which would arise if address $FFF0 had to contain RAM, an indirect vector must be stored in RAM at address VECTOR. Figure 8.27 shows an interrupt handler for the serial communications interface.

```
            ORG    $FFF0
            FDB    IRQSCI         Vector points to interrupt handler

            ORG    $F800
     IRQSCI LDX    VECTOR         get VECTOR and
            JMP    0,X            go to that address
```

Figure 8.27 SCI interrupt handler

Because the SCIs interrupt vector can be changed, it must be initialised when the system is reset. When the communications software is waiting for a message to arrive, the interrupt vector must point to RXDEST, the section of code which compares the byte at the start of the message with the address of this microcomputer. Before the receiver can function, it too must be initialised by setting up its control register TRCS. The transmitter and receiver are connected to the serial bus by means of an RS-422 transceiver chip as shown in Figure 8.28, and this must be switched to 'receive' mode to allow the receiver to pick up messages.

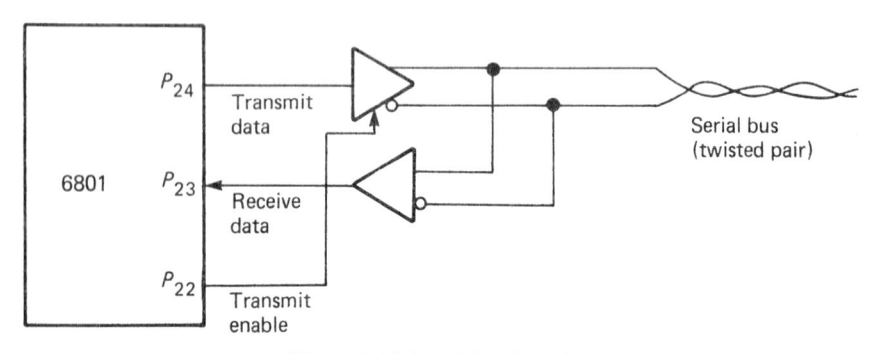

Figure 8.28 Serial bus interface

The initialisation subroutine which carries out these functions may be seen in Figure 8.29. Note that the lines associated with setting up the input/output port are incomplete because this port is used for several different functions.

```
INIT    LDA    #DRIVER+...  Set up data direction
        STA    DDR1         register
        LDA    #...         and
        STA    PORT1        port.
        LDX    #RXDEST      Set up SCI vector
        STX    VECTOR
        LDA    #RE+RIE+WU   Switch on receiver, Rx
        STA    TRCS         interrupt and 'wakeup' mask.
        RTS                 Return to main program.
```

Figure 8.29 Initialising the serial interface

Now the interrupt vector has been defined and the serial port has been initialised. However, interrupts cannot occur immediately because the 'wakeup' mask bit has been set, and it will only be cleared when ten successive logic ones have been received. The first byte to be received after the wakeup bit has been cleared by the SCI hardware will cause an interrupt.

The receiver interrupt software consists of a number of routines, each associated with a different aspect of the message format. While the receiver is waiting for a message, the interrupt vector points to RXDEST, which compares the byte received with the address of this microcomputer. If they do not match, the 'wakeup' mask is set again so that the rest of the message is ignored. However, if the byte does match the address, the SCI interrupt vector is changed to point to RXSEND, so that the next interrupt from the SCI will cause RXSEND to run instead of RXDEST. RXSEND picks up the identity of the sender and stores it in FROM, and then changes the vector to point to RCOUNT. RCOUNT picks up the number of data bytes to be received and then changes the vector to point to RXDATA, which receives a number of data bytes as indicated by RCOUNT and stores them in a buffer with the somewhat predictable name of BUFFER. Finally the vector is changed to RCHECK which checks that the checksum CKSM is zero. At the completion of a message, bit RXFULL of register STATUS is set to logic 1 to indicate that BUFFER contains a new message. The main program can find out if a message has been received by reading the contents of STATUS and it can find out whether the message contains errors by testing to see if CKSM is zero. The receiver interrupt handler can be seen in Figures 8.30 and 8.31.

Transmitted messages have the same format as those received, and the transmission of messages is handled in a similar fashion. When the program wishes to send a message, it must put the address of the recipient in register TO, the number of bytes to be sent in COUNT, and the message in the buffer which starts at address BUFFER. Next the transmitter must be switched on

```
RXDEST  LDA   RXBUF      Read receiver data register.
        CMPA  ADDRES     Is this our address?
        BNE   RDEST1
        STA   CKSM       Yes; initialise checksum
        LDX   #RXSEND    Update vector to point
        STX   VECTOR     to next section
        RTI              and return from interrupt...
RDEST1  LDA   TRCS       Not us, so set 'wakeup'
        ORA   #WU        bit to prevent rest of
        STA   TRCS       this message from interrupting.
        RTI
RXSEND  LDA   RXBUF      Read receiver data register.
        STA   FROM       Note identity of sender
        ADDA  CKSM       Update checksum
        STA   CKSM
        LDX   #RCOUNT    and update pointer
        STX   VECTOR     ready for next interrupt
        RTI
RCOUNT  LDA   RXBUF      Read receiver data register.
        STA   BYTES      Set up byte count and keep
        STA   COUNT      a copy for main program.
        ADDA  CKSM       Update checksum
        STA   CKSM
        LDX   #BUFFER    Set up buffer pointer
        STX   POINT1
        LDX   #RXDATA    and update pointer
        STX   VECTOR     ready for next interrupt
        RTI
```

Figure 8.30 Receiver interrupt handler

and the receiver switched off, to start the transmission process. This is done by setting and clearing the appropriate bits in the control register TRCS. The transmitter buffer amplifier must also be switched on and the receiver amplifier switched off, which is achieved by switching a control line (controlled by bit DRIVER of port PORT2) low. A signal will now appear on the bus, but because the transient at the start of transmission may confuse some receivers into treating this transient as a character, the transmitter must wait at least two character times before starting the message proper. This time delay can be produced either by software or by using a system timer. At the end of this time delay the first byte is sent to the transmitter and the interrupt vector is set to point to the second step (TXSEND) in the sequence, the sending of the transmitter's identity. The transmit subroutine may be seen in Figure 8.32; once the transmitter has been started, it will

RXDATA	LDX	POINT1	Get pointer
	LDA	RXBUF	When an interrupt occurs,
	STA	0,X	and put byte into buffer
	ADDA	CKSM	Update checksum
	STA	CKSM	and
	INX		increment pointer.
	STX	POINT1	
	DEC	BYTES	Decrement byte count
	BNE	RDATA1	and if a count of zero has
	LDX	#RCHECK	been reached, update vector
	STX	VECTOR	for next section
RDATA1	RTI		
RCHECK	LDA	RXBUF	Get checksum byte
	ADDA	CKSM	and update checksum
	STA	CKSM	to give 0 if all is OK.
	LDA	STATUS	Indicate that
	ORA	#RXFULL	that the buffer is
	STA	STATUS	full once again
	LDX	#RXDEST	Put vector back
	STX	VECTOR	to beginning
	RTI		and return...

Figure 8.31 Receiver interrupt handler (continued)

raise interrupts whenever it requires servicing, causing the corresponding interrupt service routine to be run.

Once the transmission of a message has been started, the process can continue automatically without the intervention of the main program. When the message has been sent, bit TEMPTY will be set in register STATUS; the main program should reset this flag bit before sending any more messages. At this point the transmitter will have been switched off and the receiver switched on again. Note that the message protocol as defined thus far does not provide for acknowledgement of messages received; this is left to the main program. In practice the microcomputer would probably start a timer after sending a message, and if no reply were taken within a specified time, another attempt at transmission might be made.

8.7.2 Interfacing to the IEEE-488 bus

This example provides a contrast to the first one not simply because it is a parallel bus system, but because the use of a standard bus means that LSI circuits are available to handle the interfacing task. This allows functions such as address comparison, which would otherwise have needed software, to be carried out directly by the hardware. As before, the interface can be

```
TRANSM  LDA   TRCS              Switch to transmit mode:
        ORAA  #TE               Transmitter on
        ANDA  #$FF-RE-RIE       Receiver and Rx interrupt off
        STA   TRCS
        LDA   PORT2             Get current data on port 2
        ANDA  #$FF-DRIVER       and switch on transmitter
        STA   PORT2             driver.
        JSR   AWAIT             Wait for 2 character times
        LDA   TO                Output address of destination
        STA   TXBUF             to transmitter
        STA   CKSM              Initialise checksum
        LDX   #TXSEND           Update vector to point
        STX   VECTOR            to next section.
        LDA   TRCS              Switch on
        ORA   #TIE              transmitter
        STA   TRCS              interrupt
        RTS                     and return from subroutine...
```

Figure 8.32 Transmit subroutine

divided up into a section which handles the protocols used by the bus and one which converts the TTL signals used by the microcomputer into the signal levels used on the bus, as shown in Figure 8.33.

Programming an interface device for the IEEE-488 bus is similar in principle to the programming of any other communications interface. In this example the interfacing function is performed by the 68488, which as the device's number might suggest, is intended to interface microprocessors of the 6800, 6801, and 6809 families to the IEEE-488 bus. This has eight registers, of which the functions are listed in Table 8.1, together with the arbitrary names and addresses assigned to them for this example.

The 68488 can be used as either a 'talker' or a 'listener' under program control, but it cannot be programmed to act as a controller for the IEEE-488 bus without external hardware. This interface device would be suitable for an intelligent instrument controlled by an external microcomputer, but less suitable for a microcomputer which would be required to act as a system controller.

Before the 68488 can be used, it must be configured for the system in which it is to be used, by appropriately initialising its registers. Although this device can be programmed to show quite complex behaviour, it is not difficult to configure it for 'talking' and 'listening'. Figure 8.34 shows a simple configuration subroutine which will allow the interface to be used for input and output under program control. Note that AUXCOM, STATUS and ADSTMD are initialised to the all '0's state to disable interrupts and the more sophisticated functions of which the circuit is capable. The addresses of the

8.7.2

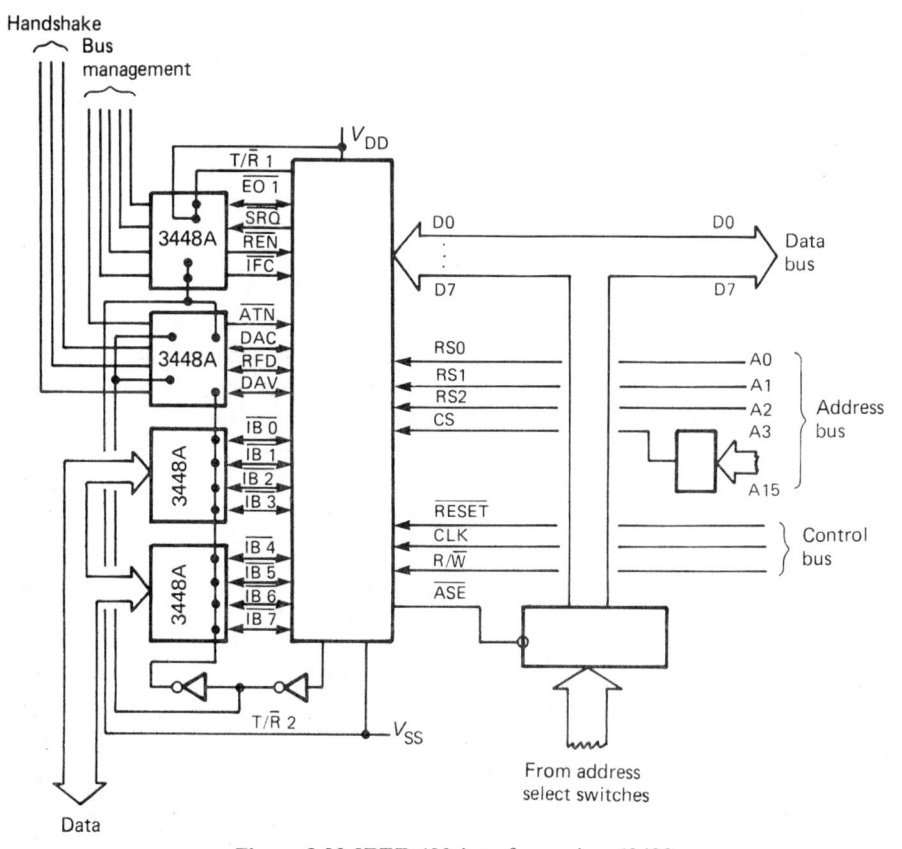

Figure 8.33 IEEE-488 interface using 68488

registers within the 68488 are defined by means of EQU statements; defining the addresses of the device's registers relative to that of the first one allows the device to be relocated in the address map of the computer with minimal changes to the program. The address of the instrument or other piece of

Table 8.1 Registers of 68488

Address	Name	Function	
		read	write
$C000	STATUS	Interrupt status	Interrupt Mask
($C001)		Command status	—
$C002	ADSTMD	Address status	Address mode
$C003	AUXCOM	Auxiliary command	Auxiliary command
$C004	ADDRES	Address switch	Address
($C005)		Serial Poll	Serial Poll
($C006)		Command passthrough	Parallel Poll
$C007	DATA	Data in	Data out

```
        STATUS  EQU    $C000        Interrupt status, Interrupt Mask
        ADSTMD  EQU    STATUS+2     Address status, Address mode
        AUXCOM  EQU    STATUS+3     Auxiliary command
        ADDRES  EQU    STATUS+4     Address switch, Address on 488 bus
        DATA    EQU    STATUS+7     Data in, Data out

        CONFIG  LDA    ADDRES       Read address from switches
                STA    ADDRES       and write it to address register
                LDA    #$00         load accumulator with '0's
                STA    AUXCOM       Clear reset bit in command reg'r
                STA    STATUS       Mask all interrupts
                STA    ADSTMD       select no special features
                RTS                 Back to main program...
```

Figure 8.34 Configuring the 68488

equipment containing the 68488 is defined by a set of switches which are connected to the data bus by means of three-state drivers. When an attempt is made to read register ADDRES, the 68488 enables the three-state drivers so that the setting of the switches is read instead. When data is written into address ADDRES, it is stored in a register within the 68488. Whenever a command is received from the controller via the IEEE-488 bus, the address of the command is compared with that in register ADDRES and the 68488 responds by taking appropriate action only if the two are the same.

Subroutine TALK causes the 68488 to send a message as a sequence of bytes stored in memory starting at the address in index register X when the subroutine is called. The number of bytes to be sent is passed in accumulator B when the subroutine is called. Note that there is no need to send a byte count or a special 'end of message' code using special data patterns; instead, the 68488 is programmed to send the 'EOI' signal on a separate wire after the end of the message. The auxiliary command register is programmed to do this immediately before the last byte is sent, to avoid delay. Similarly, the short communication path and the use of special cables means that the likelihood of errors occurring is very small, so there is no need to include error control information in the signal. If error detection or error correction is necessary, it can be included in the software.

In the listing of subroutine TALK shown in Figure 8.35, symbols BO and EOI are used instead of hexadecimal numbers in order to make the program more understandable. When the subroutine is called, it will wait until the 'byte out' flag BO goes to logic 1. This will indicate that the controller has requested this device to 'talk'. Subroutine TALK will then read a succession of bytes from memory, using accumulator B as a counter.

Figure 8.36 shows the listing of subroutine LISTEN This performs the complementary operation of receiving a message from the controller and

8.7.2

```
BO      EQU     $40             'Byte Output' bit in STATUS
EOI     EQU     $20             Writing this to AUXCOM causes EOI signal
*                               to be sent after next data byte

TALK    LDA     STATUS          Get interrupt status
        BITA    #BO             Wait for flag
        BEQ     TALK

        LDA     0,X+            Get data byte from buffer
        STA     DATA            and output to bus
        DECB                    and decrement count
        BNE     TALK            until done...

        LDA     #EOI            get EOI code, which will be
        STA     AUXCOM          asserted after next & last byte
        LDA     0,X             Get last byte from buffer
        STA     DATA            and send it...

        RTS                     Return to calling program...
```

Figure 8.35 68488 'Talker' subroutine for 6809

```
BI      EQU     $02             'Byte input' flag

LISTEN  CLR     COUNT           Reset the byte count

LOOP    LDB     STATUS          Get interrupt status
        BITB    #BI             Wait for input flag
        BEQ     LOOP

        LDA     DATA            Get data byte from bus
        STA     0,X+            and output to buffer
        INC     COUNT           and byte count.

        BITB    #END            Has controller sent EOI?
        BEQ     LOOP            If so, message has finished...

        LDB     COUNT           Recover byte count
        RTS                     and return to calling program...
```

Figure 8.36 68488 'Listener' subroutine for 6809

storing it in memory starting at the location pointed to by index register X when the subroutine is called. The contents of accumulator B when LISTEN returns to the calling program indicate the number of bytes which were received.

The speed at which data will be transfered between the subroutines TALK and LISTEN and the controllers will depend upon the speed of the controller and that of the microcomputer connected to the 68488; the transfer of data can proceed only as fast as the slower of the two. The speed of the 68488 itself is sufficiently faster than that of the 6809 microprocessor running the software that it does not limit the speed of data transfer.

8.8 Questions

1. Write a program in Pascal or BASIC to work out the (odd) parity of a 7-bit (bits 0 to 6) number and place the parity bit in bit 7 ('128' bit) of the number. Repeat this exercise using assembly language.

2. Write an interrupt handler to allow messages to be sent using the serial protocol of Figure 8.25.

3. An experiment controller based upon a microcomputer has two serial ports: one is connected to a computer terminal and the other to another computer. Anything typed at the terminal is to be relayed to the computer, while the program operating the computer must 'package' messages to the controller in such a way as to allow the controller to distinguish between messages intended for display at the terminal and commands for the controller. Typical controller commands are to read an input port and send the result to the computer, and to allow a value sent from the computer to be sent to an output port. Devise a format for the messages to be used.

4. The cables which connect together the various transducers, instruments, switches, lamps, and other electrical equipment within a car are expensive to manufacture and install. Discuss the ways in which a single signal cable could carry digitally coded information from one part of a car to another. A second, thicker, power cable could be used to distribute the power required.

5. Describe the CSMA/CD (Carrier Sense Multiple Access with Collision Detection) and Token Passing techniques for sharing the transmission medium of a baseband Local Area Network.

6. Discuss the relative merits of using either an Ethernet or ring type network to provide a distributed computing (or control) environment within the bounds of your place of study or work.

7. Fibre optics has been called the ideal communications medium for ring type LANs. Give reasons which either support or oppose this statement.

8. List the key issues which affect the security of a computer network, and discuss the various automated control methods by which security can be maintained.

Chapter 9 SYSTEM DESIGN AND DEVELOPMENT

The last chapter of this book will concentrate upon the problems encountered when a system is to be implemented using one or more microcomputers. In most respects the building of a microcomputer-based system passes through the same phases that are encountered in other kinds of engineering although the emphasis upon certain aspects may be different.

9.1 Designing microcomputer-based systems

First the requirements of the system must be specified, which is not always as easy as it might seem because of the complexity of systems of this type. However, this is a very important part of the overall engineering process, because these requirements provide the foundation for the remaining phases to follow. Once the requirements are accurately known, a design can be produced. In doing so, the designer must take account of many factors, some of them fairly obvious, such as the need to minimise cost and development time, and others which become more important as the complexity of systems increases, such as the need to design for easier testing during commissioning and maintenance. Next the design must be implemented; that is, prototype hardware must be built, and the software coded. There will almost certainly be problems with both hardware and software once the time comes to test the system, and these hardware faults and software 'bugs' will need to be removed before the system will perform as intended.

The project is still not finished at this point, because the final documentation for the system must be written. There are several types of this documentation needed, but it can be classified loosely as information relating to the manufacture of the system, a 'user's manual' for the user of the system, and a service manual containing information such as circuit diagrams and software explanations as well as suggested procedures to be followed when the system fails.

The steps involved with the hardware and software aspects of the project are very similar, which is not surprising in view of the fact that they are inextricably linked in the same system. In view of this it is quite meaningful — and customary — to speak of 'software engineering' when referring to the production of software. The importance of software engineering becomes apparent when it is borne in mind that in most microcomputer designs it is software which accounts for the greater part of the overall development cost.

9.1.1 Requirements

It is unfortunate that many 'customers' for whom microcomputer-based systems are developed do not state their requirements completely, despite the fact that they are so important. One problem with the specification of requirements is that everyday 'natural' language is not always precise or unambiguous enough for the purpose.

The result of discussions between the parties involved in a project should be a document which sets out the agreed requirements for the system. This document will then form the basis for the remainder of the project, although allowance must be made for minor changes in the requirements as the project proceeds. Heniger (1980) gives six requirements for the document itself, and although his paper referred to software projects, the same points are applicable to any engineering exercise.

- It should specify only external system behaviour. There is no reason why the requirements should specify internal behaviour, and by not doing so the designers are allowed greater freedom in performing their job.
- It should specify constraints on the implementation. The constraints may be physical, such as size or weight, or otherwise, such as cost or country of origin of components.
- It should be easy to change. A careful document control system is needed to ensure that everyone involved with a project knows the current version of the requirements.
- It should serve as a reference tool for those involved in system maintenance. Understanding exactly what a system has to do can be important in trying to find out what is wrong when it malfunctions!
- It should record forethought about the life-cycle of the system. This includes the way in which it is to be designed, implemented, commissioned, manufactured, tested and maintained.
- It should characterise acceptable responses to undesired events. When things go wrong, it should behave in a predictable (and hopefully responsible) way.

9.1.2 Design philosophy

Any project needs management if it is to proceed smoothly to completion, and because microcomputer-based systems tend to be complicated, even greater care is needed to ensure that the project will be completed on time. The approach normally taken to the problem of designing microcomputer-based equipment is the so-called **top-down approach**. Here the designer breaks down the problem into smaller and smaller sections, and once it has been analysed in enough detail a solution can be proposed to each part of the problem. The design process proceeds from the overall general view of the problem into greater and greater detail until the design has been fully described. The specification of the subsystem needed to handle each part of the problem must be carefully written and the interaction between the subsystems carefully defined. Once all the sections have been specified, the designer can decide which of the sections are to be constructed in hardware and which are to be based on software methods.

A careful approach to design, with consideration being given to tolerancing of components and to the ways in which hardware or software failures can affect the system, will be repaid later in the 'life-cycle' of the project. A reliable design will have a long **MTBF (Mean Time Between Failures)**, while if the problem of repairability is taken into consideration from the outset the **MTTR (Mean Time To Repair)** will be kept short. Both these factors affect the important **availability** of a system. All this must be carefully documented so that reference can be made to it later.

9.1.3 Scheduling

Many technical factors have to be taken into consideration when any new project is undertaken. The size, weight, power consumption, performance, and a whole host of other factors will need to be specified, but most importantly it will have to be completed within a defined budget and by a certain date. Once the specification of a project has been completed, it should be possible to schedule the remaining phases. The necessary staff must be found, hardware must be bought or constructed, and software must be written.

The design and implementation of the various sections of the hardware and software will require time, and the various tasks involved will be interrelated so that some tasks cannot begin until others have been completed. Some sort of control is needed to make sure that the whole project will not be held up simply because one part is behind schedule. Estimating how long each part of a project is going to take is best left to the individuals overseeing that project, but there is one important rule: be realistic, not optimistic. It must not be assumed that all will go well; allowance must be made for staff shortages due to illness and leave and for the unforeseen problems that experience tells us will always appear!

The relationship between the tasks which make up a project and the time which each task will take to complete can be shown diagrammatically using a PERT chart (Figure 9.1). Each task is represented by a line against which is written its name and the time which it is anticipated that the task will take to complete. At any point in the progress of a project the PERT chart will show the staff needed and the task or tasks which are critical for the project to remain on schedule. A task of this sort is spoken of as being in the **critical path** and if this task takes longer than anticipated then the whole project will run late. If a task which is not in the critical path falls behind schedule, the overall timing of the project will not be affected until that task falls so far behind that it in turn becomes critical. This chart provides a simple tool for estimating the staff requirements and the time which

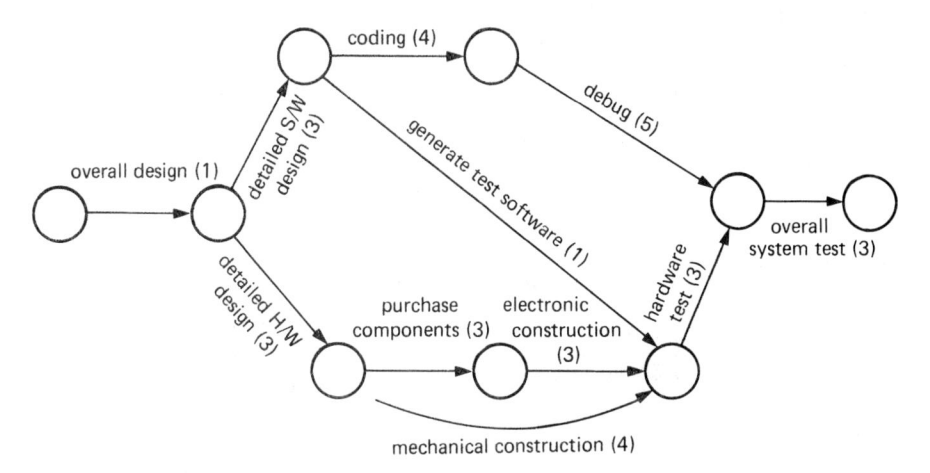

Figure 9.1 PERT chart

a project will take and it also enables the effects of overruns in timing or other problems to be examined as the project progresses.

The next maxim to be observed is not to assume that the amount of effort that is applied to a project will necessarily be reflected in the amount of progress which is made. The PERT chart of Figure 9.1 shows that there is only a certain number of tasks which can be carried out at any time, and extra effort may not be able to make these progress any faster. Brooks in his very readable book on software engineering, *The Mythical Man-Month* (Brooks, 1975) demonstrates that labour and overall time cannot be traded while keeping the overall number of man-months of effort the same. The writing of software is a creative exercise, and it may not be productive, or it might even be counterproductive, to use two people on a programming task instead of one. The same is true of the hardware design process, where there is a limit to the extent to which the various tasks involved can be spread over a large number of designers.

9.2 Hardware

Once a system has been specified by its designers, a decision must be made concerning the methods of construction: in some cases the same function could be performed using hardware or software. The microcomputer appears in an enormous range of forms, from single integrated circuits to boxed units with very considerable computing power. The designer of a system must also determine whether the project should use an 'off the shelf' computer or, if some electronic construction is to be undertaken, the necessary approach.

9.2.1 Choosing the integration level

A number of factors must be weighed when deciding how a system is to be built. Designing a computer is not a task to be undertaken lightly, and for this reason it is almost always better to make use of 'standard' equipment wherever possible. For many applications where only one or a few of a design are going to be produced the best solution will be to make use of a boxed system from a microcomputer manufacturer.

Where more specialised applications are involved it may prove necessary to assemble a system out of standard printed circuit boards (PCBs) which can be wired together to produce a microcomputer-based system with the necessary functions. Here the problem arises that there is not a single standard for interconnecting microcomputer PCBs; in fact there is a considerable number of standards, and for practical reasons of construction, the PCBs used in a system must all conform to a single standard. Unfortunately the situation is even worse than it may already appear at this point because different manufacturers interpret some of the standard specifications differently. Happily some independent bodies, notably the Institution of Electrical and Electronic Engineers (IEEE) have produced more rigorous interpretations of many of these standards (IEEE-488, 1978; IEEE-796, 1980), and PCBs which are designed to conform to the same IEEE standard should be compatible.

Standard chassis and other subassemblies are available from a number of manufacturers for each of the widely used standard sizes of printed circuit board, and this allows a system to be built up with a suitable mixture of processor, memory, and input/output for most applications.

The very low cost of integrated circuits makes the idea of designing one's own microcomputer-based system quite attractive. The hidden costs in this approach can, however, be considerable. We shall see later in this chapter that the detection of faulty operation in a microcomputer-based system can present very considerable problems, and the equipment needed to tackle these problems is generally expensive. Designing with large-scale integrated circuits requires a considerable amount of expertise, and once a design has been produced it must still be tested and made to work. Once the prototype system has been commissioned, a printed circuit board must be designed, which is also a time-consuming and expensive process for circuits of the complexity commonly found within microcomputers. Indeed it is not uncommon for the PCB used in a microcomputer to cost more than the microprocessor circuit itself! To summarise, therefore, designing a special purpose circuit rather than buying standard PCBs is quite possible, but not an exercise to be undertaken lightly unless a large number of the boards are to be made.

In very large scale production another possibility presents itself: the design of special integrated circuits. It is quite possible to incorporate special purpose analog and/or digital interface circuitry on the same chip as the microprocessor, ROM and RAM. This provides a low-cost solution for applications such as controlling the frequency of a television set or sensing the temperature within a microwave cooker. In this way the unit cost can be minimised, but the high design cost means that a very large production run is needed to make this approach cost effective.

The person responsible for the 'buying in' of material for the project should make sure that it will be available in the necessary quantities and that it can be delivered by the date when it will be needed. This is normal procedure, of course, but it is especially pertinent in this high technology field because there may be 'teething troubles' with new equipment designs. If possible, a second source of supply should be located in case supplies from the normal source 'dry up'. These points are relevant no matter whether the project is to be based upon 'bought in' components or computers, because at this point any manufacturer becomes dependent upon others. Sometimes 'famines' of particular types of component arise due to sudden increases in demand elsewhere or because of manufacturing problems. The production of integrated circuits requires very pure materials, very close control of temperature to within about a degree at a temperature of more than 1000°C, and very careful positioning of the 'masks' used to tolerances comparable with the wavelength of light. Consequently the 'yield', the proportion of working components from the production process is much lower than that of almost any other industry, and if any parameter in the process changes appreciably, the yield rate may easily fall to zero!

9.2.2 Specifying functions

The hardware components must interact to form a composite system, and the exact nature of all these interactions must be set out in the specification. The precise relationship between the inputs and outputs of each function must be specified. Once this has been done, hardware must be bought or built to produce modules which carry out these functions. This can be done at the level of integrated circuit components, printed circuit boards, or boxed systems, depending upon the application.

All the voltages and timings, and the logic of the signals passing between the functions must be listed and checked for compatibility. Wherever possible, the electronic modules of the system will communicate by a system bus, because this makes the design of the system more 'regular' and provides a uniform standard which can be applied throughout the design.

9.2.3 Circuit and signal diagrams

The documentation for the hardware will contain diagrams showing the interconnections between the modules used as well as descriptions of their functions. When systems were much simpler, circuit diagrams were simpler, but now several conventions are used to ensure that the diagrams are not unnecessarily cluttered with wires, as mentioned in Chapter 4. Power supply leads are not normally shown, and buses made up of several wires are normally shown as one thick line rather than as a cluster of individual connections. Against each connection must be written the name of the signal, so that it can be referred to in the documentation. Signals are usually given short names rather like the labels used in assembly language, not surprisingly, because the signal names are simply labels identifying points in the circuit in the same way that the labels in assembly language identify points in the program. An asterisk at the end of a signal name indicates that it is active low, so that the signal name 'IRQ*' identifies an interrupt request line which causes an interrupt request when a logic '0' is present on the line. Sometimes a description of the signal is included for test purposes, for example a typical voltage reading, a waveform, or a waveform 'signature'. Signatures will be discussed later in section 9.6.7.

9.2.4 Hardware/software tradeoffs

There are two main aspects to the design of a microcomputer-based system: the hardware and the software. When a system is being designed decisions often have to be made as to whether a function can best be performed using hardware or software.

As an example, consider the problem of counting external events which appear to the microcomputer as a series of pulses from some kind of transducer. If these pulses occur relatively infrequently, perhaps due to the movement of a mechanical component in the system, they can be counted in software by means of a suitable subroutine, possibly using an interrupt. Each time the count is incremented a fixed amount of processor time is used, so that the faster the rate of arrival of pulses, the greater the proportion of the processor time is taken up by the relatively simple operation of counting pulses. Thus another method must be used to count pulses if the pulse frequency is expected to be high, and this will mean the inclusion of extra electronic hardware to perform the counting operation. The designer is faced with a choice between increasing the complexity of the hardware to count the pulses and increasing the program complexity — and possibly increasing the problems in 'debugging' the software. Any function performed by software could in principle be carried out using hardware, but it is not always possible to replace hardware with software.

As a further example, consider the case where the microcomputer is to be used to control the position of an actuator, and it is known that there is some 'slack' in the mechanical coupling so that hysteresis results; here the position of the actuator depends upon the direction of approach to the desired position. One approach would be to use higher quality mechanical components so that the effect of the hysteresis was negligible. Another might be to ensure in software that the actuator always approached the required position from the same direction, so that the effect of the hysteresis was always a slight change in position in the same direction.

There is an understandable temptation to make use of the microcomputer to correct for various small deficiencies in the hardware with which it is used, and provided that this can be done without too great an input of effort it is usually well worth doing, since the result will be cheaper production. However, there is a pitfall in trying to make 100% use of the processor time to correct for other deficiencies. The software will be more difficult to get working, because it will be more complicated, and this will increase the development cost. Moreover, if it turns out during development that the program needs to be extended for any reason, the processor may not be able to spare enough time for the correction software. Finally, the diagnosis of problems in the system may prove to be more complicated, thus increasing maintenance costs.

9.3 Software

The exact proportion of time spent in the various phases of a software project will vary from project to project, but not by much. Table 9.1 shows figures from two sources (Brooks, 1975), (Zelkowitz et al., 1979), from which it will be seen that the programming effort represents at most one-fifth of the overall effort needed for the project. Put another way, the overhead on the programming effort is a factor of four to five. However, Table 9.1 shows only the effort required to implement a working piece of software, and no account is taken of the need for **software maintenance**, the need to fix programs when they are later found to be wrong, or when minor changes to the specification are made.

Table 9.1 Analysis of effort in software projects

Project Phase	Zelkowitz	Brooks
Requirements	10%	
Specification	10%	33%
Design	15%	
Programming	20%	17%
Module Test	25%	25%
System Test	20%	25%

These figures show that the programming must be finished within the first half of the time allotted for the project, otherwise there is little hope that it will be finished on time. Once a project falls behind in this way it is unlikely to catch up with the schedule again even if extra effort is made available. This is because the extra staff would need to become acquainted with the project before they could contribute to it, and they would almost certainly require assistance from the programmers and others already working on it. To quote **Brook's Law**, 'adding manpower to a late project makes it later'.

When assessing the amount of effort that will be needed in a project, an important statistic is the productivity of the people working on it. There are different ways in which this can be measured. For example, is it more meaningful to quote the amount of source code written by the programmer per day or the amount of machine code produced? Although the amount of machine code gives a measure of the 'size' of the program, it appears that programmers produce much the same amount of source code per working day irrespective of the programming language which they are using, provided that they are reasonably 'fluent' in that language. However, some types of programming are more difficult than others, and Table 9.2 shows that the average number of program lines produced per day varies considerably according to the type of program being written.

Table 9.2 Programmer productivity

Lines of debugged and documented code per day	
Commercial applications	10-100
Scientific and technical	8-50
System and related real-time	5-10

One of the reasons for the variation of the figures in Table 9.2 is the dependence upon the size of the program being written. Larger programs, being more complicated tend to have more interactions between the sections, and they require more care in writing and checking. An empirical law has been obtained relating effort to the size of large programs: the effort required is proportional to the size of the program raised to the power 1.5. This observation is another point in the favour of high level languages. Because programs written in high level languages require fewer lines of source code than their equivalents written in assembly code, they normally require much less effort to implement and maintain. Reducing the size of a program reduces the space in which software 'bugs' can hide!

9.3.1 Software design

Because the software is designed according to the same 'top-down' principles that are used with the hardware, it is not surprising that the same need to specify functions and document their interactions applies to the software.

Specifying functions
The software which results from a top-down approach to design will consist of a main program that consists largely of calls to subroutines, which will in turn reference other subroutines, and so on, down to the lowest level of subroutines and the instructions or statements which make them up. Thus the subroutines are the functions which comprise the overall software, and it is these subroutines which must first be specified in terms of what they do and how they interact with the rest of the system.

Each subroutine must be called with the parameters being presented in a defined way, and the manner in which any parameters are returned to the calling program must also be defined. The subroutine will carry out a set of operations on the input parameters to produce the output parameters, and these operations must be closely specified.

The specification for the subroutine should contain

- the name of the subroutine.

- the number of parameters passed to the subroutine, the number of parameters returned, and the manner in which they are passed between the calling program and the subroutine.
- the specification of the subroutine, in terms of the operations which it carries out upon the input parameters and how the output parameters are generated. If the subroutine handles any input/output functions, then the way in which these relate to the parameters must be specified.
- the time taken to carry out its function, and any restrictions upon values of the input parameters. Although average timings are useful, worst-case timings are needed because these are used to ensure that there will always be enough time for all the subroutines to run in the application envisaged.
- any unusual action which the subroutine may take in the case of an 'exception' condition such as an attempt to divide by zero, or as the result of being presented with 'illegal' data.
- the amount of memory used by the subroutine, for code storage, data storage, and stack usage. Again, worst-case figures must be specified for the stack usage if this is variable.
- the names of any other subroutines which are called in turn by this subroutine, so that any interaction between them can be checked.

Interaction of subroutines.
If subroutines are specified in this way, enough information should be available to allow the software designer to check that there will be enough memory for the subroutines to operate together and there will be enough time for them all to perform their tasks. The interactions between the subroutines should only ever take place via the parameters. If global variables are used which are shared between two or more subroutines, then this fact should also be recorded in the specification of the subroutines. Where possible, the avoidance of global variables makes for simpler interactions between subroutines and for easier checking of overall operation.

Within subroutines, programming constructs such as FOR...NEXT and **repeat...until** are used to control the flow of the program's execution. In each case these constructs should be designed to have just one entry point and one exit point. The same is true of subroutines, which should be designed to have one starting point and one return statement. This rule of having one entry and one exit makes programming much more systematic and makes debugging easier since there are always places where a single 'breakpoint' can be used to stop a program to allow registers to be examined. Correctness arguments are difficult to carry through when GOTO statements are used in a program and they should be avoided for this reason (Dijkstra, 1968). Structured programming languages need no GOTO statements but in older languages such as FORTRAN and BASIC it is not usually too difficult to simulate constructs using GOTO in a controlled way.

However, there is one circumstance where the GOTO statement is useful in its own right. This is where exceptions to normal program operation arise, either due to incorrect data causing an overflow or an attempt to divide by zero, or due to some external influence such as an interrupt indicating that there is a malfunction. In such cases the normal program is abandoned and the microcomputer takes special action ('exception processing') instead.

Structure diagrams

This philosophy of designing each section of the program with an entry point and exit point requires a different design technique from the flowcharting method described in Chapter 2. Flowcharting places no constraint upon the sequences of statements which can be used in constructing a program, while the principles of top-down design dictate that the program should be designed as a set of modules each with an entry and an exit. For this reason, other graphical techniques have been developed which also allow the logic of a program to be drawn as a diagram (this is often important for an overall understanding of its operation) but which restrict the programmer to the use of structures. A useful introduction to the principles of structured software development is given by Mühlbacher (1982), who provides examples of the use of graphical units, or structograms, which were introduced by Nassi and Schneiderman (1973).

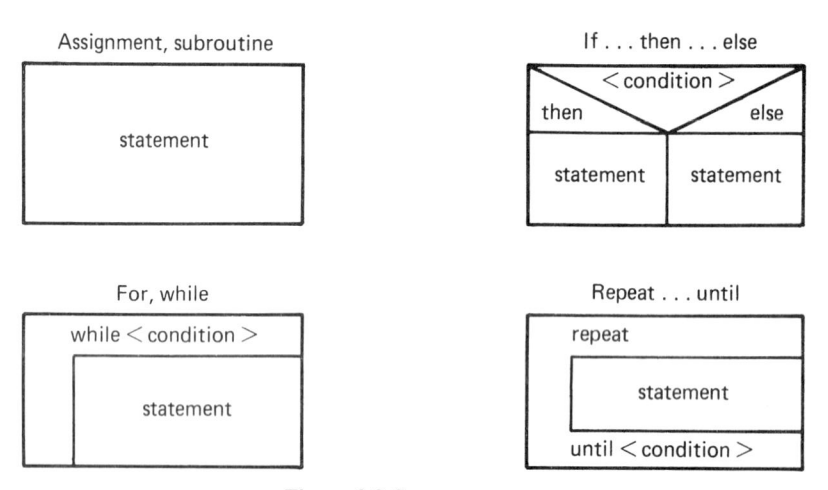

Figure 9.2 Structograms

In this system of drawing the structure of programs, the various modules used are drawn as rectangular boxes. The simplest type of box contains a list of statements which do not affect the flow of a program, and is just a rectangular box with the statements inside it. The **if..then..else** construct is represented as a rectangular box containing divisions which indicate the condition which is to be tested by the **if** statement, and two sections for the **then** and **else** cases. If there is no **else** section in the statement, one box will remain empty. Loops are shown with the group of statements within an inner box surrounded by a section containing the specification for the loop.

The more powerful **case** statement contains a number of inner boxes, each containing the code to be executed in that case and preceded by the condition for that piece of code to be executed. Figure 9.2 shows the various types of structogram used in this system.

When designing a system, the program requirement can be repeatedly broken down into smaller structogram 'boxes', a process known as 'successive refinement'. Once the structure has been fully constructed in terms of these boxes, the program itself can be coded.

Listings
The listing produced by an assembler or a compiler provides the final documentation of the code which makes up the program. For this reason it should be presented in a manner which makes it as easy to read as possible. In the case of assembly language programming the best that usually can be done is to tabulate the material into separate columns for the labels, opcodes, operands, and comments.

If a program has been designed using a structured high level language, however, it makes sense to try and indicate the structure as far as possible in the way that the listing is laid out on the printed page. This can be done quite neatly by indenting the start of a statement along the line when they are within a construct such as a loop. If they are within two 'nested' constructs, they are indented by twice the amount, and so on. This provides a useful manual method of checking that each block of program is started and finished correctly when programming, and provides an 'at-a-glance' method of looking at the way in which the program is constructed.

Because Pascal was designed to encourage structured programming, programs written in Pascal are conventionally shown with these indents. The same is true of similar languages such as C, which is widely used in systems programming. Other languages such as FORTRAN and BASIC, which were developed before the ideas of structured programming were discovered, tend to be written without this indentation, although there is no reason why listings of such programs should not be indented in this way.

In addition to showing the statements making up the program, the listing should show something of the way in which the program is represented in machine code. In the case of assembly language, as has been seen, this can be done simply by listing the corresponding machine code and the addresses at which it is stored, to the side of the source program. When relocatable modules are written in this way the addresses of the program and data locations used start from zero; after the modules have been linked, the actual addresses can be found by adding the addresses shown in the listing to the actual starting address (the **base address**) of the corresponding module. The listings produced by compilers do not show all the code generated, but often list the address where the code corresponding to each statement starts in memory, together with the address corresponding to each label used in the program.

9.3.2 The development process

The development of the hardware and software to be used in a project will require a range of facilities. Specialised tools are needed to construct and test the software just as they are for the hardware, the only difference being that the tools used for software development are usually more complicated and less familiar than their hardware counterparts.

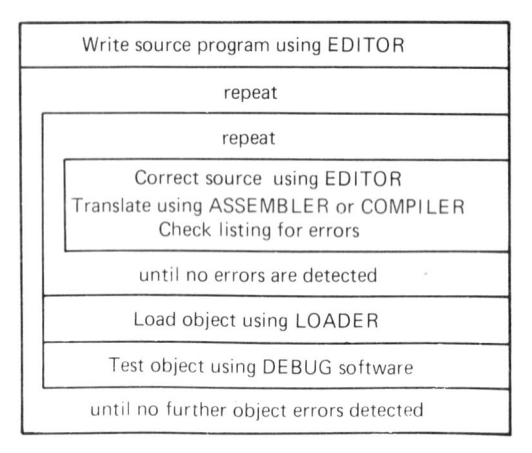

```
┌─────────────────────────────────────────────────────┐
│         Write source program using EDITOR           │
├─────────────────────────────────────────────────────┤
│ repeat                                              │
│  ┌────────────────────────────────────────────────┐ │
│  │ repeat                                         │ │
│  │  ┌───────────────────────────────────────────┐ │ │
│  │  │     Correct source using EDITOR           │ │ │
│  │  │ Translate using ASSEMBLER or COMPILER     │ │ │
│  │  │     Check listing for errors              │ │ │
│  │  └───────────────────────────────────────────┘ │ │
│  │        until no errors are detected            │ │
│  └────────────────────────────────────────────────┘ │
│         Load object using LOADER                    │
│       Test object using DEBUG software              │
│      until no further object errors detected        │
└─────────────────────────────────────────────────────┘
```

Figure 9.3 Software development

Figure 9.3 shows a structogram illustrating the sequence of operations which need to be carried out as software is being developed. First the program must be written, which requires the use of an editor program to allow corrections to be made. Then it must be translated by means of a compiler or assembler, which will check for the existence of errors in the 'grammar' of the programming language being used, called **syntactic errors**. When these have been corrected using the editor the resulting machine code must be loaded into the target system to allow it to be tested with the aid of 'debugging' software. Because nobody is a perfect programmer, there will usually be **semantic errors** — mistakes in the logic of the program — which will become apparent at this point in the proceedings. This means that more corrections will be needed before the program is finished.

The first and most obvious requirement for software development is for some kind of computer which will run the support software that the programmer needs. Computers come in all shapes and sizes (and at all sorts of prices) and it is worth determining what sort of computer is best suited to this application.

The equipment used for software development is called a **development system**. A good development system will support the programmer at each stage as programs are being developed: during writing, translation, and testing. In addition to being the vehicle for the software 'toolkit', the development system should provide facilities for testing the hardware and the relationship between hardware and software. The effectiveness of the software tools depends to a large extent upon the ease with which they can be used. This is affected by the way in which information can be stored and presented by the system using its input/output devices, and the ease with which the system can be operated.

The importance of a good development system which provides an efficient programming environment cannot be over-emphasised. The development system must also be powerful enough to allow it to run software at an adequate speed. Sophisticated development systems allow more than one terminal to be used so that the facilities offered by the system can be shared between more than one programmer at the same time. This places greater demands upon the processor. The amount of memory available for the support software is also important. An editor, for example, can operate with quite a small amount of memory provided that it can copy those parts of the file not currently being edited back to a storage device to save space. The problem with using any type of 'backing store' to hold information not currently needed in memory is that they are relatively slow when compared with the speed of access of the memory itself. Thus if more memory is available the computer needs to swap information between the memory and storage device less often and performance will be markedly improved.

So far, the development system has been considered simply as a vehicle for the software needed for developing a program. Anyone who has ever used a 'hobby' computer and a BASIC interpreter will wonder what all the fuss is about, because in this case the editing, testing, and running are all carried out on the same machine using the same software. BASIC interpreters are easy to use, and in many applications such as controlling experiments in a laboratory the use of a microcomputer running a BASIC interpreter may provide the most flexible and cost effective solution. When a microcomputer is going to be used to control a piece of equipment on a more permanent basis, it may prove preferable to use one machine for software development and a second, more robust, one to actually run the program once it has been finished. Often the microcomputer will be 'embedded' in the final system in such a way that its presence is not immediately apparent.

The machine on which the software 'tools' run is called the 'vehicle' or **host machine**, while the one on which the finished program will run is called the **target machine**. The need to develop software on a different type of computer from the one which is to run them has been in existence throughout the history of computing. Whenever a new computer is designed and developed, the software for that machine is developed as a parallel exercise. The only difference with microcomputers is that some microcomputer-based systems are not capable of being used for developing their own programs even when they have been built. In practice it is very common to use **cross-assemblers** and **cross-compilers**, which run on one type of computer or microcomputer but which generate code for another type.

9.3.3 Development systems

Many types of computer are available which can be used to develop programs. In this section the main advantages and disadvantages of each type will be briefly discussed.

Hobby computers

The huge number of microcomputers which have been manufactured for home use during the last few years means that a simple computer can be bought very cheaply, and it is tempting to believe that they could be used to produce a low-cost development system. The disadvantage of using the cheapest microcomputers is that they tend to use BASIC which is best suited to 'desk-top calculator' jobs rather than the specialised tasks required of a development system. However, they are perfectly adequate when the 'host' and 'target' machines are one and the same.

Minicomputers and mainframes

The computing power, operating system, and peripherals of a 'minicomputer' or 'mainframe' computer make them attractive as development systems. They usually have powerful editors, fast printers, and flexible operating systems, and their computing power allow them to compile and link programs quickly provided that the appropriate software is available. Unfortunately they are also expensive to run, and the inefficiency of most emulator software makes the testing of the machine code produced in this way very expensive. The problems increase when input/output software is to be tested because the computer manager is unlikely to look favourably on the suggestion that the expensive computer is to be connected directly to 'real-world' electronic circuits and interfaces. The compromise is to use a microcomputer with a monitor program which can be loaded from the mainframe and then controlled from the terminal.

Dedicated systems

The most effective development system is one which has been designed expressly for that purpose. At first, development systems were produced only by the manufacturers of microprocessors because they needed a means for their customers to develop products based upon microprocessors if they were to be able to sell the integrated circuits. The manufacturer is able to design and develop the development system hardware and software for a new microprocessor at the same time as the microprocessor itself is being developed. The result is that development systems are usually announced for new microprocessor and microcomputer integrated circuits at more or less the same time that the circuit itself becomes available.

The problem with using a development system produced by a microprocessor manufacturer is that the manufacturer is only interested in equipment which will help sell the products, not those of competing manufacturers. Consequently, once a development system has been purchased, there is an increased tendency to stay with that manufacturer's products, which is exactly what the manufacturer wants! What is more, these development systems are often appreciably more expensive than ordinary computers of the same power because they are sold in much smaller quantities and do not benefit from the same economy of scale.

More recently, instrument manufacturers have started producing development systems. These systems are usually designed to support a wide range of microprocessors, although special software and equipment such as in-circuit emulators may be needed for each different type of device.

9.4 Software development tools

When a computer is used to edit and translate programs, the programmer needs some method of communicating with the machine. For example, when a program is to be assembled, the computer must be told that the assembler is to be used, and the name of the source program must be specified. The software which allows other programs to be run is called the system's **operating system**. It provides the user with a range of software facilities which combine to make the computer usable. The various software tools can be invoked using the operating system, and the availability of an effective operating system is an important factor in improving the effectiveness of a programmer.

The operating system also manages the use of peripheral devices. It allows programs to be copied to a printer for 'listing', sent for storage, or sent via a communications link to another computer. 'Standard' input and output from a program is handled by the operating system which provides the I/O subroutines. Usually the operating system allows for the redirection of input and output, so that a program which normally sends its output to the screen of a terminal could be run with all its output being send to a disc file instead.

The main types of system software provided by a good development system are:

- an **editor** to allow programs to be entered and corrected.
- **compilers** to translate programs into machine code. Sometimes compilers translate high level source programs into assembly language rather than directly into machine code. If the development system is to be used to develop programs for types of microprocessor other than that used in the development system itself, **cross-compilers** will be needed; these are compilers which run on one type of machine but which produce code for another kind of machine. Sometimes compilers translate high level source programs into assembly language rather than directly into machine code.
- an **assembler** to translate programs written in assembly language into machine code. One or more **cross-assemblers** may be available if the final program is to run on a machine of a different type from the development system.

- a **linker** to allow modules to be linked together as a single program and to allow the use of libraries.
- a **loader** to allow machine code programs to be loaded into the computer's memory.
- **debugging** programs to allow the code to be tested. Many types of debugging aid are available, for example to trace the flow of the program one instruction or one subroutine at a time, or to measure the proportion of time spent in each subroutine.
- an **emulator** which may be provided to allow the operation of the target machine to be developed if the final program is to run on another machine. This topic of emulation is an important one, and it will receive a considerable amount of attention in section 9.4.7.

9.4.1 Operating systems

The basic function of an operating system is to allow a human operator to control a computer. At its simplest it may just allow programs to be loaded and run; the sophistication of the operating system reflects the sophistication of the machine on which it is running. The facilities which an operating system may offer also depend upon the application for which the system is intended.

Multitasking operating systems
During the software development process most of the programmer's time is spent using the editor, since the speed of the operation is limited by the typing speed of the programmer. During this period the computer spends much of its time in waiting loops, waiting for the next character to be typed in at the keyboard. When a program is being compiled, the programmer has to wait for the compiler, since the speed of operation is now limited by the speed of the processor and disc unit.

This wastage of both the programmer's and the computer's time can be minimised by using a **multitasking operating system**. This type of system allows more than one program or 'task' to appear to be active at any time. It accomplishes this trick simply by arranging for each program to proceed until it is held up by a 'resource', for example, it is waiting for new material to be typed in or for the disc unit to complete the reading or writing of a block of information. Once a task has been held up in this way, the operating system temporarily suspends it and continues with another task until that one is held up. The operating system also ensures that each task continues at a reasonable rate if one or more of them is able to continue without being limited by input/output resources by switching between tasks several times per second.

With a multitasking operating system the user is able to start editing a second program while the computer is still compiling a first one, and much more efficient use is made of the computer's resources.

9.4.1

Multiuser operating systems
A large microcomputer may have a **multiuser operating system** which allows several users to share the facilities of the same machine, apparently making simultaneous use of the processor, memory, and disc units. The multiuser system is simply a development of the multitasking system in which the tasks current at any time can interact with different terminals. Each user can use the editor, compiler, and other programs at the same time without any undesirable interactions taking place. The files held in the disc unit can be specified as being private to one user or available to all, so that groups of programmers working on the same project can access the same information and pass information to each other if they wish.

An example of a multiuser operating system is UNIX (the name is a trademark of AT&T Bell Laboratories), which provides a range of facilities to support the writing of documentation, since this is such an important part of any project. In addition to screen editors, programs are available to check spelling and even writing style (Bourne, 1982). However, systems with this sort of complexity consist of several megabytes of code and require the availability of large storage devices such as Winchester discs to support the software. They also require large amounts of random access memory, and in practice a microcomputer with a wordlength of sixteen or greater is needed to support them.

Standardisation of operating systems
One important factor with large operating systems is that they take a long time for the programmer to become really familiar with them, and this makes the idea of standardising operating systems very attractive. 'Standard' operating systems are attractive for other reasons. A computer's operating system looks after the input and output of information, and provides the 'environment' in which other programs run. The hardware is 'hidden' from the user and changes can be made to the hardware without the user needing to be aware. A program to run on a large computer is written to run on a particular type of operating system, rather than on a computer of a specified type, it is then compiled to run on the appropriate type of processor and makes use of the input/output subroutines provided by the operating system. The availability of a widely used operating system creates a widespread market for 'software products': programs which will run on that operating system. There is no doubt that operating systems will become more standardised in future.

On 8-bit microcomputers, many single-user operating systems have appeared, but the most successful of these has without doubt been CP/M. This requires a Z80 or an 8085 microprocessor, tens rather than hundreds of kilobytes of RAM, and a floppy rather than a rigid disc.

Basic programming environment

Some hobby computers are designed in such a way that the BASIC interpreter starts running as soon as the machine is switched on. In this case the command interpreter used by BASIC is also used to interpret commands which reference peripheral equipment such as the printer and disc unit, if any. Thus programs can be loaded from disc or saved back on disc by means of the commands LOAD and SAVE; there is no need to specify that the program is a BASIC program because that is the only type of program which is available.

Monitors

In microcomputers which do not have a disc store, some sort of operating system program is still needed if programs are to be loaded and run. In this case the program is usually called a **monitor** program. The facilities provided by the monitor vary from microcomputer to microcomputer, but typically they include the following:

- a **loader** to allow programs to be loaded from a larger computer
- a **dumper** to allow programs to be transmitted back to the larger machine. Sometimes the loading and dumping facilities can also be used with a simple storage system using an audio cassette recorder.
- a facility to **execute** programs once they have been loaded.
- a **breakpoint** facility to allow programs to be stopped at specified points.
- a **register examine/modify** facility to allow the contents of the registers within the microprocessor itself to be displayed, for example after a breakpoint has been encountered. This also allows the contents of a register to be changed, for example before testing a section of code in which the contents of the accumulator must be defined before that piece of program runs.
- a **memory examine/modify** facility to allow the contents of memory locations to be examined and modified if desired. The contents are usually treated as hexadecimal numbers, but sometimes the ability to handle ASCII characters or even a direct assembler is incorporated into this facility.

Typically, a microcomputer program might be edited and translated using a larger microcomputer with disc storage and then loaded into the smaller machine using a communication link such as an RS-232-C connection of the sort normally used for computer terminals. The monitor in the smaller microcomputer can then be used to test the software. This process is made easier if the small microcomputer (the 'target' of the software) has two RS-232-C connections; one of these can be used for the terminal and the other one for the connection to the larger machine (the 'host'). The target microcomputer can then communicate with the computer terminal and be loaded from another machine without the need to change connections. If the

9.4.1

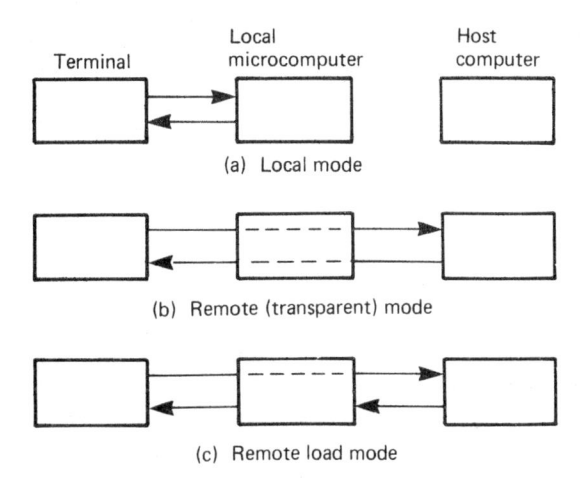

```
           Local           Host
Terminal   microcomputer   computer
  □ ⇄ □                    □
     (a) Local mode

  □ ⇄ □----□ → □
     (b) Remote (transparent) mode

  □ ⇄ □----□ → □
               ←
     (c) Remote load mode
```

Figure 9.4 Using two microcomputers from one terminal

monitor software provides for a 'transparent' mode in which characters from the keyboard travel via the target to the host and responses from the host are passed to the terminal, the user can communicate with both the host machine which is being used for software development and the target machine on which it is being tested (Figure 9.4).

9.4.2 Assemblers

An **assembler** typically acts upon a source file written in assembly language to produce two sets of output information. The first of these is a copy of the original assembly language program, to which is added the corresponding machine code information. This is the **listing** of the program and because it shows both the source information and its machine code equivalent it is an important piece of documentation concerning the software. The second is a copy of the machine code alone, stored in some form which can be read by the microcomputer being programmed. This is known as **object code** and it is often stored on cassette or floppy disc and loaded into the microcomputer when required by a special program called a **loader**. Almost all microcomputers have a loader program incorporated into their monitor software because alternative methods of entering programs into computers are, at best, difficult!

Direct assemblers

If the program being assembled is to be run on the same machine as the assembler itself, the object code can be loaded directly into memory. This is how a **direct assembler** works; it translates each line of assembly language into machine code as it is typed into the machine.

Many commercial microcomputers have simple direct assemblers built into their monitor programs, and they are very useful for making minor changes to programs that are already in memory. However, these assemblers have many shortcomings when used for writing programs. The most significant of these is that it is more difficult to correct typing errors unless they are spotted before the end of the line because each line that is typed in to the machine is translated when the 'return' key is pressed.

Macroassemblers

The need is often felt when programming in assembly language for instructions that are not available for the microcomputer in question. For example, many types of microprocessor have an 'increment memory location' instruction, INC, which can be used to increment the contents of a memory location by one. If we wanted to increment the memory location by some other number such as 5, we could use a sequence of INC instructions or we could load the contents of the memory location into one of the accumulators, add 5 to it, and store the result into the original memory locations. It would be convenient in many cases to be able to invent a new instruction called something like INCR with two operands so that the instructions

```
        INCR    MEMLOC,5
        INCR    MEMLOC,9
```

would increment memory location MEMLOC by 5 and 9 respectively.

Some assemblers allow us to do this by means of 'macroinstructions' or **macros**. At the beginning of the source program each macro is defined in terms of existing instructions for the type of microprocessor in question. This macro definition is used to relate the name of the macro to a 'macro body' in which special symbols are used to mark where the 'arguments' or operands of the macro are to be inserted. When the assembler reads the source file it replaces each occurrence of a macro name with the macro body in the definition and inserts the appropriate arguments into the macro body. The macro INCR might be defined as follows:

```
    INCR    MACR                Define the macro INCR
            PSHS    A           Save the contents of the accumulator
            LDA     ?0          Load acc-A from the memory location
            ADDA    #?1         Add the amount in the second argument
            STA     ?0          and return the result to memory
            PULS    A           Get back contents of accumulator A
            ENDM                End of macro definition
```

Now if a 'macro call' of the form

```
        INCR    MEMLOC,5
```

appears in the program, the assembler will use the macro body from the definition and insert the appropriate values from the macro call. The result is the same as if the programmer had written

```
        PSHS    A               Save the contents of the accumulator
        LDA     MEMLOC          Load acc-A from the memory location
        ADDA    #5              Add the amount in the second argument
        STA     MEMLOC          and return the result to memory
        PULS    A               Get back contents of accumulator A
```

Macros provide the ability to modify an assembly language to suit the particular programming task in hand and to fit in with any preferences or eccentricities that a particular programmer may have. Assemblers which provide this facility are called **macroassemblers**.

Conditional assembly
The macro definition used in the last example will produce inefficient code if the argument of the macro call is 0 or 1 because more efficient ways of coding this already exist; incrementing by 1 can be done using the INC instruction and incrementing by 0 has no effect apart from modifying the flag bits in the processor status register. Some assemblers provide the ability to include conditional IF clauses which determine whether a specified block of code is to be assembled or omitted from the output. For example, the fragment

```
        IFEQ    MODE-TEST       Only assemble this if MODE=TEST
        JSR     CHECK           Display check information if testing
        ENDC                    End of conditional section
```

will be assembled only when the program is being tested. When the final version of the software is produced the assembler can be run with symbol MODE set to a value different from TEST to produce a version without the test sections.

This conditional assembly technique is very useful when macros are being used because it allows the programmer to specify the best code to be used in each eventuality. Thus the definition of INCR could be expanded to include sections to deal with the cases when the second argument is 0,1, or negative, as shown in Figure 9.5.

Using subroutine libraries
Because programming is expensive in terms of labour, it makes sense to use 'off the shelf' solutions to programming problems wherever appropriate. For example, the programmer may be able to use sections of existing programs when writing a new one simply by copying them into the source material of the program. This saves the difficulty of designing the software from scratch, but it still leaves plenty of scope for typing errors, and obviously it is much easier if the program sections can simply be copied from one computer file to another one.

One way to 'streamline' this approach to programming is to have a 'library' of subroutines each with a well-defined objective. The programmer

```
INCR    MACR                    Define the macro INCR

        IFGT    ?1-1            If greater than 1
        PSHS    A               save the contents of the accumulator
        LDA     ?0              Load acc-A from memory
        ADDA    #?1             Add the amount in the second argument
        STA     ?0              and return the result to memory
        PULS    A               Get back contents of acc-A
        ENDC                    End of conditional section

        IFEQ    ?1-1            If argument is 1
        INC     ?0              simply increment memory location
        ENDC

        IFEQ    ?1              If argument is 0
        TST     ?0              test value to set flag bits
        ENDC

        IFEQ    ?1+1            If argument is −1
        DEC     ?0              decrement location
        ENDC

        IFLT    ?1+1            If less than −1
        PSHS    A               save the contents of acc-A
        LDA     ?0              load acc-A with the memory location
        ADDA    #?1             add the amount in the second argument
        STA     ?0              and return the result to memory
        PULS    A               Get back contents of accumulator A
        ENDC

        ENDM                    End of macro definition
```

Figure 9.5 Using conditional assembly in a macroassembler

can then look up which subroutines are available in a catalogue and select those which are usable in the current program. In this way the task of programming becomes one of writing a main program which calls other subroutines, some of which may be from the library and the remainder of which will be from other sources, usually the same programmer who wrote the main program.

Great care must be taken when making use of assembly code subroutines in this way that the various subroutines and the main program each use different addresses for their data storage and for the storage of the programs themselves. Suppose, for example, that two subroutines each used the same label `COUNT`, but for different purposes. If both subroutines were included in the same program and then assembled together, the assembler would not be able to distinguish between the two identical labels, although it would produce an error message if `COUNT` were defined twice.

In fact two types of label are encountered in this type of programming; those which are used only within a single subroutine or group of subroutines, and those which are used throughout a program including all its subroutines. The first type is referred to as a **local label** while the second is a **global label**. Local labels are of no interest outside their own subroutine, and there is no need for any other subroutine or the main program itself to be 'aware' of their existence.

The factors of interest to the user of the subroutine are the amount of memory needed for program storage, the amount needed for data storage, how parameters are passed between the calling program and the subroutine, and the time taken for the subroutine to carry out its function. In fact, because each library subroutine is called to carry out a function defined by a specification in the 'catalogue' of subroutines there is no need for the programmer to be aware of the actual code that makes up the subroutine.

9.4.3 Compilers

Each statement in a high level language corresponds to a sequence of instructions in machine code. Sometimes the relationship between a statement and the corresponding code is very simple, for example a `GOTO` statement is essentially the same thing as a `JMP` in machine code, although it may not be translated directly as such. In other cases a statement may produce a relatively large amount of code; a `FOR...NEXT` loop requires code to be generated which increments a variable each time a loop is traversed. However, the code used for a `FOR...NEXT` loop is more or less the same for each such loop in a program. Consequently a translation process can use a set of rules which show what machine code should be used for each type of statement in the high level language. Sometimes this code will be produced as a sequence of instructions in the main program, while in other cases a call to a subroutine containing appropriate code will be produced.

A program which translates source material written in a high level language into machine code according to rules in this way is called a **compiler**. The compiler normally produces an output which consists of coded instructions including calls to relevant subroutines in a library. This contains subroutines for handling input/output and for carrying out standard operations such as mathematical functions. Usually compilers are thought of as producing less efficient machine code than programmers writing in assembler language, but although this is still the case the difference in efficiency is not as great with modern microcomputer designs as it was with earlier machines. In fact compilers can produce more efficient code than inexperienced assembly language programmers because often they are capable of optimising the machine code produced. They can also exploit programming tricks such as using 'increment memory' instructions when statements such as

    FRED = FRED + 1

are encountered. The compiler converts programs written in a high level language into machine code which is then stored in a form which can be loaded into a microcomputer and executed when required. This need not be the same machine that was used to run the compiler, if a cross-compiler were used.

9.4.4 Linkers and loaders

When a compiler is used with a run-time library the machine code produced by the compiler must be **linked** with the material to be included from the library. This is carried out by a program called a **linker** or **link editor**. The ability to link material from different sources to produce a composite program is a very useful one. For example, it is convenient to be able to write a program in several separate sections called **modules** and test them separately. These can then be linked together to produce the final program.

The modules could be produced in quite different ways; some could be written in assembly language, others could be written in one or more high level languages, and others could be taken from libraries of subroutines such as the run-time library of the compiler. In each case the linker must ensure that the addresses used in the various modules refer to appropriate memory locations. This is normally made possible by arranging for the compilers and assemblers to translate into a special **relocatable object code** which is not quite the same as machine code. The difference is that the addresses in each module are coded in an intermediate form which specifies each address relative to the starting address of the module rather than in the absolute form used by the microprocessor itself. Each relocatable address also includes information as to whether it refers to data or program and in some cases, where a reference is made in one module to an address in another module, the corresponding label is included in the relocatable code.

The linker reads in each module and checks that each cross-reference between modules is satisfied. In this way it makes sure that the program does not have any modules or variables missing. When the linker reads a library, it looks for any modules in that library to which reference has been made in another module. For example, if a program contained a reference to the sine function, the corresponding relocatable code might contain a reference to a module called SIN, and when the linker read through the libraries it would check for the existence of SIN. If the module was found it would be translated and included in the resulting machine code program. However, if no module were found with this name, an error message would be produced by the linker.

Once the program has been translated and linked it must be loaded into the machine to be tested. Sometimes the program will be run on the same microcomputer used to run the editor, compiler (or assembler) and linker, but often another machine is used. In either case a program called a **loader** must be used to load the machine code into the memory of the microcomputer. The machine code is transferred in some standard format, either as a sequence of bytes or as a stream of information coded using ASCII characters, usually as hexadecimal digits.

9.4.5 Editors

The function of an editor is quite simply to allow text to be entered into a computer and to allow modifications to be made to that text whenever required. Most editors do not make any assumptions about what the text means; it could be in a low or a high level language, or indeed it could be in a natural language such as English. The importance of documentation in any project has already been pointed out, and a good editor is an invaluable aid in maintaining good documentation.

The very simplest editors are **line editors** which allow the programmer to insert or delete lines or change blocks of lines. They were the first type to be developed and are relatively small, fast-running programs. Unfortunately they are not very convenient to use, and **text editors** are more commonly used. These allow groups of characters to be changed within a line as well as the manipulation of whole lines of text, which makes them much more flexible.

Both these types of editor are suitable for use with printers because they do not attempt to show the programmer what is in the 'file' of text at any time. The third type of editor, the **screen editor**, displays a 'page' of text on the screen all the time which is changed as the program is edited. This allows the programmer to see exactly what is going on, and this makes editing very much simpler. The price paid for this ease of use is a relatively large editor program which sometimes has to output a whole new page of information when a simple correction is made, and in general screen editors require a very fast display system compared with the other types. Despite their greater complexity, there has been a very marked move to screen editors over the last few years, and most hobby microcomputers, for example, now use screen editors.

Some editors can be used to check the material automatically as it is typed into the machine. If an assembly program is being entered, for example, the labels, opcodes, operands, and comments can be arranged into columns without the programmer having to go to the trouble of laying out each line. It is very useful when text is being typed in, for example when documentation is being written, if the editor ensures that words are not broken at the end of lines. Some editors even check programs as they are typed in for errors in syntax so that mistakes can be identified as early as possible.

9.4.6 Interpreters

Instead of translating the whole source program into machine language, **interpreters** store it in more or less the same form in which it was typed. When the user tells the computer to start running the program, usually by typing RUN in BASIC, the machine takes the first line of the program and reads the sequence of characters on that line. This sequence of characters is analysed according to a set of rules in much the same way as a compiler translates each line of source code, but instead of producing machine code for later execution, the computer immediately carries out the action specified by the statement. Once that action has been finished, the machine reads the next line and continues as before until the end of the program. If a statement is encountered which changes the flow of the program such as an IF or a GOTO, the machine will continue at the point in the program specified. This process of interpretation is slow, but not simply because each line must be translated before it is acted upon. If, as is almost always the case, the program contains one or more loops in which the same statement is repeated several times, that statement will need to be interpreted afresh each time it is encountered.

The versions of BASIC found in microcomputers have associated translators which are nearly always interpreters because they are easier to use than most compilers. When microcomputers are used in applications in which the program does not change, the relative ease of programming is no longer an advantage and other factors become important. The fact that interpreters are slow in operation, typically 10 to 100 times slower than compiled or assembled code, may not be important in some applications where the remainder of the system is slowly acting, because of mechanical inertia or other delays. If a large number of systems are to be produced to the same design, the manufacturing cost becomes important and the more compact programs produced by a compiler are more attractive. On the other hand, the ease of programming offered by interpreters can reduce the development cost, a factor to be borne in mind if only a small number of systems are to be produced. If high speed operation is required, interpreters cannot be used and the choice lies between compilers and assemblers.

9.4.7 Emulators

As an alternative to loading the program into a microcomputer for testing, the microcomputer in question can be 'emulated' by means of a program running on another computer. Programs of this type, which are called **emulators**, simulate the registers and instruction set of the microprocessor. They also contain sufficient memory for program and data storage to allow the program being tested to be run on the emulator. Unfortunately emulators tend to be very inefficient in the way that they use computer time, with the result that they are used rather sparingly in software testing. However, they do provide a useful way to check the operation of each subroutine against its specification with different data.

Real-time emulation techniques
A major problem arises when developing systems in which the microcomputer is 'embedded' to the extent that the microcomputer cannot be used for software 'debugging'. This is especially true in relatively simple systems such as the controller for a washing machine or microwave cooker where there is no monitor or similar program. In cases like this the functions of the microprocessor and the memory used by the target system must be emulated by means of circuits which appear to the target system to have the same function as the circuits which they replace. The emulating system must be capable of behaving in exactly the same way as the system being emulated, at the full operating speed.

Use of ROM emulators
The ROM emulator, as its name implies, is a device which allows a read-only memory to be emulated. It is connected to the system being developed by means of a plug which fits in the socket normally used for the ROM, and using a ROM emulator avoids the need to program successive versions of EPROMs as a program is developed.

It consists of a read-write memory together with some logic which allows it to be loaded from the development system via a cable. While the RAM is being loaded from the development system its address lines are under the control of the loading system, but once the program is in the RAM, control of the address lines is returned to the address pins of the ROM emulator

Some single-chip microcomputers are available in development versions which have a socket on top of the package to allow a read-only memory to be plugged in instead of using the ROM on the chip itself. These can be used with a ROM emulator to simplify the development process, and an EPROM can be used once the program appears to be functioning correctly.

Microprocessor emulation
An alternative to emulating the function of the ROM is to emulate the microprocessor's functions within the target system using the development system. The overall form of the microprocessor emulator is the same as that of the ROM emulator with a plug to fit the microprocessor socket and a cable linking it to the development system. This approach allows a program to be run a section at a time by using hardware to cause breakpoints at designated addresses. In this case the program is defined by the contents of a ROM or a ROM emulator.

The microprocessor emulator allows programs to run at full speed, unfettered by the inefficiencies of software emulation on another kind of processor, but the hardware used is specific to one type of microprocessor and represents a considerable extra expense when developing circuits based upon microprocessor integrated circuits.

Figure 9.6 An in-circuit emulator in use

In-circuit emulation
In fact the best of both worlds is possible by using an emulator which emulates both the microprocessor and its associated ROM. This allows a program to be started, stopped, modified, and rerun as many times as desired under the control of a monitor program. This process of **in-circuit emulation (ICE)**, shown in Figure 9.6, is virtually essential during the development of target systems in which the final system does not have a terminal and debugging software available. With in-circuit emulation the programmer has the same degree of control over the execution of programs enjoyed by the programmer who works with larger computers and microcomputers.

To see how the in-circuit emulation of a microprocessor or microcomputer can be so valuable in the development of a system, consider in general terms the development of a 'black box' in which the input is by switches and push buttons and the output is by means of lamps. Inside the unit is a microcomputer consisting of a microprocessor, input/output devices, RAM and ROM.

Initially the subroutines can be tested where possible using the development system alone, and the hardware can be checked out as far as possible using conventional electronic equipment such as oscilloscopes and test meters. The difficulties arise when the software and hardware are combined; there may be errors in the input/output subroutines, faults in the microcomputer hardware, or complicated interactions between the software and the system to which the unit is connected. To ease the commissioning of the system, we must proceed cautiously.

First the in-circuit emulator is connected into the 'target system', and the development system is then used to run some test programs to check out

9.4.7

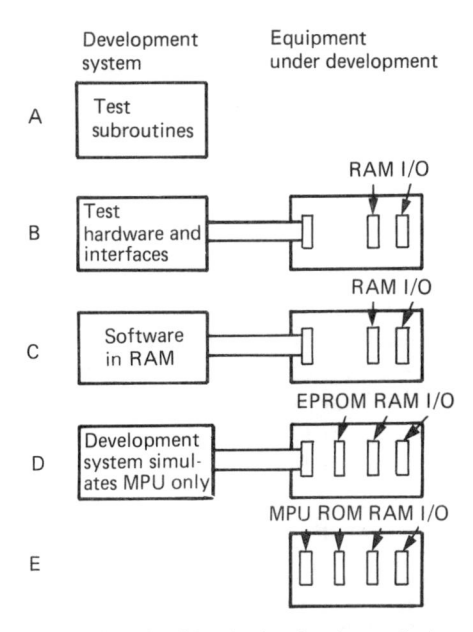

Figure 9.7 Steps involved in the in-circuit emulation process

the RAM and the I/O circuitry (Figure 9.7). Once any problems in the hardware have been sorted out, the I/O subroutines can be tested to ensure that the inputs can be read and the lamps switched on and off as required. Now all the subroutines making up the software should have been tested individually, and next more complicated combinations of subroutines can be checked until the whole of the system software is running. At this point the software appears to be correct, although there may be some 'hidden bugs' which may not become apparent for some time, and the greater part of the hardware must be functioning since no problems are being observed.

The next step is to copy the software into an erasable programmable read-only memory, mercifully abbreviated as EPROM. The system can be run again with the development system emulating only the microprocessor, and if the equipment continues to function correctly, the in-circuit emulator can be replaced with the microprocessor itself. At this stage the project appears to be complete, but there may be problems with the software under some circumstances, and it is good practice to test a few examples of the system for some time to make sure as far as possible that no 'bugs' remain. Once confidence has been established in the system, production can be begun with ROMs being used instead of EPROMs. In fact there is not a great difference in price between the two sorts of memory, and it may be worth while to use EPROMs in production machines until supplies of ROMs arrive from the chip manufacturers.

9.5 Test equipment

The complexity of microcomputer-based equipment requires complicated test equipment to diagnose faults when things go wrong, and the development of the microcomputer has been accompanied by the evolution of increasingly sophisticated test equipment. Some of these techniques will be discussed in the next few pages.

9.5.1 Oscilloscopes

For a long time, one of the most widely used instruments in electronics has been the oscilloscope, which shows an image of a waveform on a screen. While this is extremely useful in analog and simpler digital circuits, however, it is of rather limited usefulness in microprocessor-based circuits.

The first problem is that most oscilloscopes usually show only two waveforms. This is not enough to give any real insight into a system where typically there might be up to thirty bus waveforms which must be monitored simultaneously. Secondly, the waveform shown on the screens of conventional oscilloscopes is not the result of monitoring a single section of the waveform but that of superimposing many successive identical waves as the signal waveform repeats. Unfortunately, there is no guarantee that a waveform in a microcomputer will ever repeat. The third shortcoming of the oscilloscope centres on the fact that the sweep of the electron beam across the tube face which draws the waveform is triggered by the appearance of a particular voltage level on a triggering input. In a digital system there are only two voltage levels, and each time a signal changes state the voltage will pass through the triggering level.

Nevertheless, oscilloscopes are useful for monitoring 'non-digital' signals such as noise, and for examining the exact form of waveforms such as clock signals. The bandwidth of an oscilloscope — the highest frequency component which can be displayed — depends largely upon price. For logic circuits such as those found within microcomputers a bandwidth of 50-100 MHz is most useful because of the need to observe details of waveforms without distortion.

9.5.2 Word recognisers

Although the principle of an oscilloscope is applicable to microcomputer circuits, in practice many modifications to the way in which it works are needed before it can be used effectively. The first improvement is to use a logic circuit called a **word recogniser**. This compares the binary word formed by the signals being monitored with a 'trigger word' stored in a register and generates a pulse when a match occurs which can be used to trigger an oscilloscope. For instance, if the user wanted to trigger an oscilloscope when the microcomputer generated the address $F123, the word recogniser would be loaded with the pattern '1111 0001 0010 0011'. The comparison logic of the word recogniser consists of a set of exclusive-OR gates with their outputs gated together. A further refinement is to gate the signals from the microcomputer so that 'don't care' conditions can be introduced. This requires AND gates and a second 'mask' register; each bit in the mask register is ANDed with the correspnding input bit, leaving '0's where there was a '0' in the mask register, corresponding to a 'don't care'.

For example, if the word recogniser is to trigger on the pattern '1111 0001 xxxx xxxx', where 'x' denotes 'don't care', the mask and comparison registers would need to be loaded with data as follows

Test pattern	1111	0001	xxxx	xxxx
Mask register	1111	1111	0000	0000
Compare register	1111	0001	0000	0000
Result	0000	0000	0000	0000

If the inputs to the word recogniser were connected to the lines of an address bus the word stored in the word recogniser would cause the oscilloscope to be triggered at a certain point in the program.

9.5.3 Storing waveforms digitally

The next improvement which can be made to the oscilloscope principle is to include some storage mechanism in the system so that a waveform which might only occur once can be 'captured'. Storage oscilloscopes using special storage tubes have been available for many years, but they are expensive. However, the availability of cheap memory circuits and ADCs and DACs now means that waveforms can be stored digitally. In the case of equipment designed expressly for displaying logic waveforms the converters are not necessary and storing waveforms becomes even easier. In a system using digital storage the waveform which preceded the triggering event can be shown just as easily as that following the triggering, a trick which is almost impossible with more conventional methods.

9.5.4 Timing analysers

We now have the main elements of an instrument suitable for displaying the waveforms which appear inside a microcomputer, although practical systems have a number of sophistications. To be useful in analysing the operation of a microcomputer the screen must be capable of displaying a relatively large number of waveforms compared with an ordinary oscilloscope, but this can be done without any great difficulty. The logic associated with the storage mechanism can be programmed to sample the input signals at different speeds, allowing an overview of the waveform or a more detailed look at part of it.

Once a waveform has been captured by a timing analyser and stored in digital form, it can be analysed by means of suitable software. For example, the analyser can be programmed to search for 'glitches' — spurious pulses — of less than a prescribed duration. Glitches usually appear due to timing errors in logic circuits, often as a result of inadequate design. The effect of a glitch can be to cause a register to be loaded at a time when its contents should not alter, with profound effects upon the running of the microcomputer. Because the waveform can be held indefinitely once it has been captured, analysers of this type allow quite rare events to be observed; it is simply a question of leaving the system running with the analyser connected until the analyser is triggered.

9.5.5 State analysers

When microcomputers are running correctly their waveforms should change only at times specified by the system clock. The changing information flowing on the data bus and the information on the address and control buses provide a means of observing the activity of the microprocessor as it runs the program. All that is necessary is to capture the states on the buses during each cycle.

The essential difference between this approach and that of the timing analyser is that in the latter case the information captured is a function of time, while in this case it is a function of the changing state of the microcomputer. The sequence of states through which the microcomputer passes as a program runs should be the same irrespective of the speed of the processor, subject of course to any speed limitations imposed by the electronics of the integrated circuits.

The sequence of states observed on the various buses could be displayed as a set of waveforms in the same way as with the timing analyser, but this would be confusing. Another, scarcely less confusing, method would be to display the information as a sequence of binary numbers, one digit representing the state of each line being monitored. Because the analyser is displaying information on the buses of a microcomputer, it makes sense to arrange to code the information in the form of hexadecimal numbers so that the sequence of addresses and data words can be compared with a listing of the program.

The triggering specification in a state analyser can be made very complicated. The triggering word can be made to contain 'don't care' sections which need not match the input word, so that for example the analyser could be set to trigger on any address of the form 'Exxx' where 'x' could have any hexadecimal value. The triggering circuit could be programmed to recognise an input data pattern in a certain range, or to cause the analyser to capture information after a certain number of triggering events had occurred rather than the first time. Another powerful addition is to include 'trigger enable' and 'trigger disable' word recognisers which determine whether the triggering event will be recognised or not. The triggering circuit can only operate after it has been enabled by the 'trigger enable' word recogniser and has not been disabled by the 'trigger disable' recogniser.

9.5.6 Microprocessor analysers

Further sophistication is possible. The analyser can determine which of the states correspond to the microprocessor fetching instructions and which correspond to executing them. The mnemonic codes for any given type of

9.5.6

microprocessor can be substituted for the hexadecimal opcodes read by the analyser. Now each instruction that is fetched by the microprocessor as it runs can be observed by the analyser and displayed as mnemonic instructions on the screen. The information read from or written to memory during each instruction can also be displayed, giving a deep insight into the internal operation of the microcomputer. Specialised state analysers of this type are called **microprocessor analysers**, and one is shown in Figure 9.8.

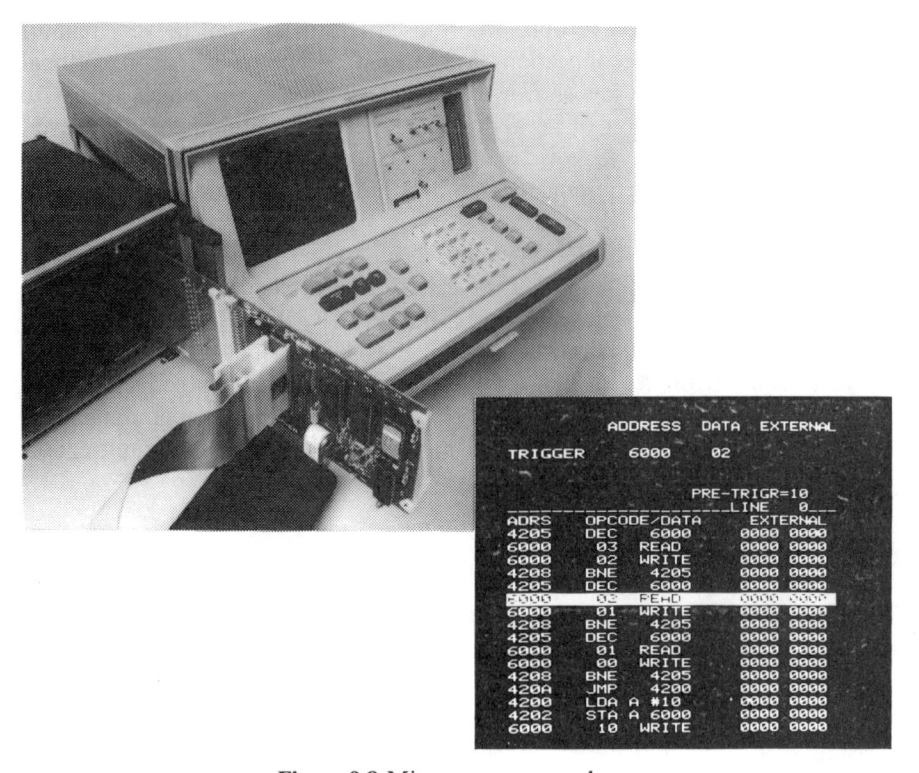

Figure 9.8 Microprocessor analyser

The usefulness of such an instrument can be seen rather more clearly by considering an actual example of fault finding using a microprocessor analyser. A microcomputer system was found to be completely inoperable, yet preliminary tests with an oscilloscope had failed to reveal any obvious problem. Connecting the analyser and resetting the microprocessor showed that it was fetching the reset vector correctly and that the program started to run correctly. Further investigation showed that after a few dozen instructions had been executed it began to malfunction and by trying various triggering addresses it was found that the fault appeared immediately after the program returned from the first subroutine call. The analyser was then set to record each microprocessor cycle which involved the stack, since it was suspected that return addresses from subroutine were not being recovered correctly. This indeed proved to be the case; the correct return address was being pushed on to the stack by writing it to memory, but when the microprocessor attempted to pull the return address from the stack, different data appeared. The fault was rapidly traced to a faulty read/write connection to the memory circuit used to store information on the stack.

9.6 Design for testability and fault localisation

Despite its complexity, microprocessor-based equipment can be very reliable because relatively few connections are needed compared with conventional equipment. Experience has shown that it is the connections which appear to cause a large proportion of breakdowns in electronic equipment. However, it can be difficult to discover the cause of any malfunction in microprocessor-based equipment once it is in the field because it is so complicated.

In the past a useful approach has been to partition equipment into a number of separate printed-circuit boards so that faults can be localised by a 'board-swapping' approach. This is not usually the best approach in microprocessor-based equipment, since it may be possible to fit all the electronics required on to a single PCB, so that deliberately introducing more connections can only reduce reliability. With microcomputer equipment of this complexity it becomes especially important to consider how fault conditions are going to be diagnosed.

9.6.1 Self-testing

One technique is to use the microprocessor to run a self-test routine as the equipment is switched on, and possibly during operation. The microprocessor compares the results of various operations on its interfaces and memories with 'correct' values which are stored in memory. If the results obtained in this way are the same as the 'correct' values, the microprocessor assumes that the remainder of the system is functioning normally. The equipment must have some means of indicating the existence of a fault and, preferably, what is wrong.

9.6.2 Testing read-only memory

A read-only memory can be tested by including in the ROM contents an extra word or words of redundant information calculated using one of the techniques described in the last chapter. For example, the word can be chosen so that the checksum of the ROM contents (the sum of all the ROM contents excluding carry) is zero. The self-test routine for the ROM then consists of adding up all the words in that ROM and checking the result. If the result is non-zero then there is a fault of some description, while if the result is zero it is likely (to better than 99% confidence) that the ROM is functioning correctly. More sophisticated error checking, using multiple words or cyclic redundancy codes, can give very high levels of confidence if required.

9.6.3 Typical RAM faults

There are many ways in which a memory may fail to perform satisfactorily; some of these are listed below.

- **Cell Opens and Shorts.** These are the simplest failures to detect. A cell may remain stuck in a fixed state, either zero or one, but this is usually due to manufacturing faults and should be discovered during initial testing of the component.
- **Address Nonuniqueness.** This is due to a failure in the address decoding circuitry, either within the memory circuit itself or somewhere in the associated circuitry.
- **Disturb Sensitivity.** Capacitative coupling between cells can cause the contents of a memory cell to be affected when an adjacent cell is accessed.
- **Sense Amplifier Interaction.** The tiny signals stored in the storage cells of the amplifier must be amplified on the memory chip before being connected to the external circuitry. Sometimes the level in one of the amplifiers can affect that in an adjacent one.

9.6.4 RAM test methods

A large number of methods have evolved for testing random-access memories, and a few of them are listed below. Again, the possibilities both of defective connections and defective storage must be checked. The basic methods really test only a single cell at a time, but they can be extended to test whole words simply by using a 'sliding' bit. Here a test pattern in the data word and its complement are used instead of zero and one bits, and after each test the pattern is shifted until the memory has been tested exhaustively. Some tests take an amount of time which is proportional to the size of the memory, while the more thorough ones, which check every memory location after each change of memory contents, take an amount which is proportional to the square of the memory size.

Zeros and ones
This consists of writing a zero into each location in memory and then testing to see that the data was stored correctly. It is very simple to program and it runs quickly, but it is limited to detecting 'stuck' cells, open cells, and the inability to store zeros. The same test can be carried out writing ones instead.

Checkerboard read/write
This program writes a 'checkerboard' of alternate zeros and ones throughout memory, and then waits some time before checking to make sure that the pattern has been retained correctly.

Walking pattern
This method tests for cell 'opens', 'shorts', and address nonuniqueness. The following procedure is repeated for each memory address in turn:

- set all memory contents to zero.
- complement contents of location being addressed.
- check contents of all other locations.

Marching pattern
This test begins by writing zeros to all locations. The program then reads the first location, and if it still contains a zero it writes a one to that address. The next address is then tested in turn and the process repeated until the memory is filled with ones. Then the whole sequence is repeated, reading to test for ones and writing zeros if the result is satisfactory.

Galloping pattern (GALPAT)
This is a very thorough method which is widely used in testing, and it is capable of detecting 'open' and 'shorted' cells as well as address uniqueness and sense amplifier interactions. The procedure is as follows:

- complement contents of cell.
- read contents of this memory cell alternately with those of each other memory location and check that everything is as expected.
- repeat steps 1 and 2 for each memory location in turn.

Using GALPAT with a sliding bit
The procedure below shows how the single-bit GALPAT test can be combined with a sliding bit to test eight bit wide memories such as those used in microcomputers:

- store the pattern 00...01 at every location in the RAM.
- starting at the lowest address, check that the pattern is stored correctly in each location, and if so, replace the pattern with its one's complement (11...10). Otherwise, abandon the test and report the fault.
- rotate the bit pattern one place to the left (to give 00...10) and repeat steps 1 and 2. Rotate and repeat steps 1 and 2 until each bit position has been tested for every word.

This method can check for many types of open- and short-circuit connections and for malfunctioning and interacting memory bit 'cells'. Unfortunately, the time taken to test a memory exhaustively rises rapidly as the size of the memory increases to the extent that it becomes prohibitive. If a memory device is tested in the microcomputer by means of a test program, that program will take in the order of tens of microseconds to carry out each test. Since the number of tests rises as the square of the number of memory addresses and proportionally to the wordlength, the time taken to test memories with several kilobytes' storage can be minutes or even hours. In effect the problem of testing is that of being able to control the logic level at each point in the circuit being tested and to be able to observe the result. These problems of **controllability** and **observability** are even worse for other types of circuit.

9.6.5 Testing input/output devices

Input/output devices are more difficult to test because the techniques used depend upon the type of device. For example, the data direction registers of an I/O device can be checked by writing a pattern into them and then reading back the result, but a correct result does not guarantee that the system is working correctly. Input and output registers are more difficult to test unless the input state is known at the time of testing. This is best done by connecting the microcomputer to a special test jig which allows the input states to be controlled and the output states to be monitored, using another microcomputer if necessary.

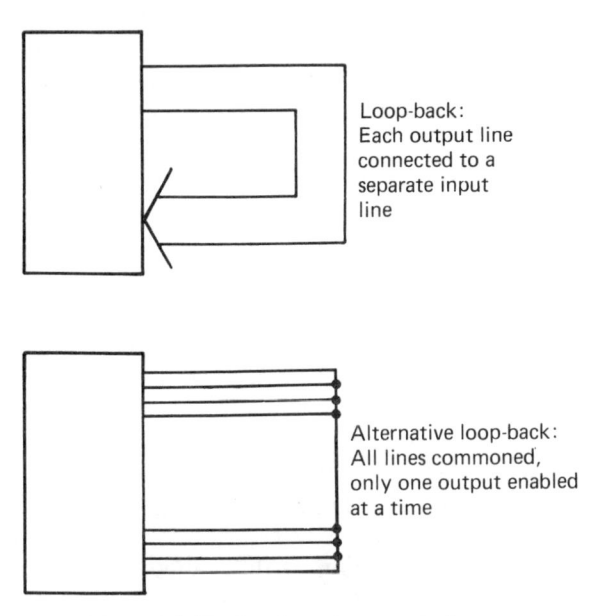

Loop-back: Each output line connected to a separate input line

Alternative loop-back: All lines commoned, only one output enabled at a time

Figure 9.9 Testing input/output devices

Often it is possible to test many of the input/output lines, and sometimes sections of the remainder of the system, by connecting some of the outputs back to the inputs in 'loop-back' arrangements, as shown in Figure 9.9. For example, if an actuator has limit switches, the microcomputer might be able to drive the actuator between the limits and check that the limits are indeed reached. If it does not sense a limit, there must be a fault in the output line, in the servo amplifier, in the actuator, or in the sensing circuit or input. The important point is that the microcomputer senses that a fault has occurred and can indicate the fact. A competent technician should then be able to ascertain the precise nature of the fault.

9.6.6 Testing the microprocessor

Testing the microprocessor itself is very difficult. If the circuit is to be checked thoroughly, not only must every instruction be tested with a large range of data patterns, but different sequences of instructions must be checked. This is because of the difficulty of controlling and observing all the points in the circuit, which requires a very detailed knowledge of the microprocessor design. Normally a relatively simple series of tests is carried out to check that the microprocessor performs a test program properly, and the most that can usually be done in practice is to make sure that all the registers are present and that the instructions function correctly. However, if the fault prevents the microcomputer from executing any programs at all, it may still be possible to use the microprocessor to generate test signals within the circuit, as described in the next section.

9.6.7 Forced instruction and signature analysis methods

A powerful technique for localising a fault is to use the microprocessor to generate repetitive waveforms which can be analysed by other equipment. This can be done by breaking the data bus near the microprocessor and substituting a connector wired so that it forces a bit pattern corresponding to a single byte instruction on to the microprocessor's data bus, as shown in Figure 9.10. The actual instruction used is not very important so long as it does not load the program counter with a new value; the instruction most commonly used is the No-OPeration (NOP) code which does not affect any of the processor registers.

The instruction is executed by the microprocessor and then the next instruction is fetched in the normal way. Now, however, the microprocessor will simply read the same opcode once more and the action will be repeated. The microprocessor will continue generating successive addresses and fetching the same instruction which is being forced on to its data bus input. The result is that it generates a set of square wave signals on its address lines as it attempts to fetch instructions, and these waveforms can be traced using an oscilloscope to check the operation of the address decoding circuitry. Some microprocessors have a test instruction which causes them to repeatedly fetch successive opcode bytes without stopping. When this instruction is forced on to the bus momentarily it causes the microprocessor to enter the test mode. It will then remain in this mode, ignoring all external stimuli except the reset input. This instruction has been given the unofficial and somewhat figurative mnemonic code of HCF... 'Halt and Catch Fire'!

Although the fact that repetitive waveforms are being generated means

9.6.7   SYSTEM DESIGN AND DEVELOPMENT   **419**

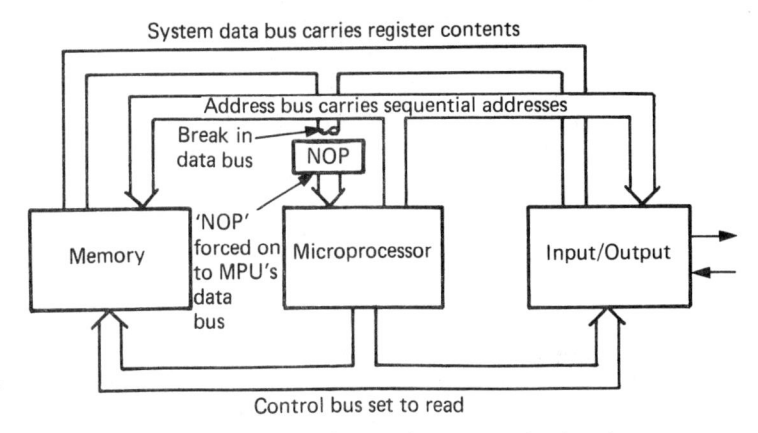

Figure 9.10 Forcing instructions on to the data bus

that an oscilloscope can be used, some of the waveforms found in the address decoding circuitry may be difficult to see distinctly on the screen. The waveforms seen on the data bus while this is happening will almost certainly be too complicated to analyse unless there is a gross fault such as a data line short-circuited to the zero volt line. In fact it is not so much the shape of the waveform which is of interest but the decision as to whether that waveform is the correct one or not. Many methods exist for 'condensing' a waveform into a simple number which is in some way characteristic of that particular waveform, for example the number of changes of logic level in a defined interval or the proportion of time which the waveform spends

Figure 9.11 Signature analyser in use

at logic '1'. One particularly effective technique is to use a feedback shift register arrangement to generate a 16-bit pattern in a register rather like the cyclic redundancy code generator of Chapter 8. This pattern is called the 'signature' of a waveform, and it can be shown that any single bit change in a waveform will change its signature; indeed, if a waveform has a correct signature, then one can say with reasonable confidence that it is correct, while if the signature is incorrect, so is the waveform.

In practice the signature of a waveform is obtained by an instrument called a 'signature analyser' which is connected to the system clock. It also has 'start' and 'stop' inputs which are connected to suitable points in the system under test to define the points in the waveform sequence at which the monitoring of the waveform signature should start and finish. The test probe can then be placed on the point in the circuit whose waveform is being monitored. Figure 9.11 shows a signature analyser being used in this way.

Signature analysis is essentially a test technique to be used when fault-finding during maintenance rather than while commissioning a new design. This is because the signatures corresponding to the correct waveforms must be known, and although in principle they could be calculated, it is much easier to obtain them from a 'healthy' piece of equipment. These signatures are printed in the service manual, with fault-tracing information and circuit diagrams showing the correct signature on each connection in the circuit. The signature is written as a four-digit hexadecimal number, with the minor difference that letters other than A-F are used to improve readability on the seven segment display used. In a typical application the 'start' and 'stop' inputs of the analyser might be connected to the most significant address line and the 'clock' input to the system clock.

As with any electronic testing, the instrument only provides information as to the waveform being observed, and the interpretation of the readings thus obtained is left to the technician. Some faults are reasonably obvious, however. If a waveform is correct at one end of a connection but not at the other, there must be a break in continuity, while if two adjacent connections share the same incorrect signature then there must be a short-circuit somewhere. In any system logic '0' always gives the signature 0000, while there is a characteristic logic '1' signature for any test connection. If this is not listed in the manual it can be found quite easily by placing the probe at some point in the circuit known to be at logic '1', for example the power supply connection to one of the integrated circuits.

Appendix I A COMPARISON OF INSTRUCTION SETS

The majority of examples of assembly language used throughout this book are in 6809 code. This is because the 6809 has a very 'regular' instruction set without any quirks or peculiarities. The problem with assembly language is that it is specific to the microprocessor for which it is written, but where interfaces are being controlled it provides a more precise and direct method of programming than either BASIC or Pascal.

It should be possible to rewrite the examples for other microcomputers in the 6300, 6500, and 6800 families without difficulty. The designs of the 8085 and the Z80 are somewhat different, but nevertheless it should not prove difficult for the reader with a microcomputer based upon one of these microprocessors to convert the examples.

In the next few pages five of the more common types of microprocessor will be compared. Three of them are 8-bit microprocessors: the 6809, which we have used as our main example and the 6502 and Z80 which between them have supplied the processing function in the majority of 8-bit personal computers. The other two types are 16-bit machines: the LSI11 which shares the instruction set of the widely used PDP11 minicomputer, and the 68000 which is a member of a family of similar microprocessors having bus widths ranging from 8 to 32 bits: 16-bit machines are capable of handling two or even four bytes in a single instruction, and have register-to-register architectures. The translation of the program examples from the accumulator-based 6809 to these types is a little more difficult, but it should not prove too daunting to the reader who is familiar with either of these machines.

I.1 Programming models

Before embarking upon a comparison of these five machines, it will be useful to compare their programming models. The simplest machines from this point of view are the 6502 and 6809 shown in Figure A.1. The 6502 is the simpler of the two, having fewer registers and a more restricted instruction set. Its index registers and stack pointer are each only eight bits long, although the index registers can be used with a 16-bit offset to produce a 16-bit address as described earlier in Chapter 3. Whenever the 6502's stack pointer is used to address memory, the eight less significant bits of the address used are the contents of the stack pointer and the more significant eight bits are the binary pattern '00000001'.

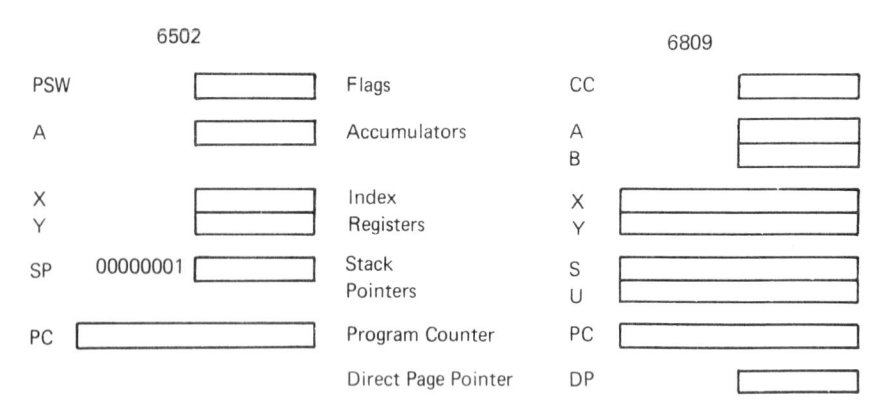

Figure A.1 Programming models of 6502 and 6809

6502
- PSW
- A
- X
- Y
- SP 00000001
- PC

Flags
Accumulators
Index Registers
Stack Pointers
Program Counter
Direct Page Pointer

6809
- CC
- A
- B
- X
- Y
- S
- U
- PC
- DP

The two accumulators of the 6809 can be used as a single 16-bit accumulator. This allows the 6809 to carry out 16-bit arithmetic directly without the need to link together two successive 8-bit additions via the carry bit. This ability to treat the accumulators as either a pair of 8-bit registers or as a single 16-bit register also makes for increased flexibility when data is being transferred between registers within the microprocessor. In the 6502, all transfers are between 8-bit registers, which makes it possible to copy the contents of the 8-bit accumulator into either of the 8-bit index registers. The 6809 allows transfers between any pair of registers of the same length (8 or 16 bits), and transfers are possible between the accumulators and registers of either length. The Direct Page Register is used to provide the high byte of the address presented to the address bus when the single-byte addressing mode is in use, allowing this 'shorthand' form of addressing to be used anywhere in the address space of the microprocessor.

The Z80 has a similar number of working registers to the 6809, except that it also boasts a second set of registers which can be used to switch context quickly in response to an interrupt. The second set of registers is shown on the right-hand side in Figure A.2.

Registers A, B, C, D, E, H, L, PC, and SP represent the programming model of the 8080 and 8085 microprocessors, and the Z80 can be regarded from the programming point of view as a development of the 8080. The interrupt (IV) register holds a byte which can be used to provide the high byte of the interrupt vector when an interrupt occurs. Register R of the Z80 is used by a feature which is almost unique to this processor. It is a counter which is used to generate the addresses used when the Z80 refreshes dynamic memories.

The programming models of the LSI11 and 68000 are more regular than those of the 8-bit microprocessors. The LSI11 has eight general purpose registers, each 16 bits long, labelled R0 to R7. Each of these registers can be used as a single 16-bit register or as an 8-bit register when handling

A COMPARISON OF INSTRUCTION SETS

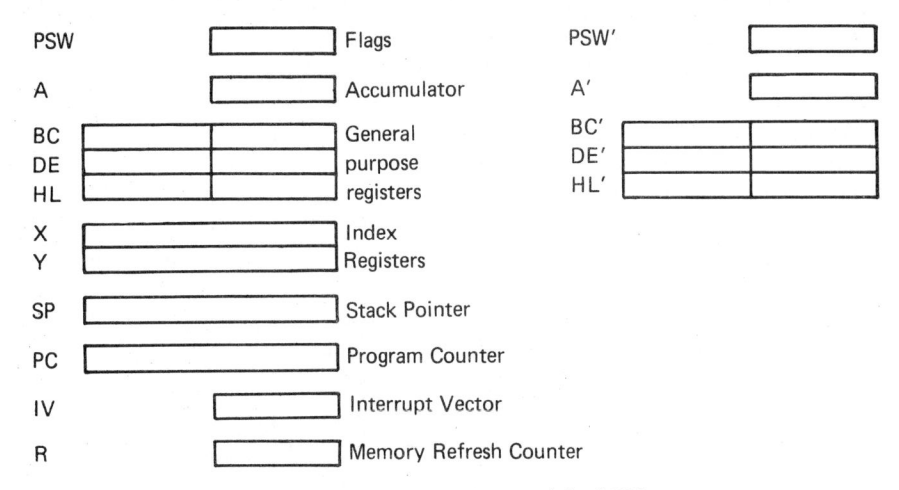

```
PSW  [        ] Flags              PSW' [        ]
A    [        ] Accumulator        A'   [        ]
BC   [        ] General            BC'  [        ]
DE   [        ] purpose            DE'  [        ]
HL   [        ] registers          HL'  [        ]
X    [        ] Index
Y    [        ] Registers
SP   [        ] Stack Pointer
PC   [        ] Program Counter
IV   [        ] Interrupt Vector
R    [        ] Memory Refresh Counter
```

Figure A.2 Programming model of Z80

single bytes. One of the registers, R7, is used as the program counter and another, R6, is used as the stack pointer. As shown in Figure A.3, the only other register is the processor status register.

```
PSW [        ]
R0  [        ]
R1  [        ]
R2  [        ]
R3  [        ]
R4  [        ]
R5  [        ]
R6  [        ] Stack Pointer
R7  [        ] Program Counter
```

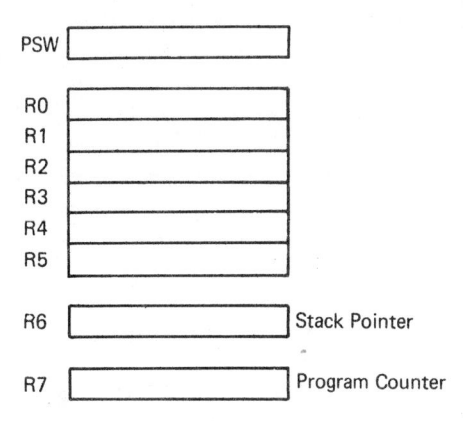

Figure A.3 Programming model of LSI11

The 68000 has more registers than the LSI11, and they are longer: 32 bits in length, although many instructions use only 8 or 16 of these bits when handling shorter data. The registers are in two groups of eight registers each; eight data registers (D0-D7) and eight address registers (A0 to A7). One of the address registers (A7) is used as a system stack pointer for subroutine linkage, rather like the LSI11. The 68000 shows a feature found in many of the newer 16-bit microprocessor designs. Programs can be run with the machine in one of two operating modes — 'supervisor mode' in which all registers and instructions are available, and a less privileged 'user' mode in which some instructions cannot be executed. This mode is selected by means of a bit in the processor status register.

The 68000 has two system stack pointers, the choice of register depending upon the mode in which the processor is operating, rather like the separate user and system stack pointers of the 6809. It has a separate program counter and as already mentioned, a status register. This is the most recent of the designs being compared and, as will be seen from the programming model in Figure A.4, the most complicated. Despite this complexity, however, the 68000 is not more difficult to program than the other types.

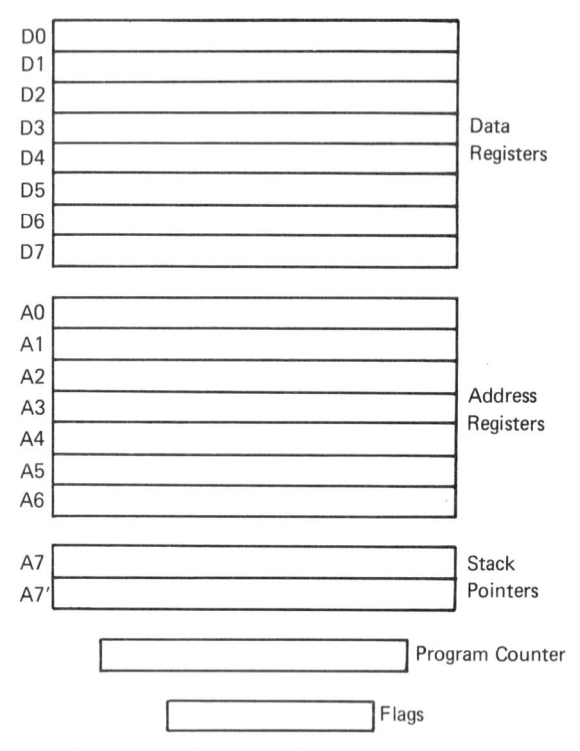

```
D0 ┌──────────────────────────────┐
D1 ├──────────────────────────────┤
D2 ├──────────────────────────────┤
D3 ├──────────────────────────────┤  Data
D4 ├──────────────────────────────┤  Registers
D5 ├──────────────────────────────┤
D6 ├──────────────────────────────┤
D7 └──────────────────────────────┘

A0 ┌──────────────────────────────┐
A1 ├──────────────────────────────┤
A2 ├──────────────────────────────┤
A3 ├──────────────────────────────┤  Address
A4 ├──────────────────────────────┤  Registers
A5 ├──────────────────────────────┤
A6 └──────────────────────────────┘

A7  ┌─────────────────────────────┐  Stack
A7' └─────────────────────────────┘  Pointers

    ┌──────────────────────┐ Program Counter

    ┌──────────────┐ Flags
```

Figure A.4 Programming model of 68000

I.2 Single operand data instructions

These instructions allow the contents of a single data register (accumulator or memory) to be manipulated. The newer microprocessor designs such as the 6809 and the 68000 have a reasonably complete set of such instructions, but as may be seen in Table A.1, other types such as the 6502 offer a more limited range. However, this does not mean that it is impossible to program the missing operations on the 6502, but simply that a sequence of instructions must be used instead of a single one.

For example, a memory location may be cleared to zero by loading the

Table A.1 Single operand data handling instructions.

Instruction	6502	6809	Z80	LSI11	68000
Clear	(STZ)*	CLR	–	CLR	CLR
Complement	–	COM	–	COM	COM
Decrement by 1	DEC	DEC	DEC	DEC	DEC
Increment by 1	INC	INC	INC	INC	INC
Negate	–	NEG	–	NEG	NEG
Test	–	TST	–	TST	TST
Arith. shift L	ASL	ASL	SLA	ASL	ASL
Arith. shift R	–	ASR	SRA	ASR	ASR
Logic. shift L	ASL	LSL	SLA	ASL	LSL
Logic. shift R	LSR	LSR	SRL	–	LSR
Rotate L	ROL	ROL	RL	ROL	ROL
Rotate R	ROR	ROR	RR	ROR	ROR

* See text

accumulator with zero and then storing the accumulator contents in that memory location. If the current contents of the accumulator are to be preserved, they can be pushed on to the stack before the two instructions are executed and recovered afterwards. Similar sequences can be used to program other 'missing' instructions on the 6502 and other microprocessors which have 'gaps' in their instruction sets, as shown in Figure A.5.

```
TST xx    CLR xx    COM xx      NEG xx    ASR xx

PHA       PHA       PHA         PHA       PHA
LDA xx    LDA #0    LDA #255    LDA #0    LDA xx
CMP #0    STA xx    EOR xx      SEC       ASL A
PLA       PLA       STA xx      SBC xx    ROR xx
                    PLA         STA xx    PLA
                                PLA
```

Figure A.5 Simulating missing instructions on the 6502

The entry (STZ) in the 6502 column of Table A.1 refers to the fact that the more recent members of the 6502 family such as the 65C02 have a larger instruction set in which some of the 'missing' instructions are included. Thus STZ, which is an abbreviation of 'store zeros' allows a location to be cleared.

I.3 Dual operand data instructions

Again, an examination of the dual operand instructions in Table A.2 will show that while the 6809, for example, has an entry in almost every row of its column, some of the other types of microprocessor have a few 'missing' instructions. In accumulator-based microprocessors (6502, 6809, Z80), the destination register, where the result of the operation is to be found, is the accumulator. The gaps in this table arise from the fact that different designs of microprocessor handle the linking of the carry bit from one addition to the next in different ways. Most microprocessors and microcomputers have an 'add with carry instruction' which allows the contents of a memory location to be added together with the contents of the carry register to the accumulator. The LSI11 differs from this norm, however, in having an 'add carry' instruction which allows the contents of the carry register to be added in a separate operation from the ordinary addition, so that two instructions are needed to program an 'add with carry'.

Table A.2 Dual operand data handling instructions

Instruction	6502	6809	Z80	LSI11	68000
Load acc.	LDA	LDA	LD A,()	MOV	MOVE
Store acc.	STA	STA	LD (),A	MOV	MOVE
Add	–	ADD	ADD	ADD	ADD
Add with carry	ADC	ADC	ADC	–	ADC
Add carry	–	–	–	ADC	–
Subtract	–	SUB	SUB	SUB	SUB
Subtract with C	SBC	SBC	SBC	–	SBC
Subtract carry	–	–	–	SBC	–
Compare	CMP	CMP	CP	CMP	CMP
Logical AND	AND	AND	AND	(BIC)*	AND
Logical OR	ORA	ORA	OR	BIS	OR
Exclusive OR	EOR	EOR	XOR	XOR	EOR
Bit test	BIT	BIT	(BIT)*	BIT	BIT

* See text

Two entries in Table A.2 are starred to indicate that they operate in a somewhat different fashion from the corresponding instructions in other microprocessors. The 'bit clear' instruction in the LSI11 (BIT) clears each bit in the destination that corresponds to a bit at logic 1 in the source register, but leaves the other bits of the destination and all the bits of the source unaffected. This differs from the AND instruction used in the other types of microprocessor being compared. In the 68000, for example, bits are cleared in the destination where there are zeros in the source register. The 'bit test' instruction of the Z80 tests a bit in the memory location specified by an argument of the instruction rather than performing a simple logic 'AND' as in the case of the other types.

The 6502 does not have an 'add without carry' instruction; the 'add with carry' instruction must be used instead, preceded with a 'clear carry'. Subtraction without carry must likewise be programmed using the sequence 'set carry, subtract with carry':

ADD *xx*   SUB *xx*

CLC   SEC
ADC *xx*   SBC *xx*

I.4 Program control instructions

The program control instructions all operate by affecting the contents of the program counter. In Table A.3, program control instructions which unconditionally affect the contents of the program counter are tabulated for the five sample microprocessor types. Branch instructions, which affect the program counter only if the correct condition appears in the processor status register, will be discussed in the next section.

Table **A.3** Program control instructions

Instruction	6502	6809	Z80	LSI11	68000
Jump	JMP	JMP	JP	JMP	JMP
Branch uncon.	(BRA)*	BRA	JR	BR	BRA
Jump to Subr.	JSR	JSR	CALL	JSR	JSR
Ret. from Subr.	RTS	RTS	RET	RTS	RTS
Software Intr.	BRK	SWI	RES	TRAP	TRAP
Ret. from Intr.	RTI	RTI	RET	RTI	RTI

* See text

All the types have instructions for unconditional jump, jump to subroutine, return from subroutine and return from interrupt. The entry marked 'Branch uncon.' refers to jump instructions which use relative rather than absolute addressing. The original 6502 did not have an unconditional branch instruction with relative addressing, but it has been included in later variants such as the 65C02.

All of these microprocessors also have instructions for software interrupts or traps, but the mechanisms used differ considerably between the different types (as do their interrupt mechanisms) and readers interested in a detailed comparison of these instructions should refer to the manufacturers' data.

I.5 Branch instructions

In this section branch instructions which use relative addressing are discussed. These cause program execution to move forward or back a defined number of bytes if the appropriate condition exists in the processor status register. Here again there are some 'missing' instructions which must be simulated where absent by means of sequences of other instructions.

Only the Z80 lacks a full set of conditional branches with relative addressing, but it does have a set of conditional branches using absolute addressing, as shown in the second of the two 'Z80' columns of Table A.4.

Table A.4 Branch instructions

Instruction	6502	6809	Z80 rel	Z80 abs	LSI11	68000
Branch if Z=0	BNE	BNE	JR NZ	JP NZ	BNE	BNE
Branch if Z=1	BEQ	BEQ	JR Z	JP Z	BEQ	BEQ
Branch if N=0	BPL	BPL	–	JP P	BPL	BPL
Branch if N=1	BMI	BMI	–	JP M	BMI	BMI
Branch if C=0	BCC	BCC	JR NC	JP NC	BCC	BCC
Branch if C=1	BCS	BCS	JR C	JP C	BCS	BCS
Branch if V=0	BVC	BVC	–	JP PO	BVC	BVC
Branch if V=1	BVS	BVS	–	JP PE	BVS	BVS

The conditional branch instructions are based upon the status of the individual flag bits, whereas when programming it is preferable to have arithmetic tests such as 'greater than' and 'less than'. These instructions can always be simulated by testing combinations of the flag bits, but the test to be carried out depends upon the number representation being used. Thus the pattern '11111111' is greater than '00000000' if the unsigned number representation is being used (because $255 > 0$) but less if two's complement representation is in use (because $0 > -1$). Two different sets of conditional tests are needed: one for unsigned comparisons and one for two's complement ones.

Table A.5 Arithmetic comparisons using flag bits

Branch if	Two's Comp.			Unsigned		
$\neq$	(Z=0)			(Z=0)		
$>$	(Z=0)	and	(N=V)	(Z=0)	and	(C=0)
$\geq$			(N=V)			(C=0)
$=$	(Z=1)			(Z=1)		
$\leq$	(Z=1)	or	(N$\neq$V)	(Z=1)	or	(C=1)
$<$			(N$\neq$V)			(C=1)

Some microprocessors have special branch instructions for all the combinations, while others require combinations to be programmed. Table A.5 shows the way in which the flag bits are tested by instructions in the 6809, 68000, and LSI11, which all provide special instructions for each of these tests, as shown in Tables A.6 and A.7.

The Z80 and 6502 can both be programmed to carry out corresponding tests using sequences of instructions, provided that note is taken of the difference in the ways in which the carry bit operates in the case of the 6502.

I.6 Manipulating the flag bits

It will be apparent that there is a strong similarity between three of the five processors being compared in this appendix — the 6809, 68000, and LSI11, despite their differences in architecture. The bits of the processor status

Table A.6 Signed arithmetic branch instructions

Instruction	6502	6809	Z80	LSI11	68000
Branch if ≠	BNE	BNE	JR NZ	BNE	BNE
Branch if >	-	BGT	-	BGT	BGT
Branch if ≥	-	BGE	-	BGE	BGE
Branch if =	BEQ	BEQ	JR Z	BEQ	BEQ
Branch if ≤	-	BLE	-	BLE	BLE
Branch if <	-	BLT	-	BLT	BLT

Table A.7 Unsigned arithmetic branch instructions

Instruction	6502	6809	Z80	LSI11	68000
Branch if ≠	BNE	BNE	JR NZ	BNE	BNE
Branch if >	-	BHI	-	BHI	BHI
Branch if ≥	BCS	BHS	JR NC	BHIS	BHS
Branch if =	BEQ	BEQ	JR Z	BEQ	BEQ
Branch if ≤	-	BLS	-	BLOS	BLS
Branch if <	BCC	BLO	JR C	BLO	BLO

registers appear in the same positions and have the same identifying letters: NZVC. The 6502 also has N, Z, V, and C bits, but they appear in a different sequence in the processor status register. By contrast, the Z80 has four main flag bits labelled C, Z, S (sign) and P/O (parity/overflow) which correspond in some circumstances to the C, Z, N, and V bits of the other types.

The various types of processor also differ in the way and the extent to which the flag bits can be controlled directly by the programmer. In the case of the 6502 the carry and overflow bits can be set and cleared by the instructions (SEC, SEV, CLC and CLV). The LSI11 has these instructions too, but it also has instructions for setting and clearing the N and Z bits (SEN, SEZ, CLN, CLZ). The 6809 and 68000 take a different approach by allowing all the bits in the flag register to be modified in a single instruction. Some assemblers for these types of microprocessor recognise instructions of the sort used by the other types of microprocessor but produce the appropriate sequence of instructions to achieve the corresponding result.

Appendix II SYMBOLS

The principal symbols used in this book are shown below together with their meanings.

Symbol	Meaning
h_{FE}	Current gain in common emitter mode
I	Current
I_B	Base current
I_C	Collector current
I_E	Emitter current
I_{IH}	Current flowing into input at logic 'high'
I_{IL}	Current flowing into input at logic 'low'
I_{OH}	Maximum current sourced by output at logic 'high'
I_{OL}	Maximum current sunk by output at logic 'low'
I_{fb}	Feedback current
I_{in}	Signal input current
K_D	Derivative gain of three-term controller
K_I	Integral gain of three-term controller
K_P	Proportional gain of three-term controller
N_{in}	Clock cycles during which input connected
N_{REF}	Clock cycles during which reference connected
N_t	Number of clock cycles
R_{fb}	Feedback resistance
R_{in}	Input resistance
R_L	Load resistance
R_o	Fixed part of transducer resistance
R_{out}	Output resistance
R_t	Transducer resistance

T_c	Duration of one clock cycle
T_{in}	Time during which input connected
T_n	Duration of noise pulse
T_{REF}	Time during which reference connected
V_+	Voltage on non-inverting input of amplifier
V_-	Voltage on inverting input of amplifier
V_1	Voltage on input 1 of differential amplifier
V_2	Voltage on input 2 of differential amplifier
V_{BB}	Base supply voltage
V_{BE}	Base-emitter voltage of bipolar transistor
V_{CC}	Collector supply voltage for bipolar transistor
V_{CE}	Collector-emitter voltage of bipolar transistor
V_{DD}	Drain supply voltage for MOS transistor
V_{fb}	Feedback voltage
V_{IH}	Minimum input voltage for logic 'high'
V_{IL}	Maximum input voltage for logic 'low'
V_{in}	Signal input voltage
V_{NH}	Noise immunity for logic 'high'
V_{NL}	Noise immunity for logic 'low'
V_{OH}	Minimum output voltage at logic 'high'
V_{OL}	Maximum output voltage at logic 'low'
V_{out}	Signal output voltage
V_{REF}	Reference voltage
V_{SS}	Source supply voltage for MOS transistor
V_T	Threshold voltage
V_{TH}	Upper threshold voltage of Schmitt trigger
V_{TL}	Lower threshold voltage of Schmitt trigger
V_t	Transducer output voltage

BIBLIOGRAPHY

Chapter 1 Introducing the microcomputer

Evans, C. 1981. *The Making of the Micro.* London: Gollancz

Chapter 2 Processing information

Aho, A.V., Hopcroft, J.E. and Ullman, J.D. 1974. *The Design and Analysis of Computer Algorithms.* Reading, Mass: Addison-Wesley

American National Standards Institute. 1978. American National Standard for Minimal BASIC, *ANSI X3.60-1978.* New York: American National standards Institute.

Atkinson, L.V. and Harley, P.J. 1983. *An Introduction to Numerical Methods with Pascal.* London: Addison-Wesley

Brown, P.J. 1982. *Pascal from BASIC.* London: Addison-Wesley

Bycer, B.B. 1975. *Flowcharting: Programming, Software Design and Computer Problem Solving.* New York: John Wiley

Froehlich, J.P. 1969. *Information Transmittal and Communicating Systems.* New York: Holt, Rinehart and Winston

Grogono, P. 1980. *Programming in PASCAL.* Reading, Mass: Addison-Wesley

Jensen, K. and Wirth, N. 1978. *PASCAL User Manual and Report.* New York: Springer-Verlag

Jury, E.I. 1964. *Theory and Application of the z-transform Method.* New York: John Wiley

Koffman, E.B. 1982. *Pascal: A Problem Solving Approach.* Reading, Mass: Addison-Wesley

Nussbaumer, H.J. 1981. *Fast Fourier Transform and Convolution Algorithms.* Berlin, Heidelberg: Springer-Verlag

Chapter 3 Inside the microcomputer

Anceau, F. 1985. *The Architecture of Microprocessors.* London: Addison-Wesley

Booth, A.D. 1951. A Signed Binary Multiplication Technique, *Q. J. Mech. Appl. Math.*, **4(2)**, 236-40.

Kernighan, B.W. and Ritchie, D.M. 1978. *The C Programming Language.* Englewood Cliffs, NJ: Prentice-Hall

Leventhal, L.A. 1978. *Introduction to Microprocessors: Software, Hardware, Programming.* Englewood Cliffs, NJ: Prentice-Hall

Lippiatt, A.G. and Wright, G.G. 1985. *The Architecture of Small Computer Systems.* 2nd. edn., London: Prentice-Hall

Magar, S., Hester, R. and Simpson, R. 1982. Signal-Processing Microcomputer Builds FFT-Based Spectrum Analyser, *Electronic Design*, 149-154

Mano, M.M. 1982. *Computer System Architecture.* Englewood Cliffs, NJ: Prentice-Hall

Sloan, M.E. 1980. *Introduction to Minicomputers and Microcomputers.* Reading, Mass: Addison-Wesley

Tocci, R.J. and Laskowski, L.P. 1982. *Microprocessors and Microcomputers.* Englewood Cliffs, NJ: Prentice-Hall

Chapter 4 Input/Output techniques

Buzen, J.P. 1975. I/O Subsystem Architecture, *Proc. IEEE.*, **63,** 871-9

Kraft, G.D. and Toy, W.N. 1979. *Mini/Microcomputer Hardware Design.* Englewood Cliffs, NJ: Prentice-Hall

Lewin, M.H. 1983. *Logic Design and Computer Organization.* Reading, Mass: Addison-Wesley

Stone, H.S. 1982. *Microcomputer Interfacing.* Reading, Mass: Addison-Wesley

Chapter 5 Digital interfacing

Artwick, B.A. 1980. *Microcomputer Interfacing.* Englewood Cliffs, NJ: Prentice-Hall

Stone, H.S. 1982. *Microcomputer Interfacing.* Reading, Mass: Addison-Wesley

Chapter 6 Analog interfacing

Artwick, B.A. 1980. *Microcomputer Interfacing.* Englewood Cliffs, NJ: Prentice-Hall

Clayton, G.B. 1982. *Data Converters.* London: Macmillan

Nyquist, H. 1928. Certain Topics in Telegraph Transmission Theory, *AIEE Trans.*, **47,** 614-44

Stone, H.S. 1982. *Microcomputer Interfacing.* Reading, Mass: Addison-Wesley

Chapter 7 Transducers: sensors and actuators

Ballard, D.H. and Brown, C.M. 1982. *Computer Vision.* Englewood Cliffs, NJ: Prentice-Hall

Beckwith, T.G., Buck, N.L. and Marangoni, R.D. 1982. *Mechanical Measurements.* 3rd. edn., Reading, Mass: Addison-Wesley

Faugeras, O.D. 1983. *Fundamentals in Computer Vision.* Cambridge: Cambridge University Press

Foley, J.D. and van Dam, A. 1982. *Fundamentals of Interactive Computer Graphics.* Reading, Mass: Addison-Wesley

Hetenyi, M. (ed.) 1950. *Handbook of Experimental Stress Analysis.* New York: John Wiley

Neubert, H.K.P. 1963. *Instrument Transducers.* Oxford: Clarendon Press

Sydenham, P.H. (ed.) 1983. *Handbook of Measurement Science. Vol.2: Practical Fundamentals.* Chichester: John Wiley

Wolf, J.J. 1980. Speech Recognition and Understanding in *Digital Pattern Recognition*, ed. Fu, K.S. Berlin: Springer-Verlag, 167-204

Chapter 8 Machine communications

American National Standards Institute. 1979. American National Standard for Advanced Data Communication Control Procedures (ADCCP), *ANSI X*3.66-1979

Arthurs, E. and Stuck, B.W. 1981. A Theoretical Performance Analysis of Polling and Carrier Sense Collision Detection Communication Systems, *Proc. Seventh Data Communications Symposium*

Chou, W. (ed.) 1983. *Computer Communications. Vol. 1: Principles.* Englewood Cliffs, NJ: Prentice-Hall

Davies, D.W., Barber, D.L.A., Price, W.L. and Solomonides, C.M. 1979. *Computer Networks and their Protocols.* Chichester: John Wiley

Denning, D.E. 1982. *Cryptography and Data Security.* Reading, Mass: Addison-Wesley

Diffie, W. and Hellman, M.E. 1976. New Directions in Cryptography, *IEEE Trans. Information Theory*, **IT-22**, 644-54

Doll, D.R. 1978. *Data Communications.* New York: John Wiley

Graube, M. June 1982. Local Area Nets: A Pair of Standards, *IEEE Spectrum*

Hamming, R.W. 1950. Error Detecting and Error Correcting Codes, *Bell Syst. Tech. J.*, **29**, 147-60

Hellman, M.E. 1980. A Cryptanalytic Time-Memory Tradeoff, *IEEE Trans. Inf. Theory*, **IT-26**, 401-406

Hopper, A., Williamson, R. and Temple, S. 1985. *Local Area Networks.* Wokingham: Addison-Wesley

IBM Corp. 1970. *General Information — Binary Synchronous Communications.* Manual No. GA27-3004-2, 3rd. edn.

IBM Corp. 1975. *Synchronous Data Link Control — General Information.* Manual No. GA27-3093-1

IEEE. 1982. IEEE Project 802 — Local Network Standards, Draft C

INMOS, 1984. *occam Programming Manual*, Englewood Cliffs, NJ: Prentice-Hall

International Standards Organization. 1979. Data Communication — High Level Data Link Control Procedures — Frame Structure, Ref. No.ISO3309-1979; Data Communication — High-Level Data Link Control Procedures — Elements of Procedures, Ref. No. ISO4335-1979

International Standards Organisation August 1981. *Open Systems Interconnection — Basic Reference Model*, Draft Proposal 7498

Martin, J. 1970. *Teleprocessing Network Organisation*. Englewood Cliffs, NJ: Prentice-Hall

McNamara, J.E. 1977. *Technical Aspects of Data Communication*, Maynard, Mass: Digital Equipment Corp.

Metcalfe, R.M. and Boggs, D.R. 1976. Ethernet: Distributed Packet Switching for Local Computer Networks, *Comm. Assoc. Comp. Mach.*, **19(7)**, 395-403

Myers, W. August 1982. Toward a Local Network Standard, *IEEE Micro*, **2(3)**, 28-45

Needham, R.M. and Herbert, A.J. 1982. *The Cambridge Distributed Computing System*. London: Addison-Wesley

Peterson, W.W. 1961. *Error Correcting Codes*. Cambridge, Mass: MIT Press

Swanson, R. 1975. Understanding Cyclic Redundancy Codes, *Computer Design*, **14(11)**, 93-9

Tanenbaum, A.S. 1981. *Computer Networks*. Englewood Cliffs, NJ: Prentice-Hall

Chapter 9 System design and development

Bourne, S.R. 1982. *The UNIX System*. London: Addison-Wesley

Breuer, M.A. and Friedman, A.D. 1976. *Diagnosis and Reliable Design of Digital Systems*. Cal: Computer Science Press Inc.

Brooks, F.P. 1975. *The Mythical Man-Month*. Reading, Mass: Addison-Wesley

Dijkstra, E.W. 1968. Goto Statement Considered Harmful, *Comm. Assoc. Comp. Mach.* **11(3)**, 147-8

Heniger, K.L. 1980. Specifying Software Requirements for Complex Systems — New Techniques and their Applications, *IEEE Trans. Software Engineering*, **SE-6**, 2-13

IEEE. 1978. IEEE Standard Digital Interface for Programmable Instrumentation, IEEE Std. 488-1978

IEEE 796 Bus Working Group. 1980. *Computer*, **13(10)**, 89-105

Levine, L. and Meyers, W. 1976. Semiconductor Memory Reliability with Error Detecting and Correcting Codes, *IEEE Computer*, **9**, 43-50

Mühlbacher, J.R. 1982. A Tutorial Introduction into Structured Software Development, *Journal of Microcomputer Applications*, **5**, 67-86

Nassi, I. and Schneiderman, D. 1973. Flow Chart Techniques for Structured Programming, *SIGPLAN Notices*, **8(5)** 12-26

Sommerville, I. 1982. *Software Engineering.* London: Addison-Wesley

Zelkowitz *et al.* 1979. *Principles of Software Engineering and Design.* Englewood Cliffs, NJ: Prentice-Hall

INDEX

Absolute pressure, 304
Access control, 366
Access matrix, 366
Accumulator, 83
Accuracy, 237
ACIA, 337
Acoustic pad, 317
Acoustic sensing, 315
Acquisition time, 244
ACR, 182
ADC, 234
Address, 27
  decoder, 148
  register, 100
  space, 27, 150
Addressing mode, 88,
  autodecrement, 122
  autoincrement, 122
  direct, 88
  immediate, 89
  indirect, 89
  register direct, 101
  register indirect, 101
  relative, 125
Alarm point, 67
Amplitude, 236
Analog, 10
Analog interface, 234
Analog-to-digital converter, 234
AND, 20, 57
Anode, 217
Arbitration, 348
Architecture, 5
Arithmetic shift left, 87
Array, 43, 89
ASCII, 19
Assembler, 80, 395, 399
  directive, 80, 123
Assembly language, 31, 80

Assembly time, 82
Assignment statement, 35
Assignment symbol, 54
Asynchronous system, 331
Auxiliary control register, 171, 182
Availability, 382

Back e.m.f., 296
Base
  address, 391
  bipolar transistor, 186
Baseband system, 344
Base-emitter voltage, 187
Baud rate generator, 337
Baud rate, 331
BCD, 18, 97
Bias current, 242
Bidirectional bus, 26
Bidirectional I/O port, 151, 194
Binary
  arithmetic mode, 103
  coded decimal, 18, 97
  search, 260
  weighted resistor DAC, 250
Bit, 13
  oriented protocol, 356
  period, 342
  plane, 313
  reversed order, 73
  test, 109
Block, program, 53, 63, 357
Boolean algebra, 22
Boolean type, 57
BOP, 356
Branch, 107
Breakpoint, 398
Broadband system, 344
Brook's law, 387

437

Buffer insertion, 363
Buffer, 335
Burst firing, 258
Burst mode DMA, 177
Bus, 26
    address, 27
    bidirectional, 26
    control, 27
    data, 27
    global, 348
    local, 348
    unidirectional, 26
Bus-driving device, 196
Byte, 14

Cambridge ring, 362
Candela, 284
Capacitance sensing, 315
Carry register, 94, 103
Case statement, 115
Cathode ray tube, 310-311
Cathode, 217
CCD, 263
Central processing unit, 4
Character, 19
    printer, 309
Characteristic, transfer, 190
Charge-coupled array, 319
Charge-coupled device, 263
Checksum, 329
Cipher text, 367
CMOS logic, 197
Coding, error-control, 328
Collector, 186
    power dissipation, 190
    -AND connection, 194
    -emitter voltage, 188
    -OR connection, 194
Common mode, 268
Communication standard, 327
Commutation, 217
Compiler, 395, 403
Compound statement, 53
Concatenation, 50
Concurrency, 349
Condition code register, 103
Constant, 35
Construct, 60
Contact bounce, 206

Contention technique, 360
Control bus, 27
Control line, 157
Controllability, 417
Correlation, 306
CPU, 4
Critical path, 382
Cross-assembler, 393, 395
Cross-compiler, 393, 395
CSMA/CD, 361
CTR, 199
Current
    balance, 300
    loop, 339
    rating, 194
    reference source, 252
    transfer ratio, 199
    -limiting resistor, 216
Cursor, 312
Cut-off, 189
Cycle-stealing mode DMA, 177
Cyclic redundancy coding, 331

DC positive feedback, 201
DAC, 236, 250, 252
Daisychaining, 165
Daisywheel, 309
DAK, 155
Darlington connection, 215
Darlington phototransistor, 200
Data
    acknowledge, 155, 157
    available, 155, 157
    direction register, 151
    encryption standard, 367
    register, 100
    type, 10
Datum point, 288
DAV, 155
DDR, 151
Debouncing, 206
Debugging, 396
Decimal number system, 12
Development system, 392
DFT, 71
Differential input voltage, 242
Differential mode, 268
Digital-to-analog converter, 236
Direct addressing, 88

Direct assembler, 399
Direct memory access, 175, 263
Directive, assembler, 80, 123
Discrete Fourier transform, 71
Disc, 307
Displays, 221
DMA, 175-7, 263
Doppler effect, 296
Drain, 189
Droop, 245
Dumper, 398
Duty cycle, 257
Dynamic RAM, 28

EAROM, 28
Earth connection, 191
Earth loop, 267
Eavesdropping, 366
Edge-sensitive input line, 208
Editor, 395, 405
EEPROM, 28
Effective address, 91
Elasticity, 301
Electromagnetic grid, 316
Element, 43
Embedded system, 9
Emitter, 186
Emulator, 396, 406
Encryption, 367
EPROM, 28
Error, 237
  roundoff, 58
  semantic, 392
  syntactic, 392
Error signal, 65
Error-control coding, 328
Ethernet, 361
Exception, 134, 168
Exclusive OR, 21
Executable statement, 34
Exponent, 18
Extrinsic function, 43

Fanout, 195
Fast Fourier transform, 73
Fast interrupt request, 163
FDM, 345
Feedback, 66, 242
FFT, 73

FIFO, 161, 263
File, 307
Firmware, 7
FIRQ, 163
First-in, first-out, 161, 263
Fixed point, 18
Flags, 103-5
Flash converter, 258
Floating point, 18
Floppy disc, 308
Flowchart, 32
Fluorescent display, 221
Flywheel current, 217
Forward bias, 188
Frequency synthesiser, 256
Full scale output, 239
Full-duplex, 334
Full-scale value, 237
Function
  extrinsic, 43
  intrinsic, 40
  standard, 65

Gas discharge display, 222
Gate turnoff thyristor, 219
Gate, 22, 189, 217
Gateway unit, 360
Gauge factor, 303
Gauge pressure, 305
Ghosting, 222
Global bus, 348
Global label, 403
Golfball, 309
Ground connection, 191

Half-duplex, 334
Handshake signal, 155
Hard disc, 307
Hardware, 1
  software tradeoff, 235
  timer, 209
Heat sink, 190
Hexadecimal, 15
High logic level, 13, 192
High level language, 31
High speed local network, 358
Host machine, 393
Hybrid integrated circuit, 252
Hysteresis loop, 201

I/O, 5
I/O address space, 150
IACK, 166
ICR, 180
IER, 165
IFR, 157
Illuminance, 284
Immediate addressing, 89
Impact printer, 309
In-circuit emulation, 408
Incandescent lamp, 221
Incremental encoder, 290
Index, 55
   register, 90
Indirect addressing, 89
Indivisible test and set, 352
Inductosyn, 293
Input
   capture register, 180
   interface, 197
   register, 149
   subroutine, 234
   transducer, 205
Input/output, 5
   memory mapped, 135
   section, 146
Instruction, machine, 79
Instrumentation amplifier, 281
Integer, 11
Integrated circuit, 2
Intercommunication register, 209
Interlacing, 310
Interpreter, 406
Interrupt flag register, 157
Interrupts, 162-168
Intrinsic function, 40
Inverting amplifier, 246
IRQ, 162
Irradiance, 285
ISR, 162

Joystick, 316
Jump to subroutine, 128
Jump, 107

Keyboard, 205

Label, 81, 403
LAN, 358

Laser printer, 310
Last-in, first-out, 126
Latch, 25
Layers, ISO, 347-348, 364
LCD, 221
Least significant digit, 13
LED, 199
Level-sensitive input line, 208
LIFO, 126
Light emitting diode, 199
Light pen, 312
Light-sensitive transistor, 199
Limit switch, 205
Line editor, 405
Line printer, 309
Linear array, 319
Linearity, 240
Link bit, 103
Link editor, 396, 404
Linker, 396, 404
Liquid crystal display, 222
Listing, 399
Literal, 47
Live connection, 191
Load cell, 303
Load enable, 148
Loader, 396, 398, 399, 405
Local area network, 358
Local bus, 348
Local label, 403
Local variable, 63
Logic level, 194
Logic, negative, 192
Logic, positive, 192
Loop, 37
Low level language, 31
Low logic level, 13, 192
Low-pass filter, 203
LSD, 13
Lumen, 284
Luminance, 284
Luminous intensity, 284
Luminous power, 284
Lux, 284

Machine instruction, 79
Machine language, 78
Machine-independent, 78
Macro, 400

INDEX   **441**

Macroassembler, 401
Magnetostrictive pad, 317
Maintainability, 8
Manchester code, 343
Mantissa, 18
Mask, 49
Masked ROM, 28
Matrix printer, 309
Mean time between failures, 382
Mean time to repair, 382
Memory, 5
  address space, 150
  examine/modify, 398
  location, 27
  mapped connection, 150
  mapped input/output, 135
Microprocessor, 4, 29
Minicomputer, 3
Mnemonic, 79
Modem, 339, 345
Module, 404
Moiré effect, 291
Monitor, 398
MOS transistor, 186
Most significant digit, 13
Mouse, 316
MPU, 4
MSD, 13
MTBF, 382
MTTR, 382
Multiplex rate, 222
Multiplexing, 221, 344
Multitasking system, 396
Multiuser system, 397

NAND, 24
Negative logic, 192
Nested interrupt, 168
Nested loop, 39
Nested subroutine, 129
Network, 358
Network security, 366
Neutralising, 269
Nibble, 14
NMI, 165
Noise margin, 196
Noise, 196, 327
Nonlinear resistor, 199
Nonmaskable interrupt, 165

NOR, 25
NOT, 20, 57
NRZ, 342
NRZI, 343
Number base, 12
N-key rollover, 210

Object code, 399
Observability, 417
occam, 350
OCR, 180
Octal, 15
Offset, 91, 125
One's complement, 17
Opcode, 79
Open collector, 194
Open drain, 194
Open loop system, 65
Operand, 79
  destination, 100
  source, 100
Operating system, 9, 395
Operation code, 79
Optical touch screen, 315
Optoisolator, 199
OR, 20, 57
  exclusive, 21
Orifice plate, 306
Output compare register, 180
Output register, 148

Palette, 314
Parallel converter, 258
Parallel input/output, 151
Parity, 328
Passing parameters, 130
Passive pull-up, 194
Peak-to-peak range, 69
Pel, 313
Peripheral control register, 157
Peripheral interface adapter, 151
Phase comparator, 255
Phase-lock loop, 255
Phoneme, 320
Photodiode, 287, 319
Phototransistor, 199, 200
PIA, 151
PIC, 124

Piezoelectric effect, 300
PIO, 151
Pixel, 313
Plain text, 367
PLL, 255
Pointer, 56, 89
　stack, 128
Polling, 159, 166
Popping, 126
Position independent code, 124
Positional notation, 12
Positive logic, 192
Precision, 237
Processor status register, 103
Program, 31
　block, 53, 63, 357
　counter, 106
Programmable ROM, 28
Proportional control, 66
Protocol, 352
Public key encryption, 367
Pulling, 126, 128
Pushing, 126, 128

Quantisation, 238
Quasi-bidirectional port, 194

Radiance, 285
Radiant intensity, 285
Radiant power, 285
Radix, 12
RAM, 28
Random-access memory, 28
Range, 237
Raster, 310
Ratiometric conversion, 277
Read-only memory, 28
Real number, 12
Real time, 9, 78, 173
Real time clock, 173
Record, 55
Recursion, 129
Redundant information, 328
Refreshing, 28
Register, 13, 25
　address, 100
　auxiliary control, 171, 182
　carry, 94

condition code, 103
destination, 84
DMA address, 176
DMA control/status, 176
DMA data, 176
DMA word count, 176
flag, 103
index, 90
input, 149
input capture, 180
intercommunication, 209
output, 148
output compare, 180
processor status, 103
shift, 174
source, 84
successive approx'n, 261
timer control/status, 180
Relational operator, 38
Relative addressing mode, 125
Relative pressure, 304
Relocatable object code, 404
Resistive membrane, 315
Resolution, 237
Return from interrupt, 163
Return from subroutine, 129
Ringing, 198
Rollover, 210
ROM, 28
Rotate left, 87
Rotating interrupt priority, 168
Roundoff error, 58
RS-232-C standard, 338
RS-422 standard, 340
RS-423 standard, 340
R-2R ladder network, 251

Sample and hold circuit, 244
SAR, 261
Saturation, 188
Scale factor, 240
Schmitt trigger, 201
SCR, 217
Screen editor, 405
Scrolling, 312
Section, input/output, 146
Sector, 307
Self heating, 279
Semantic error, 392

INDEX

Series connection, 211
Series feedback, 242
Set point, 65
Settling time, 252
Set-Reset latch, 208
Shift register, 174
Shunt connection, 211
Shunt feedback, 242
Signal
    conditioning, 146
    error, 65
    flow graph, 143
Sign-magnitude representation, 17
Silicon
    chip, 1
    controlled rectifier, 217
    software, 125
Simplex, 334
Single chip microcomputer, 6
Slew rate, 201
Slotted ring, 362
Snubbing circuit, 218
Software, 1
    engineering, 5, 32
    interrupt, 169
    maintenance, 387
Solid state relay, 220
Source
    MOS transistor, 189, 192
    operand, 100
    register, 84
SR latch, 207
SSR, 220
Stack, 126
    pointer, 128
Standard function, 65
Start bit, 333
State, 10, 192
    transition diagram, 33
Statement, 34
    compound, 53
Static electricity, 197
Static RAM, 28
Statistical TDM, 345
Stepper motor, 298
Stop bit, 334
Strain, 301
    gauge, 302
Stress, 302
String, 19, 47

Subroutine, 39
Subscripted array, 89
Subscripted variable, 43
Symbol, assignment, 54
Symbolic value, 81
Synchronising pattern, 332
Synchronous system, 331
Syntactic error, 392

Tachometer, 296
Target machine, 393
TCSR, 180
TDM, 344-5
Text editor, 405
Thermocouple, 282
Three term controller, 67
Three-state output, 149, 193
Time constant, 204
Timer control/status register, 180
Time-division multiplexing, 344
Topology, 358
Top-down approach, 53, 381
Totem-pole, 193
Trackerball, 316
Transfer characteristic, 190
Transformer, 199
Transistor, 186, 189
Transistor transistor logic, 195
Transparency, 351, 356
Transputer, 349
Transresistance amplifier, 246
Trap, 169
Triac, 218
TTL, 195
Twisted pair, 268
Two's complement representation, 17

UART, 336
Uncertainty, 237
Unidirectional bus, 26
Unipolar converter, 241

Value parameter, 64
Variable, 35, 88
    subscripted, 43
    local, 63
Variable parameter, 64

VCO, 255
Venturi meter, 306
Vernier fringe, 292
Versatile interface adapter, 151
VIA, 151
Video, 311
Virtual earth, 245
Volatile RAM, 28
Voltage
  comparator, 200
  controlled oscillator, 255
  follower, 244
  gain, 242
  offset, 242
  reference source, 252

Waveform, 154
Wheatstone bridge, 279
Winchester, 307
Wired-AND connection, 194
Wired-OR connection, 194
Word recogniser, 410
Word, 14
Wordlength, 29
Write signal, 149

Young's modulus, 302

Zener diode, 198
Zero flag, 104
Zero offset, 240